Student Solutions Manual

Differential Equations & Linear Algebra

Second Edition

Farlow   Hall   McDill   West

Upper Saddle River, NJ 07458

Senior Editor: *Holly Stark*
Senior Managing Editor: *Scott Disanno*
Production Manager: *Craig Little*
Manufacturing Buyer: *Lisa McDowell*
Cover Designer: *Daniel Sandin*

© 2008 by Pearson Education, Inc.
Pearson Prentice Hall
Pearson Education, Inc.
Upper Saddle River, NJ 07458

Printed in the United States of America

ISBN 0-13-186063-1
     978-0-13-186063-6

Pearson Education Ltd., *London*
Pearson Education Australia Pty. Ltd., *Sydney*
Pearson Education Singapore, Pte. Ltd.
Pearson Education North Asia Ltd., *Hong Kong*
Pearson Education Canada, Inc., *Toronto*
Pearson Educación de Mexico, S.A. de C.V.
Pearson Education—Japan, *Tokyo*
Pearson Education Malaysia, Pte. Ltd.
Pearson Education, Inc., *Upper Saddle River, New Jersey*

Contents

Chapter 1   **First-Order Differential Equations**   **1**

    1.1   Dynamical Systems: Modeling   1

    1.2   Solutions and Direction Fields: Qualitative Analysis   3

    1.3   Separation of Variables: Quantitative Analysis   13

    1.4   Approximation Methods: Numerical Analysis   20

    1.5   Picard's Theorem: Theoretical Analysis   28

Chapter 2   **Linearity and Nonlinearity**   **33**

    2.1   Linear Equations: The Nature of Their Solutions   33

    2.2   Solving the First-Order Linear Differential Equation   38

    2.3   Growth and Decay Phenomena   47

    2.4   Linear Models: Mixing and Cooling   52

    2.5   Nonlinear Models: Logistic Equation   58

    2.6   Systems of Differential Equations: A First Look   69

Chapter 3   **Linear Algebra**   **77**

    3.1   Matrices: Sums and Products   77

    3.2   Systems of Linear Equations   84

    3.3   The Inverse of a Matrix   92

    3.4   Determinants and Cramer's Rule   96

    3.5   Vector Spaces and Subspaces   102

    3.6   Basis and Dimension   106

Chapter 4   **Higher-Order Linear Differential Equations**   **115**

    4.1   The Harmonic Oscillator   115

    4.2   Real Characteristic Roots   123

    4.3   Complex Characteristic Roots   135

    4.4   Undetermined Coefficients   142

    4.5   Variation of Parameters   152

    4.6   Forced Oscillations   158

    4.7   Conservation and Conversion   165

Chapter 5   **Linear Transformations   173**

    5.1   Linear Transformations   173

    5.2   Properties of Linear Transformations   182

    5.3   Eigenvalues and Eigenvectors   191

    5.4   Coordinates and Diagonalization   205

Chapter 6   **Linear Systems of Differential Equations   215**

    6.1   Theory of Linear DE Systems   215

    6.2   Linear Systems with Real Eigenvalues   223

    6.3   Linear Systems with Nonreal Eigenvalues   234

    6.4   Stability and Linear Classification   246

    6.5   Decoupling a Linear DE System   249

    6.6   Matrix Exponential   253

    6.7   Nonhomogeneous Linear Systems   257

Chapter 7   **Nonlinear Systems of Differential Equations   263**

    7.1   Nonlinear Systems   263

    7.2   Linearization   271

    7.3   Numerical Solutions   278

    7.4   Chaos, Strange Attractors, and Period Doubling   284

    7.5   Chaos in Forced Nonlinear Systems   289

Chapter 8   **Laplace Transforms   295**

    8.1   The Laplace Transform and Its Inverse   295

    8.2   Solving DEs and IVPs with Laplace Transforms   299

    8.3   The Step Function and the Delta Function   302

    8.4   The Convolution Integral and the Transfer Function   306

    8.5   Laplace Transform Solution of Linear Systems   311

Chapter 9   **Discrete Dynamical Systems   315**

    9.1   Iterative Equations   315

    9.2   Linear Iterative Systems   323

    9.3   Nonlinear Iterative Equations: Chaos Again   329

Chapter 10   **Control Theory   337**

    10.1   Feedback Controls   337

    10.2   Introduction to Optimal Control   343

    10.3   Pontryagin Maximum Principle   345

Appendix CN: Complex Numbers   347

Appendix LY: Linear Transformations   349

Appendix PF: Partial Fractions   352

Acknowledgments

There are over 500 new problems in the second edition of *Differential Equations and Linear Algebra* by Farlow, Hall, McDill and West. This Solutions Manual was produced by the authors and two major contributors, Bill Hesselgrave and Mike Robertson, of California Polytechnic State University, San Luis Obispo. The timeliness and accuracy of this Solutions Manual are due in large part to these two contributors.

Kelly Barber contributed many months of precision typing to this effort. She cheerfully spent extra hours in multiple revisions as we all worked to get it right. We thank her for her intelligent questions and decisions about style and format. Leah King supplied many hours of excellent proofreading and had a sharp eye for errors and inconsistencies.

This and earlier versions of the Solutions Manual required many contributors as more than 2500 problems have been compiled. Among those who contributed significantly are

> Mary and Nancy Toscano, Document Engineers (and their employees)
> Hubert Hohn, IDE Developer
> Katrina Thomas, Revisions and Graphics
>
> Jim Delany, Al Jimenez, Francesca Fairbrother, Cal Poly, San Luis Obispo
> Martin Scharlemann, University of California at Santa Barbara
> Warwick Tucker, Cornell University and University of Uppsala
> Suzanne Hruska, State University of New York at Stony Brook
>
> Undergraduate Students:
>   University of California at Santa Barbara
>   Michael Bice, Nancy Heinschel
>   Cornell University
>   Lael Fisher, Samuel Wald, Jesse Mez,
>   Scott Kramer, Michael Savalli, Liz Springer,
>   Oren Harel, Melissa DiBella, Mansi Kanuga

We also wish to thank our many students and colleagues in mathematics at Cornell, Cal Poly and the University of Maine at Orono for their perceptive help in finding and correcting errors in solutions or problem statements.

We, the authors, however, claim the rights to all the errors remaining. If you should encounter one, please bring it to our attention as soon as possible. We hope that you find this Solutions Manual useful and informative.

Jerry Farlow   farlow@math.umaine.edu
Jean Marie McDill   jmcdill@calpoly.edu
Beverly H. West   bhw2@cornell.edu

Preface

Based on the power of the qualitative/graphics approach, you will be able to analyze many linear and nonlinear systems. As a student just beginning to study differential equations, you are beginning to learn how to examine the long-term behavior of multivariable systems and the effects of varying parameters.

It is important for you to develop *critical analysis skills,* and to consider the following questions: What are the limitations, and the possibilities for extensions, of these enabling techniques? To what extent can a numerical approximation or a picture be trusted? How far from an equilibrium point is a linearization reasonably accurate?

The cost of acquiring power at an early stage in the study of differential equations is that these exercises must be approached *carefully*, at a considerably slower pace than traditional exercises. *Open-ended problems should be assigned sparingly because they require much more time for both the students and the instructor who reads their work. A wide variety of responses and analyses is to be expected.*

The accuracy of the pictures is of prime importance.

- Solutions and trajectories <u>must</u> follow direction and vector fields.

- Directions must be clearly visible on all phase portraits.

- Nonsolutions (e.g., isoclines and nullclines) must <u>not</u> be confused with solutions.

- Solutions and trajectories must be carried over sufficiently long time intervals to convey long-term behaviors.

Questions on nonlinear systems will be more open-ended, because there are a vast variety of observations that can be made and investigated. A "solutions" manual can only be a guide – a great diversity of answers is to be expected. Ideally it is this diversity that thoroughly solves a problem. In later chapters in particular, the instructor will look for multiple approaches and perspectives on a given DE system and might ask that applicable pictures and conclusions be shared within a class. Our Solutions Manual can serve only as a reference in these chapters. We try to provide sufficient information for your assessment of your solution; whenever different statements or pictures seem not to agree, there is might an error, and the difference must be tracked down. An instructor can point to what is suspicious and leave it to you to do the tracking; you might have caught something the authors have missed.

The challenge for this Solutions Manual, especially in the later chapters, is to provide some basic ingredients for which various approaches to the problem can be compared or judged.

Important Note to Students:

The solution to every <u>third</u> problem has been selected for inclusion in the Student Solutions Manual

CHAPTER 1

First-Order Differential Equations

1.1 Dynamical Systems: Modeling

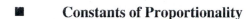

■ Constants of Proportionality

3. $\dfrac{dP}{dt} = kP(20{,}000 - P)$

■ A Walking Model

6. Because $d = vt$ where d = distance traveled, v = average velocity, and t = time elapsed, we have the model for the time elapsed as simply the equation $t = \dfrac{d}{v}$. Now, if we measure the distance traveled as 1 mile and the average velocity as 3 miles/hour, then our model predicts the time to be $t = \dfrac{d}{v} = \dfrac{1}{3}\mathrm{hr}$, or 20 minutes. If it actually takes 20 minutes to walk to the store, the model is perfectly accurate. This model is so simple we generally don't even think of it as a model.

■ Population Update

9. (a) If we assume the world's population in billions is currently following the unrestricted growth curve at a rate of 1.7% and start with the UN figure for 2000, then

$$y_0 e^{kt} = 6.056 e^{0.017t},$$

and the population in the years 2010 ($t = 10$), 2020 ($t = 20$), and 2030 ($t = 30$), would be, respectively, the values

$$6.056 e^{0.017(10)} = 7.176$$
$$6.056 e^{0.017(20)} \approx 8.509$$
$$6.056 e^{0.017(30)} \approx 10.083.$$

These values increasingly exceed the United Nations predictions so the U.N. is assuming a growth rate less than 1.7%.

(b)   2010: $6.056e^{10r} = 6.843$

$$e^{10r} = \frac{6.843}{6.056} = 1.13$$

$$10r = \ln(1.13) = 0.1222$$

$$r = 1.2\%$$

2020: $6843e^{10r} = 7568$

$$e^{10r} = \frac{7.578}{6.843} = 1.107$$

$$10r = \ln(1.107) = 0.102$$

$$r = 1.0\%$$

2030: $7.578e^{10r} = 8.199$

$$e^{10r} = \frac{8.199}{7.578} = 1.082$$

$$10r = \ln(1.082) = 0.079$$

$$r = 0.8\%$$

■   **Verhulst Model**

12. $\dfrac{dy}{dt} = y(k - cy)$. The constant k affects the initial growth of the population whereas the constant c controls the damping of the population for larger y. There is no reason to suspect the two values would be the same and so a model like this would seem to be promising if we only knew their values. From the equation $y' = y(k - cy)$, we see that for small y the population closely obeys $y' = ky$, but reaches a steady state $(y' = 0)$ when $y = \dfrac{k}{c}$.

1.2   Solutions and Direction Fields

■   **Verification**

3.   Substituting

$$y = t^2 \ln t$$
$$y' = 2t \ln t + t$$

into $y' = \dfrac{2}{t} y + t$ yields the identity

$$2t \ln t + t \equiv \frac{2}{t}\left(t^2 \ln t\right) + t.$$

■   **IVPs**

6.   Substituting $y = t + t^2 + 2e^{2t}$

$$y' = 1 + 2t + 4e^{2t}$$

into $y' = 2y + 1 - 2t^2$ yields the identity

$$1 + 2t + 4e^{2t} \equiv 2(t + t^2 + 2e^{2t}) + 1 - 2t^2$$
$$\equiv 2t + 4e^{2t} + 1$$

Note that $y(0) = 0 + 0 + 2 = 2$ so that the initial condition is satisfied.

■   **Using the Direction Field**

9. $y' = 2y$

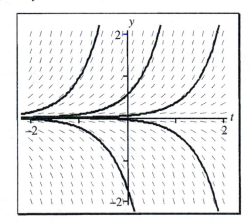

Solutions are $y = ce^{2t}$.

■   Linear Solution

12.   It appears from the direction field that there is a straight-line solution passing through $(0, -1)$ with slope 1, i.e., the line $y = t - 1$. Computing $y' = 1$, we see it satisfies the DE $y' = t - y$ because $1 \equiv t - (t - 1)$.

■   Stability

15. $y' = t^2(1 - y^2)$

Two equilibrium solutions:

$y = 1$ is stable

$y = -1$ is unstable

Between the equilibria the slopes (positive) are shallower as they are located further from the horizontal axis.

Outside the equilibria the slopes are negative and become steeper as they are found further from the horizontal axis.

All slopes become steeper as they are found further from the vertical axis.

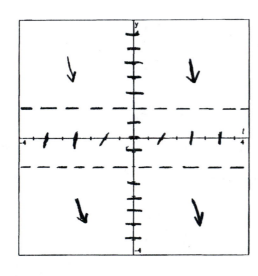   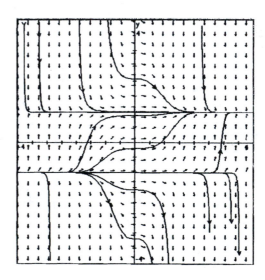

■   Match Game

18.   (F)   Because F is the only direction field that has vertical slopes when $t = 0$ and zero slopes when $y = 0$

21.   (A)   Because it is undefined when $t = 0$ and the directional field has slopes that are independent of y, with the same sign as that of t

■ Concavity

24. $y' = y^2 - t$

$y'' = 2yy' - 1$

$\quad = 2y^3 - 2yt - 1 = 0$

When $t = \dfrac{2y^3 - 1}{2y} = y^2 - \dfrac{1}{2y}$, then $y'' = 0$

and we have a locus of inflection points.

The locus of inflection points has two branches: Above the upper branch, and to the right of the lower branch, solutions are concave *up*.

Below the upper branch but outside the lower branch, solutions are concave *down*.

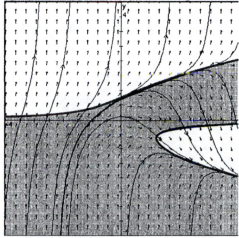

Bold curves are the locus of inflection points; shaded regions are where solutions are concave *down*.

■ Asymptotes

27. $y' = t^2$

There are *no* asymptotes.

As $t \to \infty$ (or $t \to -\infty$) slopes get steeper and steeper, but they do not actually approach vertical for any finite value of t.

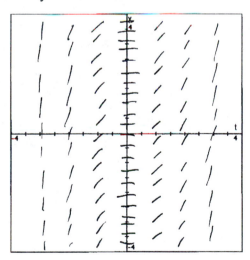

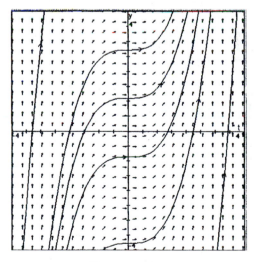

No asymptote

30. $$y' = \frac{ty}{t^2 - 1}$$

At $t = 1$ and $t = -1$ the DE is undefined. The direction field shows that as $y \to 0$ from either above or below, solutions asymptotically approach vertical slope. However, $y = 0$ is a solution to the DE, and the *other solutions do not cross* the horizontal axis for $t \neq \pm 1$. (See Picard's Theorem Sec. 1.5.)

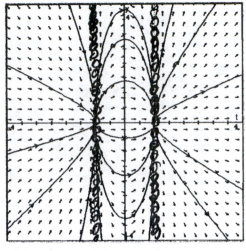

Vertical asymptotes for $t \to 1$ or $t \to -1$

■   **Isoclines**

33. $y' = y^2$.

Here the slope of the solution is always ≥ 0.

The isoclines where the slope is $c > 0$ are the horizontal lines $y = \pm\sqrt{c} \geq 0$. In other words the isoclines where the slope is 4 are $y = \pm 2$. The isoclines for $c = 0, 2,$ and 4 are shown in the figure.

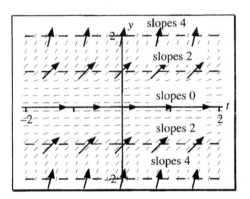

36. $y' = y^2 - t$. The isocline where $y' = c$ is a parabola that opens to the right. Three isoclines, with slopes $c = 2, 0, -2$, are shown from left to right (see figure).

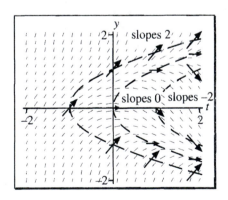

39. $y' = \cos(y - t)$

$$y' = c = \begin{cases} 0 \text{ when } y - t = -\dfrac{\pi}{2}, \dfrac{\pi}{2}, \dfrac{3\pi}{2}, \ldots \\ \qquad\qquad \text{or } y = t \pm (2n+1)\dfrac{\pi}{2} \\ 1 \text{ when } y - t = 0, 2\pi, \ldots \\ \qquad\qquad \text{or } y = t \pm 2n\pi \\ -1 \text{ when } y - t = -\pi, \pi, 3\pi, \ldots \\ \qquad\qquad \text{or } y = t \pm (2n+1)\pi \end{cases}$$

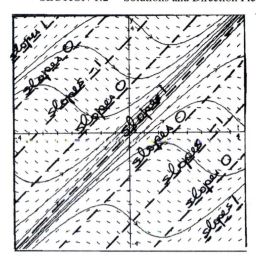

All these isoclines (dashed) have slope 1, with different y-intercepts.

The isoclines for *solution* slopes 1 are also solutions to the DE *and* act as oblique asymptotes for the other solutions between them (which, by uniqueness, do not cross. See Section 1.5).

■   ' **Periodicity**

42. $y' = -\cos y$

If $y = \pm(2n+1)\dfrac{\pi}{2}$, then $y' = 0$ and these horizontal lines are equilibrium solutions.

For $y = \pm 2n\pi$, $y' = -1$

For $y = \pm(2n+1)\pi$, $y' = 1$.

Slope y' is always between -1 and 1, and *solutions between the constant solutions cannot cross them, by uniqueness.*

To further check what happens in these cases we have added an isocline at $y = \dfrac{\pi}{4}$, where

$$y' = \cos\left(\frac{\pi}{4}\right) \approx -0.7.$$

Solutions are not periodic, but there is a periodicity to the direction field, in the vertical direction with period 2π. Furthermore, we observe that between every adjacent pair of constant solutions, the solutions are horizontal translates.

continued on the next page

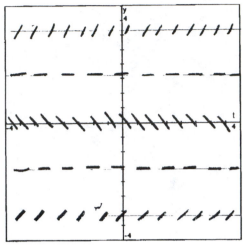
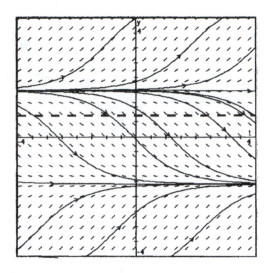

45. $y' = y(\cos t - y)$

Slopes are 0 whenever $y = \cos t$ or $y = 0$

Slopes are *negative* outside of both these isoclines;

Slopes are *positive* in the regions trapped by the two isoclines.

If you try to sketch a solution through this configuration, you will see it goes downward a lot more of the time than upward.

For $y > 0$ the solutions wiggle downward but never cross the horizontal axis—they get sent upward a bit first.

For $y < 0$ solutions eventually get out of the upward-flinging regions and go forever downward.

The solutions are *not* periodic, despite the periodic function in the DE.

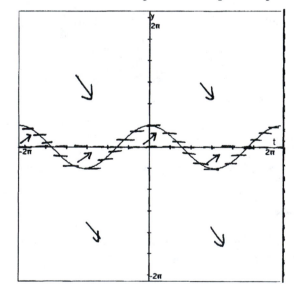
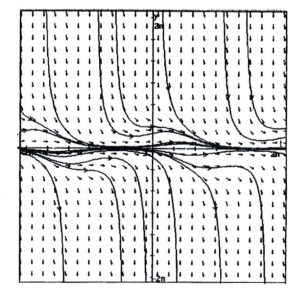

■ **Symmetry**

48. $y' = t^2$

Note that y' depends only on t, so isoclines are vertical lines.

Positive and negative values of t give the *same* slope, so the *slope values* are repeated symmetrically across the vertical axis, but the resulting direction field does *not* have visual symmetry.

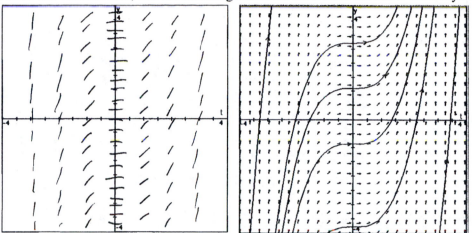

The only symmetry visible in the direction field is point symmetry through the origin (or any point on the y-axis).

51. $y' = \dfrac{1}{(t+1)^2}$

Note that y' depends only on t, so isoclines will be vertical lines.

Slopes are always positive, so they will be *repeated*, not reflected, across $t = -1$, where the DE is not defined.

If $t = 0$ or -2, slope is 1.

If $t = 1$ or -3, slope is $\dfrac{1}{4}$.

If $t = 2$ or -4, slope is $\dfrac{1}{9}$.

The direction field has *point* symmetry through the point $(-1, 0)$, or any point on the line $t = -1$.

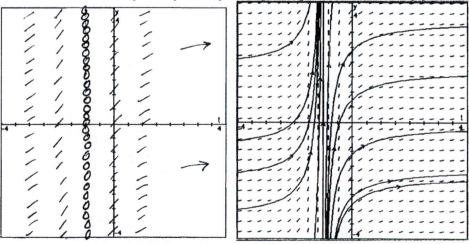

■ Long-Term Behavior

54. $y' = t + y$

(a)   There are *no* constant solutions; zero slope requires $y = -t$, which is not constant.

(b)   There are *no* points where the DE, or its solutions, are undefined.

(c)   We see one straight line solution that appears to have slope $m = -1$ and y-intercept $b = -1$. Indeed, $y = -t - 1$ satisfies the DE.

(d)   All solutions above $y = -t - 1$ are concave up; those below are concave down. This observation is confirmed by the sign of $y'' = 1 + y' = 1 + t + y$.

(e)   As $t \to \infty$, solutions above $y = -t - 1$ approach ∞; those below approach $-\infty$.

(f)   As $t \to -\infty$, going backward in time, *all* solutions are seen to emanate from ∞.

(g)   The only asymptote, which is oblique, appears if we go backward in time—then all solutions are ever closer to $y = -t - 1$.

There are *no* periodic solutions.

In shaded region, solutions are concave *down*.

57. $y' = \dfrac{1}{t - y}$

(a)   There are *no* constant solutions, nor even any point with zero slope.

(b)   The DE is undefined along $y = t$.

(c)   There appears to be one straight line solution with slope 1 and y-intercept -1; indeed $y = t - 1$ satisfies the DE.

$y' = 1$ when $y = t - 1$.   Straight line solution

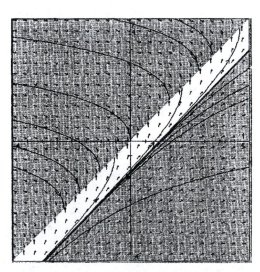

In shaded region, solutions are concave *down*.

(d) $y'' = -\dfrac{(1 - y')}{(t - y)^2} = \dfrac{y - (t - 1)}{(t - y)^3}$

   $y'' > 0$ when $y > t - 1$ and $y < t$ Solutions concave up

   $\left.\begin{array}{l} y'' < 0 \text{ when } y < t - 1 \text{ and } y > t \\ \qquad\qquad y > t - 1 \text{ and } y > t \end{array}\right\}$ Solutions concave down

(e)   As $t \to \infty$,   solutions below $y = t - 1$ approach ∞;

   solutions above $y = t - 1$ approach $y = t$ ever more vertically.

(f)   As $t \to -\infty$,   solutions above $y = t$ emanate from ∞;

   solutions below $y = t$ emanate from $-\infty$.

(g)   In *backwards* time the line $y = t - 1$ is an oblique asymptote.

   There are no periodic solutions.

◾ Logistic Population Model

60.   We find the constant solutions by setting $y' = 0$ and solving for y. This gives $ky(1 - y) = 0$, hence the constant solutions are $y(t) \equiv 0, 1$. Notice from the direction field or from the sign of the derivative that solutions starting at 0 or 1 remain at those values, and solutions starting between 0 and 1 increase asymptotically to 1, solutions starting larger than 1 decrease to 1 asymptotically. The following figure shows the direction field of $y' = y(1 - y)$ and some sample solutions.

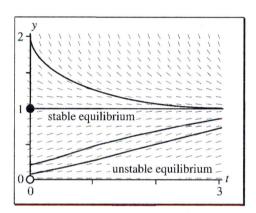

Logistic model

◾ Coloring Basins

63. $y' = y(1 - y)$. The constant solutions are found by setting $y' = 0$, giving $y(t) \equiv 0, 1$. Either by looking at the direction field or by analyzing the sign of the derivative, we conclude the constant solution $y(t) \equiv 1$ has a basin of attraction of $(0, \infty)$, and $y(t) \equiv 0$ has a basin attraction of the single value $\{0\}$. When the solutions have negative initial conditions, the solutions approach $-\infty$.

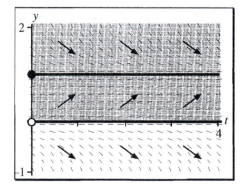

66. $y' = (1 - y)^2$. Because the derivative y' is always zero or positive, we conclude the constant solution $y(t) \equiv 1$ has basin of attraction the interval $(-\infty, 1]$.

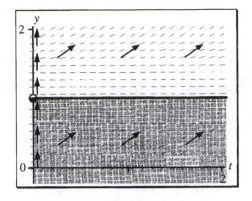

■ Computer or Calculator

69. $y' = ty$. The direction field shows one constant solution $y(t) \equiv 0$, which is unstable (see figure). For negative t solutions approach zero slope, and for positive t solutions move away from zero slope.

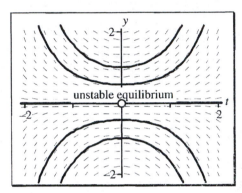

72. $y' = \sin(ty)$. We have a constant solution $y(t) \equiv 0$ and there is a symmetry between solutions above and below the t-axis. Note: This equation does not have a closed form solution.

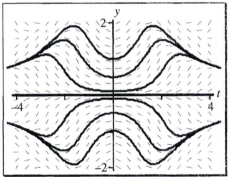

■ Suggested Journal Entry I

75.   Student Project

1.3   Separation of Variables: Quantitative Analysis

■   **Separable or Not**

3. $y' = \sin(t + y)$. Not separable; no constant solutions.

6. $y' = \dfrac{y+1}{ty} + y$. Not separable; no constant solutions.

9. $y' = \dfrac{y}{t} + \dfrac{t}{y}$. Not separable; no constant solutions.

■   **Solving by Separation**

12. $ty' = \sqrt{1 - y^2}$. The equilibrium solutions are $y = \pm 1$.

Separating variables, we get

$$\frac{dy}{\sqrt{1 - y^2}} = \frac{dt}{t}.$$

Integrating gives the implicit solution

$$\sin^{-1} y = \ln|t| + c.$$

Solving for y gives the explicit solution

$$y = \sin(\ln|t| + c).$$

15. $\dfrac{dy}{dt} = y \cos t \qquad y = 0$ is an equilibrium solution.

For $y \neq 0$, $\displaystyle\int \frac{dy}{y} = \int \cos t\ dt$

$\ln|y| = \sin t + c_1$

$e^{\ln|y|} = e^{\sin t} e^{c_1}$, so that $y = C e^{\sin t}$, where $C = \pm e^{c_1}$.

18. $y' = y^2 - 4$, $y(0) = 0$. Separating variables gives

$$\frac{dy}{y^2 - 4} = dt.$$

Rewriting this expression as a partial fraction decomposition (see Appendix PF), we get

$$\left[\frac{1}{4(y-2)} 1 - \frac{1}{4(y+2)} \right] dy = dt.$$

Integrating we get

$$\ln|y - 2| - \ln|y + 2| = 4t + c$$

or

$$\left|\frac{y-2}{y+2}\right| = e^c e^{4t}.$$

Hence, the implicit solution is

$$\frac{y-2}{y+2} = \pm e^c e^{4t} = ke^{4t}$$

where k is an arbitrary constant. Solving for y, we get the general solution

$$y(t) = \frac{2(1+ke^{4t})}{1-ke^{4t}}.$$

Substituting in the initial condition $y(0) = 0$ gives $k = -1$.

■ Integration by Parts

21. $y' = (\cos^2 y)\ln t$.   The equilibrium solutions are $y = (2n+1)\dfrac{\pi}{2}$.

Separating variables we get

$$\frac{dy}{\cos^2 y} = \ln t\, dt.$$

Integrating, we find

$$\int \frac{dy}{\cos^2 y} = \int \ln t\, dt + c$$

$$\int \sec^2 y\, dy = \int \ln t\, dt + c$$

$$\tan y = t \ln t - t + c$$

$$y = \tan^{-1}(t \ln t - t + c).$$

24. $y' = t\, ye^{-t}$.   The equilibrium solution is $y = 0$.

Separating variables we get

$$\frac{dy}{y} = te^{-t} dt.$$

Integrating, we find

$$\int \frac{dy}{y} = \int te^{-t} dt + c$$

$$\ln|y| = -te^{-t} - e^{-t} + c$$

$$y = Qe^{-(t+1)e^{-t}}.$$

■   **Equilibria and Direction Fields**

27.   (E)   **30.**   (D)

■   **Finding the Nonequilibrium Solutions**

33. $y' = y(y-1)(y+1)$

We note first that $y = 0, \pm 1$ are equilibrium solutions. To find the nonconstant solutions, we divide by $y(y-1)(y+1)$ and rewrite the equation in differential form as

$$\frac{dy}{y(y-1)(y+1)} = dt.$$

By finding a partial fraction decomposition, (see Appendix PF)

$$\frac{dy}{y(y-1)(y+1)} = -\frac{dy}{y} + \frac{dy}{2(y-1)} + \frac{dy}{2(y+1)} = dt.$$

Integrating, we find

$$-\ln|y| + \frac{1}{2}\ln|y-1| + \frac{1}{2}|y+1| = t + c$$

$$-2\ln|y| + \ln|y-1| + \ln|y+1| = 2t + 2c$$

or

$$\ln\left|\frac{(y-1)(y+1)}{y^2}\right| = 2t + 2c$$

$$\frac{(y-1)(y+1)}{y^2} = ke^{2t}.$$

Multiplying each side of the above equation by y^2 gives a quadratic equation in y, which can be solved, getting

$$y = \pm\sqrt{\frac{1}{(1+ke^{2t})}}.$$

Initial conditions will tell which branch of this solution would be used.

■   **Help from Technology**

36. $y' = \cos t$, $y(1) = 1$, $y(-1) = -1$

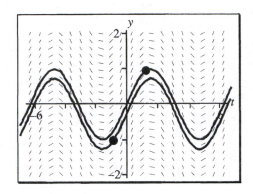

The solution of the initial-value problem

$$y' = \cos t, \ y(1) = 1$$

is $y(t) = \sin t + 1 - \sin(1)$. The solution of

$$y' = \cos t, \ y(-1) = -1$$

is $y = \sin t - 1 + \sin(-1)$. The solutions are shown
in the figure.

39. $y' = \dfrac{2t(y+1)}{y}$, $y(1) = 1$, $y(-1) = -1$

Separating variables and assuming $y \neq -1$, we
find

$$\frac{y}{y+1}\,dy = 2t\,dt$$

or

$$\int \frac{y}{y+1}\,dy = \int 2t\,dt + c.$$

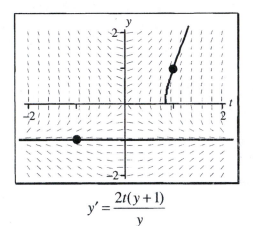

$$y' = \frac{2t(y+1)}{y}$$

Integrating, we find the implicit solution

$$y - \ln|y+1| = t^2 + c.$$

For $y(1) = 1$, we get $1 - \ln 2 = 1 + c$ or $c = -\ln 2$. For $y(-1) = -1$ we can see even more easily that
$y \equiv -1$ is the solution. These two solutions are plotted on the direction field (see figure). Note that
the implicit solution involves branching. The initial condition $y(1) = 1$ lies on the upper branch,
and the solution through that point does not cross the t-axis.

■   **Making Equations Separable**

42.   Letting $v = \dfrac{y}{t}$, we write

$$y' = \frac{y^2 + t^2}{yt} = \frac{y}{t} + \frac{t}{y} = v + \frac{1}{v}.$$

But $y = tv$ so $y' = v + tv'$. Hence, we have

$$v + tv' = v + \frac{1}{v}$$

or

$$tv' = \frac{1}{v}$$

or

$$v \, dv = \frac{dt}{t} .$$

Integrating gives the implicit solution

$$\frac{1}{2} v^2 = \ln|t| + c$$

or

$$v = \pm\sqrt{2\ln|t| + c} .$$

But $v = \dfrac{y}{t}$, so

$$y = \pm t\sqrt{2\ln|t| + c} .$$

The initial condition $y(1) = -2$ requires the negative square root and gives $c = 4$. Hence,

$$y(t) = -t\sqrt{2\ln|t| + 4} .$$

■   **Another Conversion to Separable Equations**

45. $y' = (y + t)^2$ Let $u = y + t$. Then

$$\frac{du}{dt} = \frac{dy}{dt} + 1 = u^2 + 1, \text{ and } \int \frac{du}{u^2 + 1} = \int dt, \text{ so}$$

$$\tan^{-1} u = t + c$$
$$u = \tan(t + c)$$
$$y + t = \tan(t + c) \quad \text{so} \quad y = \tan(t + c) - t$$

■   **Orthogonal Families**

48.   (a)   Starting with $f(x, y) = c$, we differentiate implicitly getting the equation

$$\frac{\partial f}{\partial x} dx + \frac{\partial f}{\partial y} dy = 0$$

Solving for $y' = \dfrac{dy}{dx}$, we have

$$\frac{dy}{dx} = -\frac{\frac{\partial f}{\partial x}}{\frac{\partial f}{\partial y}} .$$

These slopes are the slopes of the tangent lines.

(b)   Taking the negative reciprocal of the slopes of the tangents, the orthogonal curves satisfy

$$\frac{dy}{dx} = \frac{\frac{\partial f}{\partial y}}{\frac{\partial f}{\partial x}}.$$

(c)   Given $f(x, y) = x^2 + y^2$, we have

$$\frac{\partial f}{\partial y} = 2y \text{ and } \frac{\partial f}{\partial x} = 2x,$$

so our equation is $\dfrac{dy}{dx} = \dfrac{y}{x}$. Hence, from part (b) the orthogonal trajectories satisfy the differential equation

$$\frac{dy}{dx} = \frac{f_y}{f_x} = \frac{y}{x},$$

which is a separable equation having solution $y = kx$.

■   **More Orthogonal Trajectories**

51. $xy = c$. Here $f(x, y) = xy$ so $f_x = y$, $f_y = x$. The orthogonal trajectories satisfy

$$\frac{dy}{dx} = \frac{f_y}{f_x} = \frac{x}{y}$$

or, in differential form, $y\,dy = x\,dx$. Integrating, we have the solution

$$y^2 - x^2 = C.$$

Hence, the preceding family of hyperbolas are orthogonal to the hyperbolas $xy = c$. Graphs of the orthogonal families are shown.

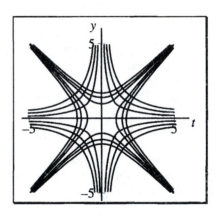

Orthogonal hyperbolas

■   **Calculator or Computer**

54. $x^2 = 4cy^3$. Here

$$f(x,\ y) = \frac{x^2}{4y^3}$$

and $f_x = \dfrac{x}{2y^3}$, $f_y = -\dfrac{3x^2}{4y^4}$. The differential equa-

tion of the orthogonal family is

$$\frac{dy}{dx} = \frac{f_y}{f_x} = \frac{-3x}{2y}$$

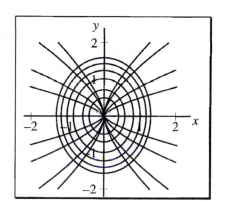

Orthogonal trajectories

or $2y\,dy = -3x\,dx$, which has the general solution $2y^2 + 3x^2 = C$, where C is any real constant. These orthogonal families are shown in the figure.

■   **Disappearing Mothball**

57.   (a)   We have $\dfrac{dV}{dt} = -kA$, where V is the volume, t is time, A is the surface area, and k is a

positive constant. Because $V = \dfrac{4}{3}\pi r^3$ and $A = 4\pi r^2$, the differential equation becomes

$$4\pi r^2 \frac{dr}{dt} = -4k\pi r^2$$

or

$$\frac{dr}{dt} = -k\ .$$

Integrating, we find $r(t) = -kt + c$. At $t = 0$, $r = \dfrac{1}{2}$; hence $c = \dfrac{1}{2}$. At $t = 6$, $r = \dfrac{1}{4}$; hence

$k = \dfrac{1}{24}$, and the solution is

$$r(t) = -\frac{1}{24}t + \frac{1}{2},$$

where t is measured in months and r in inches. Because we can't have a negative radius or time, $0 \le t \le 12$.

(b)   Solving $-\dfrac{1}{24}t + \dfrac{1}{2} = 0$ gives $t = 12$ months or one year.

■   **Suggested Journal Entry**

60.   Student Project

1.4   Euler's Method: Numerical Analysis

■   **Computer Help Advisable**

3. $y' = 3t^2 - y$, $y(0) = 1$; $[0, 1]$. Using a spreadsheet and Euler's method we obtain the following values:

<table>
<tr><td colspan="5" align="center">Spreadsheet Instructions for Euler's Method</td></tr>
<tr><td></td><td>A</td><td>B</td><td>C</td><td>D</td></tr>
<tr><td>1</td><td>n</td><td>t_n</td><td>$y_n = y_{n-1} + hy'_{n-1}$</td><td>$3t_n^2 - y_n$</td></tr>
<tr><td>2</td><td>0</td><td>0</td><td>1</td><td>$= 3*t*B2 \wedge 2 - C2$</td></tr>
<tr><td>3</td><td>$= A2 + 1$</td><td>$= B2 + .1$</td><td>$= C2 + .1*D2$</td><td></td></tr>
</table>

Using step size $h = 0.1$ and Euler's method we obtain the following results.

<table>
<tr><td colspan="4" align="center">Euler's Method $(h = 0.1)$</td></tr>
<tr><td>t</td><td>$y \approx$</td><td>t</td><td>$y \approx$</td></tr>
<tr><td>0</td><td>1</td><td>0.6</td><td>0.6822</td></tr>
<tr><td>0.1</td><td>0.9</td><td>0.7</td><td>0.7220</td></tr>
<tr><td>0.2</td><td>0.813</td><td>0.8</td><td>0.7968</td></tr>
<tr><td>0.3</td><td>0.7437</td><td>0.9</td><td>0.9091</td></tr>
<tr><td>0.4</td><td>0.6963</td><td>1.0</td><td>1.0612</td></tr>
<tr><td>0.5</td><td>0.6747</td><td></td><td></td></tr>
</table>

Smaller steps give higher approximate values $y_n(t_n)$. The DE is not separable so we have no exact solution for comparison.

6. $y' = t^2 - y^2$, $y(0) = 1$; $[0, 5]$

Using step size $h = 0.01$ and Euler's method we obtain following results. (Table shows only selected values.)

<table>
<tr><th colspan="4">Euler's Method $(h = 0.01)$</th></tr>
<tr><th>t</th><th>y</th><th>t</th><th>y</th></tr>
<tr><td>0</td><td>1</td><td>3</td><td>2.8143</td></tr>
<tr><td>0.5</td><td>0.6992</td><td>3.5</td><td>3.3464</td></tr>
<tr><td>1</td><td>0.7463</td><td>4</td><td>3.8682</td></tr>
<tr><td>1.5</td><td>1.1171</td><td>4.5</td><td>4.3843</td></tr>
<tr><td>2</td><td>1.6783</td><td>5</td><td>4.8967</td></tr>
<tr><td>2.5</td><td>2.2615</td><td></td><td></td></tr>
</table>

Smaller steps give higher approximate values $y_n(t_n)$. The DE is not separable so we have no exact solution for comparison.

9. $y' = \dfrac{\sin y}{t}$, $y(2) = 1$

Using step size $h = 0.05$ and Euler's method we obtain the following results. (Table shows only selected values.)

<table>
<tr><th colspan="4">Euler's Method $(h = 0.05)$</th></tr>
<tr><th>t</th><th>y ≈</th><th>t</th><th>y ≈</th></tr>
<tr><td>2</td><td>1</td><td>2.6</td><td>1.2366</td></tr>
<tr><td>2.1</td><td>1.0418</td><td>2.7</td><td>1.2727</td></tr>
<tr><td>2.2</td><td>1.0827</td><td>2.8</td><td>1.3079</td></tr>
<tr><td>2.3</td><td>1.1226</td><td>2.9</td><td>1.3421</td></tr>
<tr><td>2.4</td><td>1.1616</td><td>3</td><td>1.3755</td></tr>
<tr><td>2.5</td><td>1.1995</td><td></td><td></td></tr>
</table>

Smaller stepsize predicts lower value.

■   **Nasty Surprise**

12. $y' = y^2$, $y(0) = 1$

Using Euler's method with $h = 0.25$ we obtain the following values.

\multicolumn{3}{c}{**Euler's Method** $(h = 0.25)$}
t	$y \approx$	$y' = y^2$
0	**1**	1
0.25	1.25	1.5625
0.50	1.6406	2.6917
0.75	2.3135	5.3525
1.00	**3.6517**	

Euler's method estimates the solution at $t = 1$ to be 3.6517, whereas from the analytical solution $y(t) = \dfrac{1}{1-t}$, or from the direction field, we can see that the solution blows up at 1. So Euler's method gives an approximation far too small.

■   **Roundoff Problems**

15.   If a roundoff error of ε occurs in the initial condition, then the solution of the new IVP $y' = y$, $y(0) = A + \varepsilon$ is

$$y(t) = (A + \varepsilon)e^t = Ae^t + \varepsilon e^t.$$

The difference between this perturbed solution and Ae^t is εe^t. This difference at various intervals of time will be

$$t = 1 \Rightarrow \text{difference} = \varepsilon e$$
$$t = 10 \Rightarrow \text{difference} = \varepsilon e^{10} \approx 22{,}026\varepsilon$$
$$t = 20 \Rightarrow \text{difference} = \varepsilon e^{20} = 485{,}165{,}195\varepsilon.$$

Hence, the accumulate roundoff error grows at an exponential rate.

■   Runge-Kutta Method

18. $y' = t + y, \; y(0) = 0, \; h = -1$

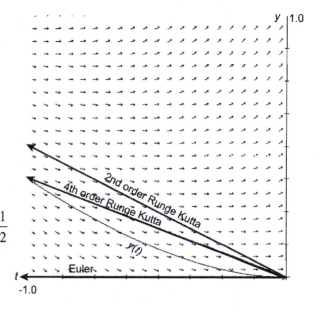

(a)   By Euler's method,

$$y_1 = y_0 + h(t_0 + y_0) = 0$$

By 2nd order Runge Kutta

$$y_1 = y_0 + hk_{02},$$

$$k_{01} = t_0 + y_0 = 0$$

$$k_{02} = \left(t_0 + \frac{h}{2}\right) + \left(y_0 + \frac{h}{2}k_{01}\right) = -\frac{1}{2}$$

$$y_1 = y_0 - 1\left(-\frac{1}{2}\right) = 0.5$$

By 4th order Runge Kutta.

$$y_1 = y_0 + \frac{h}{6}\left(k_{01} + 2k_{02} + 2k_{03} + k_{04}\right)$$

$$k_{01} = t_0 + y_0 = 0$$

$$k_{02} = \left(t_0 + \frac{h}{2}\right) + \left(y_0 + \frac{h}{2}k_{01}\right) = -\frac{1}{2} = -0.5$$

$$k_{03} = \left(t_0 + \frac{h}{2}\right) + \left(y_0 + \frac{h}{2}k_{02}\right) = -\frac{1}{2} + \left(-\frac{1}{2}\right)\left(-\frac{1}{2}\right) = -\frac{1}{4} = -0.25$$

$$k_{04} = (t_0 + h) + \left(y_0 + \frac{h}{2}k_{03}\right) = -1 + \left(-\frac{1}{2}\right)\left(-\frac{1}{4}\right) = -\frac{7}{8} = -0.875$$

$$y_1 = 0 + -\frac{1}{6}\left(0 + 2\left(-\frac{1}{2}\right) + 2\left(-\frac{1}{4}\right) + -\frac{7}{8}\right) = -\frac{1}{6}(-2.375) \approx 0.396$$

(b)   Second-order Runge Kutta is *high* though closer than Euler. Fourth order R-K is *very close*.

(c)   If $y(t) = -t - 1 + e^t,$

then $y(-1) = e^{-1} \approx 0.368.$

■ Runge-Kutta vs. Euler

21. $y' = -\dfrac{t}{y}$, $y(0) = 1$

Using the fourth-order Runge-Kutta method and $h = 0.1$ we arrive at the following table of values.

\multicolumn{4}{c}{**Runge-Kutta Method,** $y' = -\dfrac{t}{y}$, $y(0) = 1$}			
t	Y	t	y
0	1	0.6	0.8000
0.1	0.9950	0.7	0.7141
0.2	0.9798	0.8	0.6000
0.3	0.9539	0.9	0.4358
0.4	0.9165	1.0	0.04880
0.5	0.8660		

We compare this with #8 where Euler's method for step $h = 0.1$ gave $y(1) \approx 0.3994$, and the exact solution $y(t) = \sqrt{1 - t^2}$ gave $y(1) = 0$. The Runge-Kutta approximate solution is much closer to the exact solution.

■ Three-Term Taylor Series

24.　(a)　Starting with $y' = f(t, y)$, and differentiating with respect to t, we get

$$y'' = f_t(t, y) + f_y(t, y) y' = f_t(t, y) + f_y(t, y) f(t, y).$$

Hence, we have the new rule

$$y_{n+1} = y_n + h f(t_n, y_n) + \frac{1}{2} h^2 \left[f_t(t_n, y_n) + f_y(t_n, y_n) f(t_n, y_n) \right].$$

(b)　The local discretization error has order of the highest power of h in the remainder for the approximation of y_{n+1}, which in this case is 3.

(c)　For the equation $y' = f(t, y) = \dfrac{t}{y}$ we have $f_t(t, y) = \dfrac{1}{y}$, $f_y(t, y) = -\dfrac{t}{y^2}$ and so the preceding three-term Taylor series becomes

$$y_{n+1} = y_n + h \left(\frac{t_n}{y_n} \right) + \frac{1}{2} h^2 \left[\frac{1}{y_n} - \frac{t_n^2}{y_n^3} \right].$$

Using this formula and a spreadsheet we get the following results.

Taylor's Three-Term Series

Approximation of $y' = \dfrac{t}{y}$, $y(0) = 1$

t	y	t	y
0	**1**	0.6	1.1667
0.1	1.005	0.7	1.2213
0.2	1.0199	0.8	1.2314
0.3	1.0442	0.9	1.3262
0.4	1.0443	**1.0**	**1.4151**
0.5	1.1185		

The exact solution of the initial-value problem $y' = \dfrac{t}{y}$, $y(0) = 1$ is $y(t) = \sqrt{1 + t^2}$, so we have $y(1) = \sqrt{2} \approx 1.4142\ldots$. Taylor's three-term method gave the value 1.4151, which has an error of

$$\left| \sqrt{2} - 1.4151 \right| \approx 0.0009.$$

(d)   For the differential equation $y' = f(t, y) = ty$ we have $f_t(t, y) = y$, $f_y(t, y) = t$, so the Euler three-term approximation becomes

$$y_{n+1} = y_n + ht_n y_n + \frac{1}{2}h^2\left[y_n - t_n^2 y_n \right].$$

Using this formula and a spreadsheet, we arrive at the following results.

Taylor's Three-Term Series

Approximation of $y' = ty$, $y(0) = 1$

t	Y	t	y
0	**1**	0.6	1.1962
0.1	1.005	0.7	1.2761
0.2	1.0201	0.8	1.3749
0.3	1.0458	0.9	1.4962
0.4	1.1083	**1.0**	**1.6444**
0.5	1.1325		

The solution of $y' = ty$, $y(0) = 1$ is $y(t) = e^{t^2/2}$, so $y(1) = \sqrt{e} \approx 1.649\ldots$. Hence the error at $t = 1$ using Taylor's three-term method is

$$\left| \sqrt{e} - 1.6444 \right| \approx 0.0043.$$

■   **Richardson's Extrapolation**

Sharp eyes may have detected the elimination of absolute value signs when equation (7) is rewritten as equation (9). This is legitimate with no further argument if y' is positive and monotone increasing, as is the case in the suggested exercise.

27. $y' = y^2$, $y(0) = 1$.

Our calculations are listed in the following table (on the next page). Note that we use $y_R(0.1)$ as initial condition for computing $y_R(0.2)$.

	One-step Euler	Two-step Euler	Richardson approx. $y_R(t^*) =$	Exact solution
t^*	$y(t^*, h)$	$y(t^*, h)$	$2y(t^*, h) - y(t^*, h)$	$y = 1/(1-t)$
0.1	1.1	1.1051	**1.1102**	1.1111
0.2	1.2335	1.2405	**1.2476**	1.2500

■   **Computer Lab: Other Methods**

30.   **Sample study of different numerical methods.** We solve the IVP of Problem 5 $y' = \sqrt{t+y}$, $y(1) = 1$ by several different methods using step size $h = 0.1$. The table shows a printout for selected values of y using one non-Euler method.

	Fourth Order Runge-Kutta Method		
t	Y	t	y
1	1	3.5	6.8910
1.5	1.8100	4	8.5840
2	2.8144	4.5	10.4373
2.5	4.0010	5	12.4480
3	5.3618		

We can now compare the following approximations for Problem 5:

Euler's method $h = 0.1$ $y(5) \approx 12.2519$

(answer in text)

Euler's method $h = 0.01$ $y(5) \approx 12.4283$

(solution in manual)

Runge-Kutta method $h = 0.1$ $y(5) \approx 12.4480$

(above)

We have no exact solution for Problem 5, but you might use step $h = 0.1$ to approximate $y(5)$ by other methods (for example Adams-Bashforth method or Dormand-Prince method) then explain which method seems most accurate. A graph of the direction field could give insight.

1.5   Picard's Theorem: Theoretical Analysis

■   **Picard's Conditions**

3.   (a) $y' = y^{4/3}$, $y(0) = 0$

Here

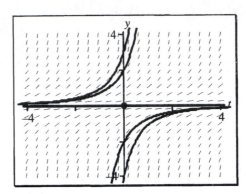

$$f(t, y) = y^{4/3}$$

$$f_y = \frac{4}{3} y^{1/3}.$$

Here f and f_y are continuous for all t and y, so by Picard's theorem we conclude that the DE has a unique solution through any initial condition $y(t_0) = y_0$. In particular, there will be a unique solution passing through $y(0) = 0$, which we know to be $y(t) \equiv 0$. The direction field of the equation is shown in the figure.

(b)   Picard's conditions hold in entire ty plane.

(c)   Not applicable - the answer to part (a) is positive.

6.   (a) $y' = \tan y$, $y(0) = \frac{\pi}{2}$

Here

$$f(t, y) = \tan y$$

$$f_y = \sec^2 y$$

are both continuous except at the points

$$y = \pm \frac{\pi}{2}, \ \pm \frac{3\pi}{2}, \dots \ .$$

Hence, there exists a unique solution passing through $y(t_0) = y_0$ except when

$$y = \pm \frac{\pi}{2}, \ \pm \frac{3\pi}{2}, \ \dots \ .$$

The IVP problem passing through $\frac{\pi}{2}$ does not have a solution. It would be useful to look at the direction field to get an idea of the behavior of solutions for nearby initial points. The direction field of the equation shows that where Picard's Theorem does not work the slope has become vertical (see figure).

(b)   Existence/uniqueness conditions are satisfied over any rectangle with y-values *between* two successive odd multiples of $\dfrac{\pi}{2}$.

(c)   There are no solutions going forward in time from any points near $\left(0, \dfrac{\pi}{2}\right)$.

■   **Linear Equations**

9. $y' + p(t)y = q(t)$

For the first-order linear equation, we can write $y' = q(t) - p(t)y$ and so

$$f(t,\ y) = q(t) - p(t)y$$
$$f_y(t,\ y) = -p(t).$$

Hence, if we assume $p(t)$ and $q(t)$ are continuous, then Picard's theorem holds at any point $y(t_0) = y_0$.

■   **Eyeballing the Flows**

For the following problems it appears from the figures given in the text that:

12.   Unique solutions exist passing through points *B* and *C* on intervals until the solution curve reaches the *t*-axis, where finite slope does not exist. Nonunique solutions at *A*; possibly unique solutions at *D* where $t = y = 0$.

15.   A unique solution will pass through each of the points *B*, *C*, and *D*. Solutions exist only for $t > t_A$ or $t < t_A$ because all solutions appear to leave from or go toward A, where there is no unique slope.

18.   A unique solution will pass through each of the points *A*, *B*, *C*, and *D*. Solutions appear to exist for all *t*.

■   **More Nonuniqueness**

21. $y' = \sqrt{y}$, $y(0) = 0$, $t_0 > 0$

For $t < t_0$, the solution is $y(t) \equiv 0$. For $t > t_0$, we have $y = \dfrac{1}{4}(t - t_0)^2$.

At $t = t_0$ the left-hand derivative of $y(t) \equiv 0$ is 0, and the right-hand derivative of $y(t) = \dfrac{1}{4}(t - t_0)^2$ is 0, so they agree.

■   **Hubbard's Leaky Bucket**

24. $\dfrac{dh}{dt} = -k\sqrt{h}$

(a) $f(t,\,h) = -k\sqrt{h}$, $\dfrac{\partial f}{\partial h} = -\dfrac{k}{2\sqrt{h}}$

Because $\dfrac{\partial f}{\partial h}$ is not continuous at $h = 0$, we cannot be sure of unique solutions passing through any points where $h(t) = 0$.

(b)   Let us assume the bucket becomes empty at $t = T < t_0$. Solving the IVP with $h(T) = 0$, we find an infinite number of solutions.

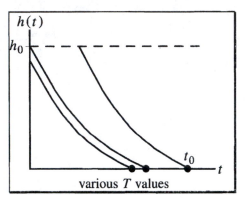

$$h(t) = \frac{1}{4}(kT - kt)^2 \quad \text{for } t < T$$
$$h(t) = 0 \quad\quad\quad\quad \text{for } t > T.$$

various T values

Each one of these functions describes the bucket emptying. Hence, we don't know when the bucket became empty. We show a few such solutions for $T < t_0$.

(c)   If we start with a full bucket when $t = 0$, then (b) gives

$$h(0) = \frac{1}{4}k^2 T^2 = h_0.$$

Hence the time to empty the bucket is

$$T = \frac{2}{k}\sqrt{h_0}\ .$$

■   **Different Translations**

27.   (a) $y' = y$ has an infinite family of solution of the form $y = Ce^t$.

(To check: $y' = (Ce^t)' = Ce^t = y$.

Note that for any real number a,
$y = e^{t-a} = Ce^t$ is a solution for every $a \in R$.

(b)   Differentiating $s(t) = \begin{cases} 0 & t < a \\ (t-a)^2 & t \geq a \end{cases}$

we obtain a continuous derivative

$$s'(t) = \begin{cases} 0 & t < a \\ 2(t-a) & t \geq a \end{cases}$$

Note that $s' = 2\sqrt{s}$ for both parts of the curve.

(c)

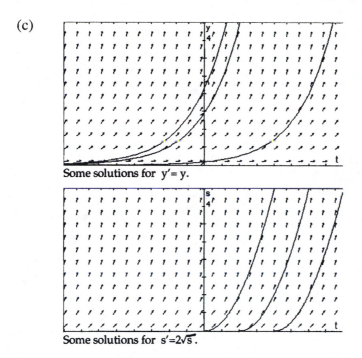

Some solutions for $y'=y$.

Some solutions for $s'=2\sqrt{s}$.

For (a), with $y' = y$, solutions $y = Ce^t$ gradually approach zero as $t \to -\infty$.

For (b), with $s' = 2\sqrt{s}$, solutions $y = \begin{cases} 0 & \text{for } t < a \\ (t-a)^2 & \text{for } t \geq a \end{cases}$ go to zero at $t \to a$.

■   **Picard Approximations**

30.

$$y_0(t) = e^{-t}$$

$$y_1(t) = 1 + \int_0^t (s - y_0)\,ds = 1 + \int_0^t (s - e^{-s})\,ds = e^{-t} + \frac{1}{2}t^2$$

$$y_2(t) = 1 + \int_0^t \left[s - \left(e^{-s} + \frac{1}{2}s^2 \right) \right]ds = e^{-t} - \frac{1}{6}t^3 + \frac{1}{2}t^2$$

$$y_3(t) = 1 + \int_0^t \left[s - \left(e^{-s} - \frac{1}{6}s^3 + \frac{1}{2}s^2 \right) \right]ds = e^{-t} + \frac{1}{24}t^4 - \frac{1}{6}t^3 + \frac{1}{2}t^2$$

■   **Calculator or Computer**

33. $y' = f(t, y) = y^{1/4}$

$\dfrac{\partial f}{\partial y} = \dfrac{1}{4} y^{-3/4}$

Note the direction field is only defined when $y \geq 0$. Picard's theorem guarantees existence through any point $y(t_0) = y_0$, but not uniqueness for points $y(t_0) = y_0$ when $y_0 = 0$. The direction field shown illustrates these ideas.

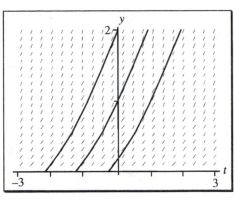

$y' = y^{1/4}$; DE does not exist for $y < 0$.

36. $y' = f(t, y) = (ty)^{1/3}$

$\dfrac{\partial f}{\partial y} = \dfrac{1}{3} (ty)^{-2/3} t = \dfrac{1}{3} t^{1/3} y^{-2/3}$

The function f is continuous for all (t, y), but f_y is not continuous when $y = 0$. Hence, we are not guaranteed uniqueness through points (t_0, y_0) when y_0 is zero. See figure for this direction field.

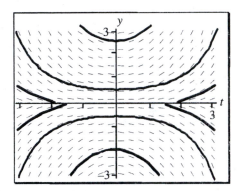

■   **Suggested Journal Entry**

39.   Student Project

2

Linearity and Nonlinearity

2.1   Linear Equations: The Nature of Their Solutions

■   **Classification**

3.   Second-order, linear, homogeneous, variable coefficients

6.   Third-order, linear, nonhomogeneous, constant coefficients

9.   Second-order, linear, homogeneous, variable coefficients

■   **Linear and Nonlinear Operations**

12. $L(y) = y' + 2y$

Suppose y_1, y_2 and y are functions of t and c is any constant. Then

$$L(y_1 + y_2) = (y_1 + y_2)' + 2(y_1 + y_2)$$
$$= (y_1' + 2y_1) + (y_2' + 2y_2)$$
$$= L(y_1 + y_2)$$
$$L(cy) = (cy)' + 2(cy) = cy' + 2cy$$
$$= c(y' + 2y) = cL(y)$$

Hence, L is a linear operator.

15. $L(y) = y' - e^t y$

Suppose y_1, y_2 and y are functions of t and c is any constant.

$$L(y_1 + y_2) = (y_1 + y_2)' - e^t (y_1 + y_2)$$
$$= (y_1' - e^t y_1) + (y_2' - e^t y_2)$$
$$= L(y_1) + L(y_2)$$
$$L(cy) = (cy)' - e^t (cy) = c(y' - e^t y)$$
$$= cL(y)$$

Hence, L is a linear operator. This problem illustrates the fact that a linear operator need not have coefficients that are linear functions of t.

■   **Pop Quiz**

18. $y' + 2y = 1 \Rightarrow y(t) = ce^{-2t} + \dfrac{1}{2}$ **21.** $y' - 3y = 5 \Rightarrow y(t) = ce^{3t} - \dfrac{5}{3}$

■   **Superposition Principle**

24.   If y_1 and y_2 are solutions of $y' + p(t) y = 0$, then

$$y_1' + p(t) y_1 = 0$$
$$y_2' + p(t) y_2 = 0.$$

Adding these equations gives

$$y_1' + y_2' + p(t) y_1 + p(t) y_2 = 0$$

or

$$(y_1 + y_2)' + p(t)(y_1 + y_2) = 0,$$

which shows that $y_1 + y_2$ is also a solution of the given equation.

If y_1 is a solution, we have

$$y_1' + p(t) y_1 = 0$$

and multiplying by c we get

$$c(y_1' + p(t) y_1) = 0$$
$$cy_1' + cp(t) y_1 = 0$$
$$(cy_1)' + p(t)(cy_1) = 0,$$

which shows that cy_1 is also a solution of the equation.

■   **Verifying Superposition**

27. $y'' + 4y = 0;$

For $y_1 = \sin 2t,$ $y_1' = 2\cos 2t$ $\Rightarrow$ $y_1'' = -4\sin 2t,$ so that $y_1'' + 4y_1 = -4\sin 2t + 4\sin 2t = 0.$

For $y_2 = \cos 2t,$ $y_2' = -2\sin 2t$ $\Rightarrow$ $y_2'' = -4\cos 2t,$ so that $y_2'' + 4y_2 = -4\cos 2t + 4\cos 2t = 0.$

Let $y_3 = c_1 \sin 2t + c_2 \cos 2t,$ then $y_3' = c_1 2\cos 2t + c_2(-2\sin 2t)$

$$y_3'' = c_1(-4\sin 2t) + c_2(-4\cos 2t)$$

Thus, $y_3'' + 4y_3 = c_1(-4\sin 2t) + c_2(-4\cos 2t) + 4(c_1 \sin 2t + c_2 \cos 2t) = 0.$

30. $y'' - y' - 6y = 0$

For $y_1 = e^{3t},$ $y_1' = 3e^{3t},$ $y_1'' = 9e^{3t}$

Substituting: $(9e^{3t}) - (3e^{3t}) - 6(e^{3t}) = 0$

For $y_2 = e^{-2t},$ $y_2' = -2e^{-2t},$ $y_2'' = 4e^{-2t}$

Substituting: $(4e^{-2t}) - (-2e^{-2t}) - 6e^{-2t} = 0$

For $y = c_1 e^{3t} + c_2 e^{-3t}$

$$y'' - y' - 6y = \left(c_1 9e^{3t} + c_2 4e^{-2t}\right) - \left(c_1 3e^{3t} - 2e^{-2t}\right) - 6\left(c_1 e^{3t} + c_2 e^{-3t}\right) = 0.$$

■   **Many from One**

33.   Because $y(t) = t^2$ is a solution of a linear homogeneous equation, we know by equation (3) that ct^2 is also a solution for any real number c.

■   **Guessing Solutions**

We can often find a particular solution of a nonhomogeneous DE by inspection (guessing). For the first-order equations given for Problems 34–37, and Problem 39 the general solutions come in two parts: solutions to the associated homogeneous equation (which could be found by separation of variables) plus a particular solution of the nonhomogeneous equation. For second-order linear equations such as Problems 38, and 40–42 we can also sometimes find solutions by inspection.

36. $y' - y = e^t \Rightarrow y_h = ce^t$ so that $y(t) = ce^t + te^t$

To check: $y' - y = (ce^t + te^t)' - (ce^t + te^t) = ce^t + te^t + e^t - ce^t - te^t = e^t$

39. $y' + \dfrac{y}{t} = t^3$

To find y_h: $y' = \dfrac{-y}{t}$ $y \neq 0$

$\displaystyle\int \dfrac{dy}{y} = -\int \dfrac{dt}{t}$ (separation of variables)

$\ln|y| = -\ln|t| + c_1$

$|y| = e^{-\ln|t| + c_1} = \left|\dfrac{1}{t}\right| e^{c_1}$

$y_h = \dfrac{c}{t}$

To find y_p: We guess that $y_p = \dfrac{t^4}{K}$ from the observation that this function would make the right

side of the DE be a function t^3 to match the left side of the DE.

$y_p' + \dfrac{y_p}{t} = \dfrac{4t^3}{K} + \dfrac{t^3}{K} = \dfrac{5t^3}{K} = t^3$, so $K = 5$, and $y_p = \dfrac{t^4}{5}$

Thus $y = y_h + y_p = \dfrac{c}{t} + \dfrac{t^4}{5}$

To check: $y' + \dfrac{y}{t} = \left(\dfrac{c}{t} + \dfrac{t^4}{5}\right)' + \dfrac{\dfrac{c}{t} + \dfrac{t^4}{5}}{5}$

$= -\dfrac{c}{t^2} + \dfrac{4t^3}{5} + \dfrac{c}{t^2} + \dfrac{t^3}{5} = t^3$

42. $y'' - y' = 0 \Rightarrow y_h = y(t) = c_1 + c_2 e^t$

■   Nonhomogeneous Principle

In these problems, the verification of y_p is a straightforward substitution. To find the rest of the solution we simply add to y_p all the homogeneous solutions y_h, which we find by inspection or separation of variables.

45. $y' - \dfrac{2}{t}y = y^2$ has general solution $y(t) = y_h + y_p = ct^2 + t^3$.

■   **Third-Order Examples**

48. $y''' + y'' - y' - y = 4\sin t + 3$

(a)   For $y_1 = e^t$, we obtain by substitution

$$y''' + y'' - y' - y = e^t + e^t - e^t - e^t = 0.$$

For $y = e^{-t}$, we obtain by substitution

$$y''' + y'' - y' - y = -e^{-t} + (e^{-t}) - (-e^{-t}) - e^{-t} = 0.$$

For $y = te^{-t}$ we obtain by substitution

$$y''' + y'' - y' - y = (-te^{-t} + 3e^{-t}) + (te^{-t} - 2e^{-t}) - (-te^{-t} + e^{-t}) - te^{-t} = 0.$$

(b) $y_h = c_1 e^t + c_2 e^{-t} + c_3 te^{-t}$

(c)   Given $y_p = \cos t - \sin t - 3$:

$$y_p' = -\sin t - \cos t$$

$$y_p'' = -\cos t + \sin t$$

$$y_p''' = \sin t + \cos t$$

To verify:

$$y_p''' + y_p'' - y_p' - y_p = (\sin t + \cos t) + (-\cos t + \sin t) - (-\sin t - \cos t) - (\cos t - \sin t - 3)$$
$$= 4\sin t + 3$$

(d) $y(t) = y_h + y_p = c_1 e^t + c_2 e^{-t} + c_3 te^{-t} + \cos t - \sin t - 3$

(e) $y' = c_1 e^t - c_2 e^{-t} + c_3(-te^{-t} + e^{-t}) - \sin t - \cos t$

$y'' = c_1 e^t + c_2 e^{-t} + c_3(te^{-t} - 2e^{-t}) - \cos t + \sin t$

$y(0)$	$= 1 = c_1 + c_2$	$+ 1 - 3$	$\Rightarrow$	$c_1 + c_2 = 3$	Equation (1)
$y'(0)$	$= 2 = c_1 + c_2 - c_3 - 1$		$\Rightarrow$	$c_1 - c_2 + c_3 = 3$	Equation (2)
$y''(0)$	$= 3 = c_1 + c_2 - 2c_3 - 1$		$\Rightarrow$	$c_1 + c_2 - 2c_3 = 4$	Equation (3)

Add Equation (2) to (1) and (3)

$$\left.\begin{array}{l} 2c_1 + c_3 = 6 \\ 2c_1 - c_3 = 7 \end{array}\right\} \Rightarrow c_1 = \frac{13}{4}, \; c_2 = -\frac{1}{4}, \; c_3 = -\frac{1}{2}.$$

$$y(t) = \frac{13}{4}e^t - \frac{1}{4}e^{-t} - \frac{1}{2}te^{-t} + \cos t - \sin t - 3.$$

2.2   Solving the First-Order Linear Differential Equation

■   **General Solutions**

The solutions for Problems 3–15 can be found using either the Euler-Lagrange method or the integrating factor method. For problems where we find a particular solution by inspection (Problem 6) we use the Euler-Lagrange method. For the other problems we find it more convenient to use the integrating factor method, which gives both the homogeneous solutions and a particular solution in one swoop. You can use the Euler-Lagrange method to get the same results.

3. $y' - y = 3e^t$

We multiply each side of the equation by the integrating factor

$$\mu(t) = e^{\int p(t)dt} = e^{\int(-1)dt} = e^{-t}$$

giving

$$e^{-t}(y' - y) = 3, \text{ or simply } \frac{d}{dt}\left(ye^{-t}\right) = 3.$$

Integrating, we find $ye^{-t} = 3t + c$,   or $y(t) = ce^t + 3te^t$.

6. $y' + 2ty = t$

In this problem we see that $y_p(t) = \dfrac{1}{2}$ is a solution of the nonhomogeneous equation (there are other single solutions, but this is the easiest to find). Hence, to find the general solution we solve the corresponding homogeneous equation, $y' + 2ty = 0$,

by separation of variables, getting

$$\frac{dy}{y} = -2tdt,$$

which has the general solution $y = ce^{-t^2}$, where c is any constant.

Adding the solutions of the homogeneous equation to the particular solution $y_p = \dfrac{1}{2}$ we get the general solution of the nonhomogeneous equation:

$$y(t) = ce^{-t^2} + \frac{1}{2}.$$

9. $ty' + y = 2t$

We rewrite the equation as $y' + \dfrac{1}{t}y = 2$, and multiply each side of the equation by the integrating

factor $\mu(t) = e^{\int dt/t} = e^{\ln t} = t$,

giving

$$t\left(y' + \frac{1}{t}y\right) = 2t, \text{ or, } \frac{d}{dt}(ty) = 2t.$$

Integrating, we find $ty = t^2 + c$.

Solving for y, we get $y(t) = \dfrac{c}{t} + t$.

12. $y' + \dfrac{3}{t}y = \dfrac{\sin t}{t^3}, \ (t \neq 0)$

We multiply each side of the equation by the integrating factor

$$\mu(t) = e^{\int (3/t)dt} = e^{3\ln t} = e^{\ln\left(t^3\right)} = t^3,$$

giving

$$t^3\left(y' + \frac{3}{t}y\right) = \sin t, \text{ or, } \frac{d}{dt}(t^3 y) = \sin t.$$

Integrating, we find $t^3 y = -\cos t + c$.

Solving for y, we get $y(t) = \dfrac{c}{t^3} - \dfrac{1}{t^3}\cos t$.

15. $y' + \left(\dfrac{2t+1}{t}\right)y = 2t, \ (t \neq 0)$

We multiply each side of the equation by the integrating factor

$$\mu(t) = \int \frac{2t+1}{t}dt = \int 2 + \frac{1}{t}dt = te^{2t},$$

giving

$$\frac{d}{dt}(te^{2t}y) = 2t^2 e^{2t}.$$

Integrating, we find $te^{2t}y = t^2 e^{2t} - te^{2t} + \dfrac{1}{2}e^{2t} + c$.

Solving for y, we have $y(t) = c\left(\dfrac{e^{-2t}}{t}\right) + \dfrac{1}{2t} + t - 1$.

■   **Initial-Value Problems**

18. $y' - \left(\dfrac{3}{t}\right)y = t^3$, $y(1) = 4$

We find the integrating factor to be

$$\mu(t) = e^{-\int (3/t)dt} = e^{-3\ln t} = e^{\ln t^{-3}} = t^{-3}.$$

Multiplying the DE by this, we get

$$\frac{d}{dt}\left(t^{-3}y\right) = 1.$$

Hence, $t^{-3}y = t + c$, or, $y(t) = ct^3 + t^4$.

Substituting $y(1) = 4$ gives $c + 1 = 4$ or $c = 3$. Hence, the solution of the IVP is

$$y(t) = 3t^3 + t^4.$$

■   **Synthesizing Facts**

21.   (a) $y(t) = \dfrac{t^2 + 2t}{t+1}$, $(t > -1)$

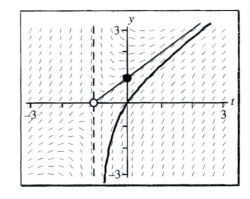

(b) $y(t) = t + 1$, $(t > -1)$

(c)   The algebraic solution given in Example 1 for $k = 1$ is

$$y(t) = \frac{t^2 + 2t + 1}{t+1} = \frac{(t+1)^2}{t+1}.$$

Hence, when $t \neq -1$ we have $y = t + 1$.

(d)   The solution passing through the origin $(0, 0)$ asymptotically approaches the line $y = t + 1$ as $t \to \infty$, which is the solution passing through $y(0) = 1$. The entire line $y = t + 1$ is *not* a solution of the DE, as the slope is not defined when $t = -1$. The segment of the line $y = t + 1$ for $t > -1$ is the solution passing through $y(0) = 1$.

On the other hand, if the initial condition were $y(-5) = -4$, then the solution would be the segment of the line $y = t + 1$ for t less than -1. Notice in the direction field the slope element is not defined at $(-1, 0)$.

■   **Using Integrating Factors**

In each of the following equations, we first write in the form $y' + p(t)y = f(t)$ and then identify $p(t)$.

24. $y' - y = e^{3t}$

Here $p(t) = -1$, therefore the integrating factor is $\mu(t) = e^{\int p(t)dt} = e^{-\int dt} = e^{-t}$.

Multiplying each side of the equation $y' - y = e^{3t}$ by e^{-t} yields

$$\frac{d}{dt}\left(ye^{-t}\right) = e^{2t}.$$

Integrating gives $ye^{-t} = \frac{1}{2}e^{2t} + c$. Solving for y gives $y(t) = ce^{t} + \frac{1}{2}e^{3t}$.

27. $y' + 2ty = t$

Here $p(t) = 2t$, therefore the integrating factor is $\mu(t) = e^{\int p(t)dt} = e^{\int 2t\,dt} = e^{t^2}$.

Multiplying each side of the equation $y' + 2ty = t$ by e^{t^2} yields

$$\frac{d}{dt}\left(ye^{t^2}\right) = te^{t^2}.$$

Integrating gives $ye^{t^2} = \frac{1}{2}e^{t^2} + c$. Solving for y gives $y(t) = ce^{-t^2} + \frac{1}{2}$.

30. $ty' + y = 2t$

Here $p(t) = \frac{1}{t}$, therefore the integrating factor is $\mu(t) = e^{\int p(t)dt} = e^{\int (1/t)dt} = e^{\ln t} = t$.

Multiplying each side of the equation $y' + \frac{y}{t} = 2$ by t yields

$$\frac{d}{dt}(ty) = 2t.$$

Integrating gives $ty = t^2 + c$. Solving for y gives $y(t) = c\frac{1}{t} + t$.

■   **A Useful Transformation**

33.   (a)   Letting $z = \ln y$, we have $y = e^z$ and $\dfrac{dy}{dt} = e^z \dfrac{dz}{dt}$.

Now the equation $\dfrac{dy}{dt} + ay = by \ln y$ can be rewritten as $e^z \dfrac{dz}{dt} + ae^z = bze^z$.

Dividing by e^z gives the simple linear equation $\dfrac{dz}{dt} - bz = -a$.

Solving yields $z = ce^{bt} + \dfrac{a}{b}$ and using $z = \ln y$, the solution becomes $y(t) = e^{(a/b) + ce^{bt}}$.

(b)   If $a = b = 1$, we have $y(t) = e^{\left(1 + ce^t\right)}$.

Note that when $c = 0$ we have the constant solution $y = e$.

■   **Bernoulli Practice**

36. $y' - y = e^t y^2$, so that $y^{-2}y' - y^{-1} = e^t$

Let $v = y^{-1}$, so $\dfrac{dv}{dt} = -y^{-2} \dfrac{dy}{dt}$.

Substituting in the DE gives $\dfrac{dv}{dt} + v = -e^t$, which is linear in v with integrating factor $\mu = e^{\int dt} = e^t$.

Thus $e^t \dfrac{dv}{dt} + e^t v = -e^{2t}$, and $e^t v = -\int e^{2t} dt = -\dfrac{e^{2t}}{2} + c$, so $v = -\dfrac{e^t}{2} + ce^{-t}$.

Substituting back for v gives $y^{-1} = -\dfrac{e^t}{2} + c_1 e^{-t}$,   or $y(t) = \dfrac{2}{-e^t + c_1 e^{-t}}$.

39. $y' + \dfrac{y}{t} = \dfrac{y^{-2}}{t}$ $y(1) = 2$

$y^2 y' + \dfrac{y^3}{t} = \dfrac{1}{t}$

Let $v = y^3$, so $\dfrac{dv}{dt} = 3y^2 \dfrac{dy}{dt}$.

Substituting in the DE gives $\dfrac{1}{3} \dfrac{dv}{dt} + \dfrac{1}{t} v = \dfrac{1}{t}$, or $\dfrac{dv}{dt} + \dfrac{3}{t} v = \dfrac{3}{t}$,

which is linear in v, with integrating factors $\mu = e^{\int 3/t \, dt} = e^{3 \ln t} = t^3$.

Thus, $t^3 \dfrac{dv}{dt} + 3t^2 v = 3t^2$, and $t^3 v = \int 3t^2 dt = t^3 + c$, so $v = 1 + ct^{-3}$.

Substituting back for v gives $y^3 = 1 + ct^{-3}$ or $y(t) = \sqrt[3]{1 + ct^{-3}}$.

For the IVP we substitute the initial condition $y(1) = 2$, which gives $2^3 = 1 + c$, so $c = 7$.

Thus, $y^3 = 1 + 7t^{-3}$ and $y(t) = \sqrt[3]{1 + 7t^{-3}}$.

■   **Computer Visuals**

42.   (a) $y' + 2y = t$

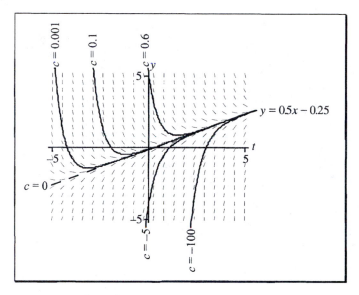

(b) $y_h(t) = ce^{-2t}$, $y_p = \dfrac{1}{2}t - \dfrac{1}{4}$

The general solution is

$$y(t) = y_h + y_p = ce^{-2t} + \frac{1}{2}t - \frac{1}{4}.$$

The curves in the figure in part (a) are labeled for different values of c.

(c)   The homogeneous solution y_h is transient because $y_h \to 0$ as $t \to \infty$. However, although all solutions are attracted to y_p, we would *not* call y_p a steady-state solution because it is neither constant nor periodic; $y_p \to \infty$ as $t \to \infty$.

45.   (a) $y' + y = \sin 2t$

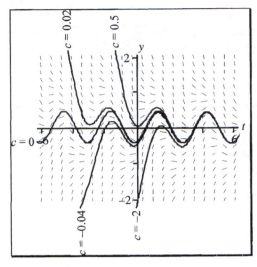

(b) $y_h(t) = ce^{-t}$, $y_p = \dfrac{1}{5}(\sin 2t - 2\cos 2t)$.

The general solution is $y(t) = \underbrace{ce^{-t}}_{y_h} + \underbrace{\dfrac{\sin 2t - 2\cos 2t}{5}}_{y_p}$.

(c)   The steady-state solution is y_p, which attracts all other solutions.

The transient solution is y_h.

■   **Computer Numerics**

48. $y' + 2y = t$, $y(0) = 1$

(a)   Using step size $h = 0.1$ and $h = 0.01$ and Euler's method, we compute the following values. In the latter case we print only selected values.

			Euler's Method			
t	$y(h = 0.1)$	$y(h = 0.01)$	T	$y(h = 0.1)$	$y(h = 0.01)$
0	1	1.0000	0.6	0.3777	0.4219
0.1	0.8	0.8213	0.7	0.3621	0.4039
0.2	0.65	0.6845	0.8	0.3597	0.3983
0.3	0.540	0.5819	0.9	0.3678	0.4029
0.4	0.4620	0.5071	1	0.3842	0.4158
0.5	0.4096	0.4552			

By Runge-Kutta (RK4) we obtain $y(1) \approx 0.4192$ for step size $h = 0.1$.

(b)   From Problem 42, we found the general solution of DE to be $y(t) = ce^{-2t} + \dfrac{1}{2}t - \dfrac{1}{4}$.

Using IC $y(0) = 1$ yields $c = \dfrac{5}{4}$. The solution of the IVP is $y(t) = \dfrac{5}{4}e^{-2t} + \dfrac{1}{2}t - \dfrac{1}{4}$,

and to 4 places, we have

$$y(1) = \frac{5}{4}e^{-2} + \frac{1}{2} - \frac{1}{4} \approx 0.4192 .$$

(c)   The error for $y(1)$ using step size $h = 0.1$ in Euler's approximation is

$$\text{ERROR} = 0.4192 - 0.3842 = 0.035$$

Using step size $h = 0.01$, Euler's method gives

$$\text{ERROR} = 0.4192 - 0.4158 = 0.0034 ,$$

which is much smaller. For step size $h = 0.1$, Runge-Kutta gives $y(1) = 0.4158$ and *zero* error to four decimal places.

(d)   The accuracy of Euler's method can be greatly improved by using a smaller step size. The Runge-Kutta method is more accurate for a given step size in most cases.

■   **Direction Field Detective**

51.   (a)   (A) is linear homogeneous, (B) is linear nonhomogeneous, (C) is nonlinear.

(b)   If y_1 and y_2 are solutions of a linear homogeneous equation,

$$y' + p(t)y = 0 ,$$

then $y_1' + p(t)y_1 = 0$, and $y_2' + p(t)y_2 = 0$. We can add these equations, to get

$$\left(y_1' + p(t)y_1 \right) + \left(y_2' + p(t)y_2 \right) = 0 .$$

Because this equation can be written in the equivalent form

$$\left(y_1 + y_2 \right)' + p(t)\left(y_1 + y_2 \right) = 0 ,$$

then $y_1 + y_2$ is also a solution of the given equation.

(c)   The sum of any two solutions follows the direction field only in (A). For the linear homogeneous equation (A) you plot any two solutions y_1 and y_2 by simply following curves in the direction field, and then add these curves, you will see that the sum $y_1 + y_2$ also follows the direction field.

continued on the next page

However, in equation (B) you can observe a straight line solution, which is

$$y_1 = \frac{1}{2}t - \frac{1}{4}.$$

If you add this to itself you get $y_1 + y_1 = 2y_1 = t - \frac{1}{2}$, which clearly does not follow the direction field and hence is not a solution. In equation (C) $y_1 = 1$ is a solution but if you add it to itself you can see from the direction field that $y_1 + y_2 = 2$ is not a solution.

2.3   Growth and Decay Phenomena

■   **Interpretation of $\dfrac{1}{k}$**

3.   If we examine the value of the decay curve

$$y(t) = y_0 e^{kt}$$

we find

$$y\left(\left|\frac{1}{k}\right|\right) = y_0 e^{k(-1/k)} = y_0 e^{-1} = y_0 (0.3678794\ldots)$$

$$\approx \frac{y_0}{3}.$$

Hence,

$$\left|\frac{1}{k}\right|$$

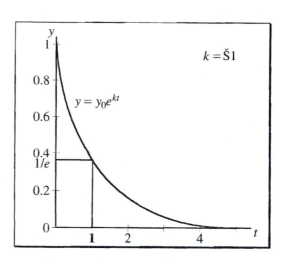

$y_0 e^{-t}$ falls from y_0 to roughly $\dfrac{y_0}{3}$

when $t = -\dfrac{1}{k}$

is a crude approximation of the third-life of a decay curve. In other words, if a substance decays and has a decay constant $k = -0.02$, and time is measured in years, then the third-life of the substance is roughly $\dfrac{1}{0.02} = 50$ years. That is, every 50 years the substance decays by $\dfrac{2}{3}$. Note that the curve in the figure falls to $\dfrac{1}{3}$ of its value in approximately $t = 1$ unit of time.

■   **Thorium-234**

6.   (a)   The general decay curve is $Q(t) = ce^{kt}$. With the initial condition $Q(0) = 1$, we have $Q(t) = e^{kt}$. We also are given $Q(1) = 0.8$ so $e^k = 0.8$, or $k = \ln(0.8) \approx -0.22$. Hence, we have

$$Q(t) = e^{-0.22t}$$

where Q is measured in grams and t is measured in weeks.

(b) $t_h = -\dfrac{\ln 2}{k} = \dfrac{\ln 2}{0.22} \approx 3.1$ weeks   (c) $Q(10) e^{-0.22(10)} \approx 0.107$ grams

■   **Radium Decay**

9.   6400 years is 4 half-lives, so that $\dfrac{1}{2^4} \approx 6.25\%$ will be present.

■   **Bombarding Plutonium**

12.   We are given $k = -\dfrac{\ln 2}{0.15} \approx -4.6209812$. The differential equation for the amount present is

$$\frac{dA}{dt} = kA + 0.00002, \; A(0) = 0.$$

Solving this initial-value problem we get the particular solution

$$A(t) = ce^{kt} - \frac{0.00002}{k}$$

where $c = \dfrac{0.00002}{k} \approx -0.000004$. Substituting in these values gives the total amount

$$A(t) \approx 0.000004\left(1 - e^{-4.6t}\right)$$

measured in micrograms.

■   **Sodium Pentathol Elimination**

15.   The half-life is 10 hours. The decay constant is

$$k = -\frac{\ln 2}{10} \approx -0.069.$$

Ed needs

$$(50 \text{ mg/kg})(100 \text{ kg}) = 5000 \text{ mg}$$

of pentathal to be anesthetized. This is the minimal amount that can be presented in his bloodstream after three hours. Hence,

$$A(3) = A_0 e^{-0.069(3)} \approx 0.813 A_0 = 5000.$$

Solving for A_0 yields $A_0 = 6{,}155.7$ milligrams or an initial dose of 6.16 grams.

■   **Extrapolating the Past**

18.   If P_0 is the initial number of bacteria present (in millions), then we are given $P_0 e^{6k} = 5$ and $P_0 e^{9k} = 8$. Dividing one equation by the other we obtain $e^{3k} = \dfrac{8}{5}$, from which we find

$$k = \frac{\ln \frac{8}{5}}{3}.$$

Substituting this value into the first equation gives $P_0 e^{2\ln(8/5)} = 5$, in which we can solve for

$$P_0 = 5e^{-2\ln(8/5)} \approx 1.95 \text{ million bacteria.}$$

▪ Growth of Tuberculosis Bacteria

21.   We are given the initial number of cells present is $P_0 = 100$, and that $P(1) = 150$ (1.5 times larger), then $100\ e^{k(1)} = 150$, which yields $k = \ln\dfrac{3}{2}$. Therefore, the population $P(t)$ at any time t is

$$P(t) = 100e^{t\ln(3/2)} \approx 100e^{0.405t} \text{ cells.}$$

▪ Rule of 70

24.   The doubling time is found in Problem 2 to be

$$t_d = \frac{\ln 2}{r} \approx \frac{0.70}{r} = \frac{70}{100r}$$

where $100r$ is an annual interest rate (expressed as a percentage). The rule of 70 makes sense.

▪ Compound Interest Thwarts Hollywood Stunt

27.   The growth rate is $A(t) = A_0 e^{rt}$. In this case $A_0 = 3$, $r = 0.08$, and $t = 320$. Thus, the total bottles of whiskey will be

$$A(320) = 3e^{(0.08)(320)} \approx 393,600,000,000.$$

That's 393.6 *billion* bottles of whiskey!

▪ Living Off Your Money

30. $A(t) = A_0 e^{rt} - \dfrac{d}{r}\left(e^{rt} - 1\right)$. Setting $A(t) = 0$ and solving for t gives

$$t = \frac{1}{r}\ln\left(\frac{d}{d - rA_0}\right)$$

Notice that when $d = rA_0$ this equation is undefined, as we have division by 0; if $d < rA_0$, this equation is undefined because we have a negative logarithm. For the physical translation of these facts, you must return to the equation for $A(t)$. If $d = rA_0$, you are only withdrawing the interest, and the amount of money in the bank remains constant. If $d < rA_0$, then you aren't even withdrawing the interest, and the amount in the bank increases and $A(t)$ never equals zero.

■   **Continuous Compounding**

33.   (a)   After one year compounded continuously the value of the account will be

$$S(1) = S_0 e^r .$$

With $r = 0.08$ (8%) interest rate, we have the value

$$S(1) = S_0 e^{0.08} \approx \$1.083287 S_0 .$$

This is equivalent to a single annual compounding at a rate $r_{eff} = 8.329\,\%$.

(b)   If we set the annual yield from a single compounding with interest r_{eff}, $S_0(1 + r_{eff})$ equal to the annual yield from continuous compounding with interest r, $S_0 e^r$, we have

$$S_0(1 + r_{eff}) = S_0 e^r .$$

Solving for r_{eff} yields $r_{eff} = e^r - 1$.

(c) $r_{daily} = \left(1 + \dfrac{0.08}{365}\right)^{365} - 1 = 0.0832775$ (i.e., 8.328%) effective annual interest rate, which is extremely close to that achieved by continuous compounding as shown in part (a).

■   **Mortgaging a House**

36.   (a)   Since the bank earns 1% monthly interest on the outstanding principle of the loan, and Kelly's group make monthly payments of \$2500 to the bank, the amount of money $A(t)$ still owed the bank at time t, where t is measured in months starting from when the loan was made, is given by the savings equation (10) with $a = -2500$.

Thus, we have

$$\frac{dA}{dt} = 0.01A - 2500, \quad A(0) = \$200,000.$$

(b)   The solution of the savings equation in (a) was seen (11) to be

$$A(t) = A(0)e^{rt} + \frac{a}{r}(e^{rt} - 1)$$

$$= 200,000e^{0.01t} - \frac{2500}{0.01}(e^{0.01t} - 1)$$

$$= -50,000e^{0.01t} + \$250,000.$$

(c)   To find the length of time for the loan to be paid off, we set $A(t) = 0$, and solve for t. Doing this, we have

$$-50{,}000e^{0.01t} = -\$250{,}000.$$

or

$$0.01t = \ln 5 \text{ or } t = 100 \ln 5 \approx 100(1.609) \approx 161 \text{ months (13 years and 5 months)}.$$

2.4   Linear Models: Mixing and Cooling

■   **Metric Brine**

3.   (a)   The salt inflow is

$$(0.1 \text{ kg/liter})(4 \text{ liters/min}) = 0.4 \text{ kg/min} .$$

The outflow is $\dfrac{4}{100}Q$ kg/min. Thus, the differential equation for the amount of salt is

$$\frac{dQ}{dt} = 0.4 - 0.04Q .$$

Solving this equation with the given initial condition $Q(0) = 50$ gives

$$Q(t) = 10 + 40e^{-0.04t} .$$

(b)   The concentration $conc(t)$ of salt is simply the amount $Q(t)$ divided by the volume (which is constant at 100). Hence the concentration at time t is given by

$$conc(t) = \frac{Q(t)}{100} = 0.1 - 0.4e^{-0.04t} .$$

(c)   As $t \to \infty$, $e^{-0.04t} \to 0$. Hence $Q(t) \to 10$ kg of salt in the tank.

(d)   Either take the limiting amount and divide by 100 or take the limit as $t \to \infty$ of $conc(t)$. The answer is 0.1 kg/liter in either case.

■   **Salty Overflow**

6.   Let x = amount of salt in tank at time t.

We have $\dfrac{dx}{dt} = \dfrac{1 \text{ lb}}{\text{gal}} \cdot \dfrac{3 \text{ gal}}{\text{min}} - \dfrac{(x \text{ lb}) \cdot 1 \text{ gal/min}}{(300 + (3-1)t)\text{gal}}$, with initial volume = 300 gal, capacity = 600 gal.

IVP: $\dfrac{dx}{dt} = 3 - \dfrac{x}{300 + 2t}$, $x(0) = 0$

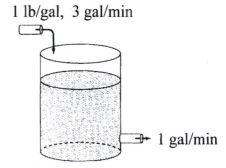

1 lb/gal, 3 gal/min

1 gal/min

The DE is linear,

$$\frac{dx}{dt} + \frac{x}{300 + 2t} = 3 ,$$

with integrating factor

$$\mu = e^{\int \frac{1}{300+2t}dt} = e^{\frac{1}{2}\ln(300+2t)} = (300 + 2t)^{1/2}$$

Thus,

$$(300+2t)^{1/2}\frac{dx}{dt}+\frac{x}{(300+2t)^{1/2}}=3(300+2t)^{1/2}$$

$$(300+2t)^{1/2}x=\int 3(300+2t)^{1/2}\,dt$$

$$=\left(\frac{3}{2}\right)\frac{(300+2t)^{3/2}}{3/2}+c,$$

so

$$x(t)=(300+2t)+c(300+2t)^{-1/2}$$

The initial condition $x(0) = 0$ implies $0 = 300 + \dfrac{c}{\sqrt{300}}$, so $c = -3000\sqrt{3}$.

The solution to the IVP is

$$x(t) = 300 + 2t - 3000\sqrt{3}(300+2t)^{-1/2}$$

The tank will be full when $300 + 2t = 600$, so $t = 150$ min.

At that time, $x(150) = 300 + 2(150) - 3000\sqrt{3}(300+2(150))^{-1/2} \approx 388$ lbs

■ **Changing Midstream**

9.   Let x = amount of salt in tank at time t.

(a)   IVP: $\dfrac{dx}{dt}=\dfrac{1\text{ lb}}{\text{gal}}\cdot\dfrac{4\text{ gal}}{\text{sec}}-\left(\dfrac{x\text{ lb}}{200\text{ gal}}\right)\dfrac{4\text{ gal}}{\text{sec}}$ $x(0)=0$

(b) $x_{eq}=\dfrac{1\text{ lb}}{\text{gal}}\cdot 200\text{ gal}=200\text{ lb}$ 1 lb/gal, 4 gal/sec

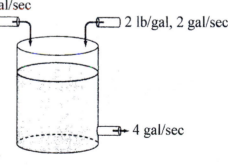

(c)   Now let x = amount of salt in tank at time t,

        but reset $t = 0$ to be when the second faucet

        is turned on. This setup gives

$$\frac{dx}{dt}=\frac{4\text{ lb}}{\text{sec}}+\frac{2\text{ lb}}{\text{gal}}\cdot\frac{2\text{ gal}}{\text{sec}}-\frac{x\text{ lb}\cdot 4\text{ gal/sec}}{(200+2t)\text{gal}},$$

        which gives a new IVP:

$$\frac{dx}{dt}=8-\frac{4x}{200+2t}\qquad x(0)=x_{eq}=200$$

(d)   To find t_f: $200 + 2t_f = 1000$ $t_f = 400$ sec

(e)   The DE in the new IVP is

$\dfrac{dx}{dt} + \dfrac{2x}{100+t} = 8$, which is linear with integrating factor

$$\mu = e^{\int \frac{2}{100+t}\,dt} = e^{\ln(100+t)^2} = (100+t)^2.$$

Thus, $(100+t)^2 \dfrac{dx}{dt} + 2(100+t)x = 8(100+t)^2,$ and

$$(100+t)^2 x = \int 8(100+t)^2\,dt = \dfrac{8}{3}(100+t)^3 + c,$$

so

$$x(t) = \dfrac{8}{3}(100+t) + c(100+t)^{-2}.$$

The initial condition $x(0) = 200$ implies $200 = \dfrac{8}{3}(100) + \dfrac{c}{(100)^2}$ or $c = -\dfrac{2}{3}\times10^6.$

Thus the solution to the new IVP is

$$x = \dfrac{8}{3}(100+t) - \dfrac{1}{3}(2\times10^6)(100+t)^{-2}.$$

When $t_f = 400$, $x(400) = \dfrac{8}{3}(500) - \dfrac{1}{3}\dfrac{(2\times10^6)}{(500)^2} \approx 1330.7$ lb.

(f)   After tank starts to overflow,

Inflow: $\underbrace{\dfrac{1\text{ lb}}{\text{gal}}\cdot\dfrac{4\text{ gal}}{\text{sec}}}_{1^{\text{st}}\text{ faucet}} + \underbrace{\dfrac{2\text{ lb}}{\text{gal}}\cdot\dfrac{2\text{ gal}}{\text{sec}}}_{2^{\text{nd}}\text{ faucet}} = \dfrac{8\text{ lbs}}{\text{sec}}$

Outflow: $\left(\underbrace{\dfrac{4\text{ gal}}{\text{sec}}}_{\text{drain}} + \underbrace{\dfrac{2\text{ gal}}{\text{sec}}\right)\cdot\dfrac{x\text{ lb}}{1000\text{ gal}}}_{\text{overflow}} = \dfrac{6x}{1000}\dfrac{\text{lbs}}{\text{sec}}$

Hence for $t > 400$ sec, the IVP now becomes

$$\dfrac{dx}{dt} = 8 - \dfrac{6x}{1000},\qquad x(400) = 1330.7 \text{ lb.}$$

■   **Three Tank Setup**

12.   Let x, y, and z be the amounts of salt in Tanks 1, 2, and 3 respectively.

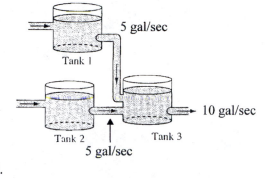

(a)   For Tank 1: $\dfrac{dx}{dt} = \dfrac{0 \text{ lbs}}{\text{gal}} \cdot \dfrac{5 \text{ gal}}{\text{sec}} - \dfrac{x \text{ lbs}}{200 \text{ gal}} \cdot \dfrac{5 \text{ gal}}{\text{sec}}$,

so the IVP for $x(t)$ is

$$\frac{dx}{dt} = \frac{-5x}{200}, \qquad x(0) = 20.$$

The IVP for the identical Tank 2 is

$$\frac{dy}{dt} = \frac{-5y}{200}, \qquad y(0) = 20.$$

(b)   For Tank 1, $\dfrac{dx}{dt} = \dfrac{-x}{40}$,   so $x = 20e^{-t/40}$.

For Tank 2, $\dfrac{dy}{dt} = \dfrac{-y}{40}$,   so $y = 20e^{-t/40}$.

(c) $\dfrac{dz}{dt} = \dfrac{x}{40} + \dfrac{y}{40} - \dfrac{z \text{ lbs}}{500 \text{ gal}} \cdot \dfrac{10 \text{ gal}}{\text{sec}}$

$\qquad = \dfrac{1}{2} e^{-t/40} + \dfrac{1}{2} e^{-t/40} - \dfrac{z}{50}.$

Again we have a linear equation, $\dfrac{dz}{dt} + \dfrac{z}{50} = e^{-t/40}$,

with integrating factor $\mu = e^{\int 1/50 dt} = e^{t/50}$.

Thus

$$e^{t/50} \frac{dz}{dt} + \frac{1}{50} e^{t/50} z = e^{-t/40 + t/50} = e^{-t/200},$$

$$e^{5/50} z = \int e^{-t/200} dt = -200 e^{-t/200} + c,$$

so $z(t) = -200 e^{-t/40} + c e^{-t/50}$.

■   **Using the Time Constant**

15.   (a) $T(t) = T_0 e^{-kt} + M(1 - e^{-kt})$, from Equation (8). In this case, $M = 95$, $T_0 = 75$, and

$k = \dfrac{1}{4}$, yielding the expression

$$T(t) = 75e^{-t/4} + 95(1 - e^{-t/4})$$

where t is time measured in hours. Substituting $t = 2$ in this case (2 hours after noon), yields $T(2) \approx 82.9\,°\text{F}$.

(b)   Setting $T(t) = 80$ and simplifying for $T(t)$ yields

$$t = -4\ln\frac{3}{4} \approx 1.15 \text{ hours,}$$

which translates to 1:09 P.M.

■   **Warm or Cold Beer?**

18.   We use

$$T(t) = M + (T_0 - M)e^{-kt}.$$

In this case, $M = 70$, $T_0 = 35$. If we measure t in minutes, we have $T(10) = 40$, giving

$$40 = 70 - 35e^{-10k}.$$

Solving for the decay constant k, we find

$$k = -\frac{\ln\left(\frac{6}{7}\right)}{10} \approx 0.0154.$$

Thus, the equation for the temperature after t minutes is

$$T(t) \approx 70 - 35e^{-0.0154t}.$$

Substituting $t = 20$ gives $T(20) \approx 44.3\,°\,\text{F}.$

■   **Case of the Cooling Corpse**

21.   (a) $T(t) = T_0 e^{-kt} + M(1 - e^{-kt})$. We know that $M = 50$ and $T_0 = 98.6\,°\text{F}$. The first

measurement takes place at unknown time t_1 so

$$T(t_1) = 70 = 50 + 48.6e^{-kt_1}$$

or $48.6e^{-kt_1} = 20$. The second measurement is taken two hours later at $t_1 + 2$, yielding

$$60 = 50 + 48.6e^{-k(t_1 + 2)}$$

or $48.6e^{-k(t_1+2)} = 10$. Dividing the second equation by the first equation gives the relationship $e^{-2k} = \dfrac{1}{2}$ from which $k = \dfrac{\ln 2}{2}$. Using this value for k the equation for $T(t_1)$ gives

$$70 = 50 + 48.6e^{-t_1 \ln 2/2}$$

from which we find $t_1 \approx 2.6$ hours. Thus, the person was killed approximately 2 hours and 36 minutes before 8 P.M., or at 5:24 P.M.

(b)   Following exactly the same steps as in part (a) but with $T_0 = 98.2°$ F, the sequence of equations is

$$T(t_1) = 70 = 50 + 48.2e^{-k(t_1)} \Rightarrow 48.2e^{-kt_1} = 20.$$

$$T(t_1 + 2) = 60 = 50 + 48.2e^{-k(t_1+2)} \Rightarrow 48.2e^{-k(t_1+2)} = 10.$$

Dividing the second equation by the first still gives the relationship $e^{-2k} = \dfrac{1}{2}$,

so $k = \dfrac{\ln 2}{2}$.

Now we have

$$T(t_1) = 70 = 50 + 48.2e^{-t_1 \ln 2/2}$$

which gives $t_1 \approx 2.54$ hours, or 2 hours and 32 minutes. This estimates the time of the murder at 5.28 PM, only 4 minutes earlier than calculated in part (a).

■   **Computer Mixing**

24. $y' + \dfrac{1}{1+t} y = 2$, $y(0) = 0$

When the inflow is greater than the outflow, the amount of dissolved substance keeps growing without end.

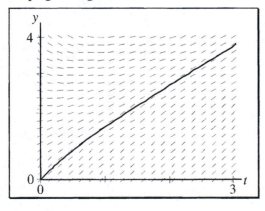

2.5   Nonlinear Models: Logistic Equation

■   **Equilibria**

Note: Problems 1–6 are all autonomous equations, so lines of constant slope (isoclines) are horizontal lines.

3. $y' = -ay + by^2$, $(a > 0,\ b > 0)$

We find the equilibrium points by solving

$$y' = -ay + by^2 = 0,$$

getting $y = 0$, $\dfrac{a}{b}$. By inspecting

$$y' = y(-a + by),$$

we see that solutions have positive slope when $y < 0$ or $y > \dfrac{a}{b}$ and negative slope for $0 < y < \dfrac{a}{b}$.

Hence, the equilibrium solution $y(t) \equiv 0$ is stable, and the equilibrium solution $y(t) \equiv \dfrac{a}{b}$ is unstable.

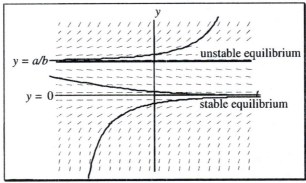

6. $y' = y - \sqrt{y}$

Setting $y' = 0$ we find equilibrium points at

$$y = 0 \text{ and } 1.$$

The equilibrium at $y = 0$ is stable; that at $y = 1$ is unstable. Note also that the DE is only defined when $y \geq 0$.

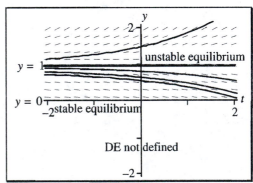

■   Nonautonomous Sketching

For nonautonomous equations, the lines of constant slope are not horizontal lines as they were in the autonomous equations in Problems 3–6.

9. $y' = \sin(yt)$

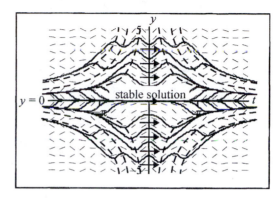

Isoclines of horizontal slopes (dashed) are hyperbolas $yt = \pm n\pi$ for $n = 0, 1, 2, \dots$. On the computer drawn graph you can sketch the hyperbolas for isoclines and verify the alternating occurrence of positive and negative slopes between them as specified by the DE.

Only $y \equiv 0$ is an equilibrium (unstable for $t < 0$, stable for $t > 0$).

■   Inflection Points

12. $y' = \cos(y - t)$

We differentiate y' with respect to t (using the chain rule), and then substitute for $\dfrac{dy}{dt}$ from the DE. This gives

$$\frac{d^2 y}{dt^2} = \frac{d}{dt}\left(\frac{dy}{dt}\right) = -\sin(y-t)\frac{dy}{dt} = -\sin(y-t)\cos(y-t).$$

Setting $\dfrac{d^2 y}{dt^2} = 0$ and solving for y yields $y - t = n\pi$, $y - t = \dfrac{n\pi}{2} + n\pi$ for $n = 0, \pm 1, \pm 2, \dots$.

Note the inflection points change with t in this nonautonomous case. See text Figure 2.5.3, graph for (2), to see that the inflection points occur only when $y = -1$, so they lie along the lines $y = t + m\pi$ where m is an odd integer.

■   Culture Growth

15.   Let y = population at time t, so $y(0) = 1000$ and $L = 100,000$.

The DE solution, from equation (10), is

$$y = \frac{100,000}{1 + \left(\dfrac{100,000}{1000} - 1\right)e^{-rt}}.$$

continued on the next page

To evaluate r, substitute the given fact that when $t = 1$, population has doubled.

$$y(1) = 2(1000) = \frac{100,000}{1 + (100 - 1)e^{-r}}$$

$$2(1 + 99e^{-r}) = 100$$

$$198e^{-r} = 98$$

$$e^{-r} = \frac{98}{198}$$

$$-r = \ln\left(\frac{98}{198}\right)$$

$$r = .703$$

Thus $y(t) = \dfrac{100,000}{1 + 99e^{-.703t}}$.

(a)   After 5 days: $y(5) = \dfrac{100,000}{1 + 99e^{-(.703)5}} = 25,348$ cells

(b)   When $y = 50,000$, find t:

$$50,000 = \frac{100,000}{1 + 99e^{-.703t}}$$

$$1 + 99e^{-.703t} = 2$$

$$t \approx 6.536 \text{ days}$$

■   **Water Rumor**

18.   Let N be the number of people who have heard the rumor at time t

(a) $\dfrac{dN}{dt} = kN(200,000 - N) = 200,000k\left(1 - \dfrac{N}{200,000}\right)N$

(b)   Yes, this is a logistic model.

(c)   Set $\dfrac{dN}{dt} = 0$. Equilibrium solutions: $N = 0$, $N = 200,000$.

(d)   Let $r = 200,000k$. Assume $N(0) = 1$.

Then

$$N = \frac{200,000}{1 + \left(\dfrac{200,000}{1} - 1\right)e^{-rt}}$$

At $t = 1$ week,

$$1000 = \frac{200,000}{1+199,999e^{-r}}$$

$$1+199,999e^{-r} = 200$$

$$e^{-r} = \frac{199}{199,999} \quad \Rightarrow \quad r - 6.913.$$

Thus

$$N(t) = \frac{200,000}{1+199,999e^{-6.913t}}.$$

To find t when $N = 100,000$:

$$100,000 = \frac{200,000}{1+199,999e^{-6.913t}}$$

$$\Downarrow$$

$$1+199,999e^{-6.913t} = 2, \qquad e^{-6.913t} = \frac{1}{199,999}, \text{ and } t = 1.77 \text{ weeks} = 12.4 \text{ days}.$$

(e)   We assume the same population. Let $t_N > 0$ be the time the article is published.

Let P = number of people who are aware of the counterrumor.

Let P_0 be the number of people who became aware of the counterrumor at time t_N.

$$\frac{dP}{dt} = aP(200,000 - P) \qquad\qquad P(t_N) = P_0, \text{ and } a \text{ is a constant of proportionality.}$$

■ Fitting the Gompertz Law

21.   (a)   From Problem 20,

$$y(t) = e^{a/b}e^{ce^{-bt}}$$

where $c = \ln y_0 - \dfrac{a}{b}$. In this case $y(0) = 1$, $y(2) = 2$. We note $y(24) \approx y(28) \approx 10$

means the limiting value $e^{a/b}$ has been reached. Thus

$$e^{a/b} = 10,$$

so

$$\frac{a}{b} = \ln 10 \approx 2.3.$$

continued on the next page

The constant $c = \ln 1 - \dfrac{a}{b} = 0 - 2.3 = -2.3$. Hence,

$$y(t) = 10e^{-2.3e^{-bt}}$$

and

$$y(2) = 10e^{-2.3e^{-2b}} = 2.$$

Solving for b:

$$-2.3e^{-2b} = \ln \frac{2}{10} \approx -1.609$$

$$e^{-2b} \approx -\frac{1.609}{2.3} \approx 0.6998$$

$$-2b = \ln(0.6998) \approx -0.357$$

$$b \approx 0.1785$$

and $a = 2.3b$ gives $a \approx 0.4105$.

(b)   The logistic equation

$$y' = ry\left(1 - \frac{y}{L}\right)$$

has solution

$$y(t) = \frac{L}{1 + \left(\dfrac{L}{y_0} - 1\right)e^{-rt}}.$$

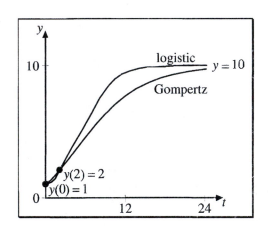

We have $L = 10$ and $y_0 = 1$, so

$$y(t) = \frac{10}{1 + 9e^{-rt}}$$

and

$$y(2) = \frac{10}{1 + 9e^{-2r}} = 2.$$

Solving for r

$$9e^{-2r} = \frac{10}{2} - 1 = 4$$

$$-2r = \ln \frac{4}{9} \approx -0.8109$$

$$r = \frac{-0.8109}{-2} \approx 0.405.$$

■ **Autonomous Analysis**

24. (a) $y' = -y\left(1 - \dfrac{y}{L}\right)\left(1 - \dfrac{y}{M}\right)$, $y' = -y(1-y)(1-0.5y)$

(b) The equilibrium points are $y = 0$, L, M. $y = 0$ is stable. $y = M$ is stable if $M > L$ and unstable if $M < L$. $y = L$ is stable if $M < L$ and unstable if $M > L$.

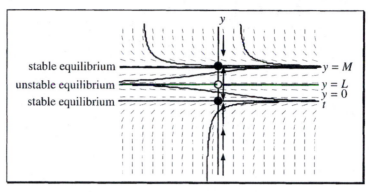

27. (a) $y' = y^2\left(4 - y^2\right)$

(b) The equilibrium solution $y(t) \equiv 2$ is stable, the solution $y(t) \equiv -2$ is unstable and the solution $y(t) \equiv 0$ is semistable.

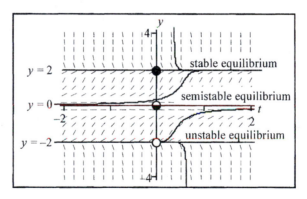

■ **Useful Transformation**

30. $y' = ky(1-y)$

Letting $z = \dfrac{y}{1-y}$ yields

$$\frac{dz}{dt} = \left(\frac{dz}{dy}\right)\left(\frac{dy}{dt}\right) = \frac{1}{(1-y)^2}\left(\frac{dy}{dt}\right).$$

Substituting for $\dfrac{dy}{dt}$ from the original DE yields a new equation

$$(1-y)^2 \frac{dz}{dt} = ky(1-y),$$

continued on the next page

which gives the result

$$\frac{dz}{dt} = \frac{ky}{1-y} \doteq kz.$$

Solving this first-order equation for $z = z(t)$, yields $z(t) = ce^{kt}$ and substituting this in the transformation $z = \frac{y}{1-y}$, we get $\frac{y}{1-y} = ce^{kt}$.

Finally, solving this for y gives $y(t) = \frac{1}{1 + \frac{1}{c}e^{-kt}} = \frac{1}{1 + c_1 e^{-kt}}$,

where $c_1 = \frac{1}{y_0} - 1$.

■ **Solving the Threshold Equation**

33. $y' = -ry\left(1 - \frac{y}{T}\right)$

Introducing backwards time $\tau = -t$, yields

$$\frac{dy}{dt} = \frac{dy}{d\tau}\frac{d\tau}{dt} = -\frac{dy}{d\tau}.$$

Hence, if we run the threshold equation

$$\frac{dy}{dt} = -ry\left(1 - \frac{y}{T}\right)$$

backwards, we get

$$-\frac{dy}{d\tau} = -ry\left(1 - \frac{y}{T}\right).$$

Equivalently it also yields the first-order equation

$$\frac{dy}{d\tau} = ry\left(1 - \frac{y}{T}\right),$$

which is the logistic equation with $L = T$ and $t = \tau$. We know the solution of this logistic equation to be

$$y(\tau) = \frac{T}{1 + \left(\frac{T}{y_0} - 1\right)e^{-r\tau}}.$$

We can now find the solution of the threshold equation by replacing τ by $-t$, yielding

$$y(t) = \frac{T}{1 + \left(\frac{T}{y_0} - 1\right)e^{rt}}.$$

■   **Another Bifurcation**

36. $y' = y^2 + by + 1$

(a)   We find the equilibrium points of the equation by setting $y' = 0$ and solving for y. Doing this we get

$$y = \frac{-b \pm \sqrt{b^2 - 4}}{2}.$$

We see that for $-2 < b < 2$ there are no (real) solutions, and thus no equilibrium solutions. For $b = -2$ we have the equilibrium solution $+1$, and for $b = +2$ we have equilibrium solution -1. For each $|b| \geq 2$ we have two equilibrium solutions.

(b)   The bifurcation points are at $b = -2$ and $b = +2$. As b passes through -2 (increasing), the number of equilibrium solutions changes from 2 to 1 to 0, and when b passes through $+2$, the number of equilibrium solutions changes from 0 to 1 to 2.

(c)   We have drawn some solution for each of the values $b = -3$, -2, -1, 0, 1, 2, and 3.

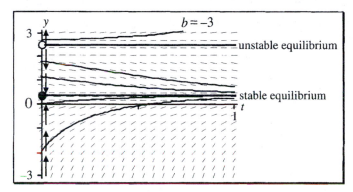

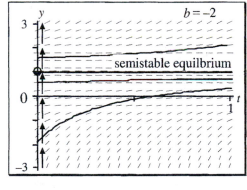
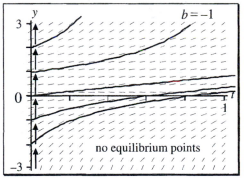

continued on the next page

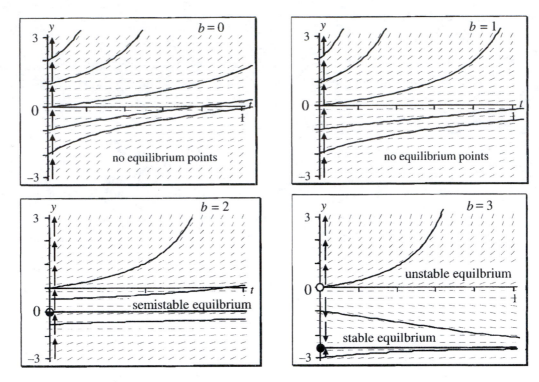

(d)   For $b = 2$ and $b = -2$ the single equilibrium is semistable. (Solutions above are repelled; those below are attracted.) For $|b| > 2$ there are two equilibria; the larger one is unstable and the smaller one is stable. For $|b| < 2$ there are *no* equilibria.

(e)   The bifurcation diagram shows the location of equilibrium points for y versus the parameter value b. Solid circles represent stable equilibria; open circles represent unstable equilibria.

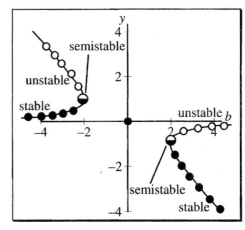

Equilibria of $y' = y^2 + by + 1$ versus b

Computer Lab: Growth Equations

39.
$$y' = ry\left(1 - \frac{y}{L}\right)$$

We graph the direction field of this equation for $L=1$, $r=0.5$, 1, 2, and 5. We keep L fixed because all it does is raise or lower the steady-state solution to $y=L$. We see that the larger the parameter r, the faster the solution approaches the steady-state L.

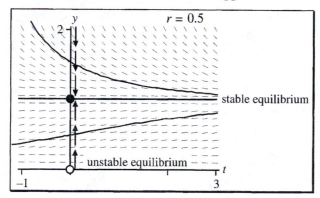

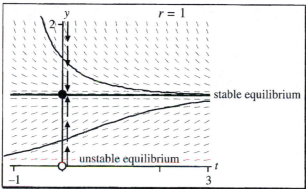

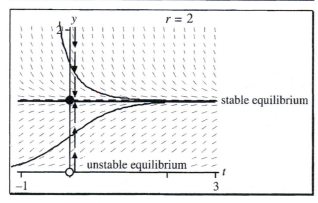

continued on next page

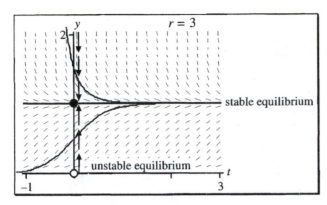

42. $y' = re^{-\beta t} y$

For larger β or for larger r, solution curves fall more steeply. Unstable equilibrium $r = 1$, $\beta = 1$

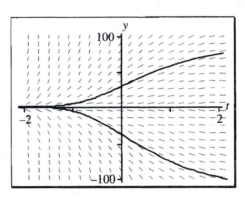

2.6   Systems of DEs: A First Look

■   **Predicting System Behavior**

3.   (a) $x' = 1 - x - y$

$y' = 1 - x^2 - y^2$

Setting $x' = 0$ and $y' = 0$ gives

h-nullcline $x^2 + y^2 = 1$

v-nullcline $x + y = 1.$

From the intersection of the two nullclines we find two equilibrium points $(0, 1)$, $(1, 0)$.

(b) 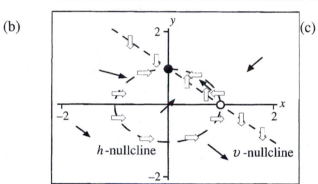   (c)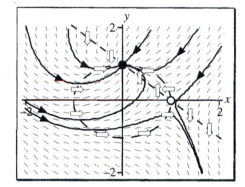

(d)   The equilibrium at $(1, 0)$ is unstable; the equilibrium at $(0, 1)$ is stable. Most trajectories seem to be attracted to the stable equilibrium, but those that approach the lower unstable equilibrium from below or form the right will turn down toward the lower right.

6.   (a) $x' = y$

$y' = 5x + 3y$

This (linear) system has one equilibrium point at $(x, y) = (0, 0)$ as do all linear systems. The 64-dollar question is: Is it stable? The v- and h-nullclines: $y = 0$, $5x + 3y = 0$, are shown following and indicate that the origin $(0, 0)$ is unstable. Hence, points starting near the origin will leave the origin. We will see later other ways for showing that $(0, 0)$ is unstable.

(b)

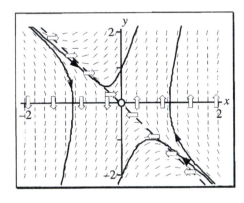

(c)   The direction field and a few solutions are drawn. Note how the solutions cross the vertical and horizontal nullclines

(d)   We see from the preceding figure that solutions come from infinity along a line (that is, not a nulllcline), and then if they are not *exactly* on the line head off either upwards and to the left or downwards and go to the left on another line. Whether they go up or down depends on whether they initially start above or below the line. It appears that points that start exactly on the line will go to $(0, 0)$. We will see later in Chapter 6 when we study linear systems using eigenvalues and eigenvectors that the solutions come from infinity on one eigenvector and go to infinity on another eigenvector.

■   Creating a Predator-Prey Model

9.   (a) $$\frac{dR}{dt} = 0.15R - 0.00015RF$$

$$\frac{dF}{dt} = -0.25F + 0.00003125RF$$

The rabbits reproduce at a natural rate of 15%; their population is diminished by meetings with foxes. The fox population is diminishing at a rate of 25%; this decline is mitigated only slightly by meeting rabbits as prey. Comparing the predator-prey rates in the two populations shows a much larger effect on the rabbit population, which is consistent with the fact that each fox needs several rabbits to survive.

(b) $\quad \dfrac{dR}{dt} = 0.15R - 0.00015RF - 0.1R = 0.05R - 0.00015RF$

$\dfrac{dF}{dt} = -0.25F + 0.00003125RF - 0.1F = -0.35F + 0.00003125RF$

Both populations are diminished by the harvesting. The equilibrium populations move from (8000, 1000) in Part (a) to (11200, 333) in Part (b), i.e., more rabbits and fewer foxes if both populations are harvested at the same rate. Figures on the next page.

In figures, x and y are measured in thousands. Note that the vertical axes have different scales from the horizontal axes.

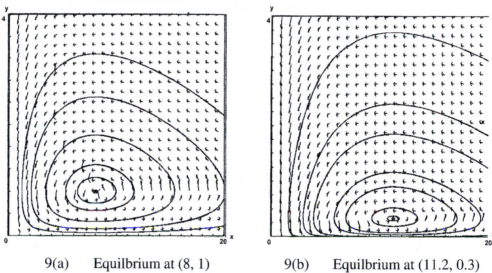

9(a)   Equilbrium at (8, 1)   9(b)   Equilbrium at (11.2, 0.3)

Analyzing Competition Models

12. $\quad \dfrac{dR}{dt} = R(1200 - 3R - 2S)$

$\dfrac{dS}{dt} = S(500 - R - S)$

The explanations of the equations are the same as those in Problem 11 except that the rabbit population is affected more by the crowding of its own population, less by the number of sheep.

Equilibria occur at $\begin{bmatrix} 0 \\ 0 \end{bmatrix}, \begin{bmatrix} 0 \\ 500 \end{bmatrix}, \begin{bmatrix} 400 \\ 0 \end{bmatrix}$, or $\begin{bmatrix} 200 \\ 300 \end{bmatrix}$.

continued on the next page

In this system the equilibria on the axes are all unstable, so the populations always head toward a coexistence equilibrium at $\begin{bmatrix} 200 \\ 300 \end{bmatrix}$. See Figure, where x and y are measured in hundreds.

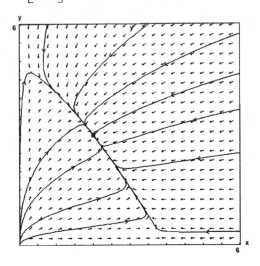

■   Finding the Model

An example of an appropriate model is as follows, with real positive coefficients.

15. $x' = ax - bx^2 - cxy - dxz$

$y' = ey - fy^2 + gxy$

$z' = -hz + kxz$

■   Competition

18.   (a) $x' = x(1 - x - y)$

$y' = y(2 - x - y)$

Setting $x' = 0$ and $y' = 0$ we find

v-nullclines $x + y = 1$, $x = 0$

h-nullclines $x + y = 2$, $y = 0$.

Equilibrium points: $(0, 0)$, $(0, 2)$, $(1, 0)$.
The directions of the solution curves are
shown in the figure.

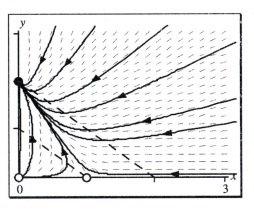

(b)   It can be seen from the figure, that the equilibrium points $(0, 0)$ and $(1, 0)$ are unstable; the point $(0, 2)$ is stable because all solution curves nearby point toward it.

(c)   Some solution curves are shown in the figure.

(d)   Because all the solution curves eventually reach the stable equilibrium at $(0, 2)$, the x species always die out and the two species described by this model cannot coexist.

■   **Simpler Competition**

21. $x' = x(a - by)$

$y' = y(c - dx)$

Setting $x' = 0$, we find the v-nullclines are the vertical line $x = 0$ and the horizontal line $y = \dfrac{a}{b}$.

Setting $y' = 0$, we find the h-nullclines are the horizontal $y = 0$ and vertical line $x = \dfrac{c}{d}$. The

equilibrium points are $(0, 0)$ and $\left(\dfrac{c}{d}, \dfrac{a}{b} \right)$. By observing the signs of x', y' we find

$$x' > 0, \; y' > 0 \; \text{ when } x < \frac{c}{d}, \; y < \frac{a}{b}$$

$$x' < 0, \; y' < 0 \; \text{ when } x > \frac{c}{d}, \; y > \frac{a}{b}$$

$$x' < 0, \; y' > 0 \; \text{ when } x > \frac{c}{d}, \; y < \frac{a}{b}$$

$$x' > 0, \; y' < 0 \; \text{ when } x < \frac{c}{d}, \; y > \frac{a}{b}$$

Hence, both equilibrium points are unstable. We can see from the following direction field (with $a = b = c = d = 1$) that one of two species, depending on the initial conditions, goes to infinity and the other toward extinction.

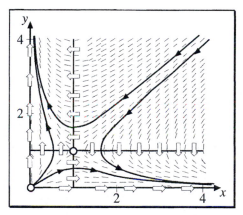

One can get the initial values for these curves directly from the graph.

■   **Nullcline Patterns**

24.   (a–e)   When the two nullclines intersect as they do in the figure, then there are four equilibrium points in the first quadrant: $(0, 0)$, $\left(\dfrac{a}{b}, 0\right)$, $\left(0, \dfrac{d}{f}\right)$, and (x_e, y_e), where (x_e, y_e) is the intersection of the lines $bx + cy = a$, $ex + fy = d$. Analyzing the sign of the derivatives in the four regions of the first quadrant, we find that (x_e, y_e) is stable and the others unstable. Hence, the two populations can coexist.

(a)–(b) 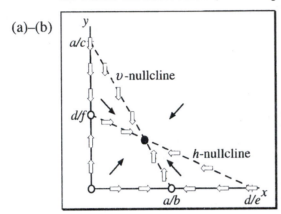   (c)–(d)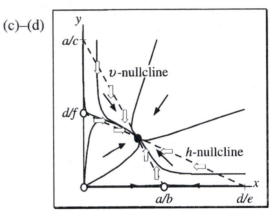

Nullclines and equilibria

Typical trajectories when the nullclines intersect and the slope of the vertical nullcline is more negative.

■   **Basins of Attraction**

27.   Adding shading to the graph obtained in Problem 2 shows the basis of the stable equilibrium at

$$\left(\frac{1}{4}\left(1 - \sqrt{5}\right)^2, \frac{1}{2}\left(\sqrt{5} - 1\right)\right) \approx (0.38, 0.60).$$

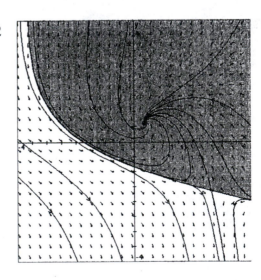

30.   The graph obtained in Problem 21 has *no* stable equilibrium, but we say that there are three basins:

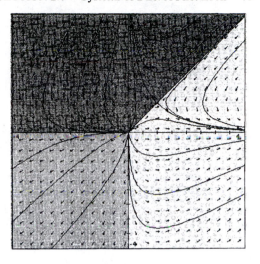

For $x > y$ and $y > 0$, trajectories are attracted to $(0, \infty)$.

For $x > 0$ and $x < y$, trajectories are attracted to $(\infty, 0)$.

For $x < 0$ and $y < 0$, trajectories are attracted to $(-\infty, -\infty)$.

■   **Suggested Journal Entry**

33.   Student Project

CHAPTER 3 | Linear Algebra

3.1 Matrices: Sums and Products

■ Do They Compute?

3. $2\mathbf{C} - \mathbf{D}$, Matrices are not compatible

6. $\mathbf{CD} = \begin{bmatrix} 3 & -1 & 0 \\ 8 & -1 & 2 \\ 9 & 2 & 6 \end{bmatrix}$

9. $\mathbf{C}^\mathrm{T}\mathbf{D}$, Matrices are not compatible

12. $\mathbf{AD}$, Matrices are not compatible

15. $\mathbf{C} - \mathbf{I}_3$, Matrices are not compatible

■ More Multiplication Practice

18. $\begin{bmatrix} a & b \\ c & d \end{bmatrix}\begin{bmatrix} d & -b \\ -c & a \end{bmatrix} = \begin{bmatrix} ad - bc & -ab + ba \\ cd - dc & -db + da \end{bmatrix} = \begin{bmatrix} ad - bc & 0 \\ 0 & ad - bc \end{bmatrix}$

21. $\begin{bmatrix} 0 & 1 \end{bmatrix}\begin{bmatrix} a & b & c \\ d & e & f \end{bmatrix}\begin{bmatrix} 1 & 1 & 0 \end{bmatrix} = \begin{bmatrix} d & e & f \end{bmatrix}\begin{bmatrix} 1 & 1 & 0 \end{bmatrix}$ not possible

■ Which Rules Work for Matrix Multiplication?
24. **Counterexample:**

$\mathbf{A} = \begin{bmatrix} 1 & 1 \\ 1 & 0 \end{bmatrix} \qquad \mathbf{B} = \begin{bmatrix} 2 & -1 \\ 0 & 1 \end{bmatrix} \qquad (\mathbf{A} + \mathbf{B})(\mathbf{A} - \mathbf{B}) = \begin{bmatrix} 3 & 0 \\ 1 & 1 \end{bmatrix}\begin{bmatrix} -1 & 2 \\ 1 & -1 \end{bmatrix} = \begin{bmatrix} -3 & 6 \\ 0 & 1 \end{bmatrix}$

$\mathbf{A}^2 - \mathbf{B}^2 = \begin{bmatrix} 2 & 1 \\ 1 & 1 \end{bmatrix} - \begin{bmatrix} 4 & -3 \\ 0 & 1 \end{bmatrix} = \begin{bmatrix} -2 & 4 \\ 1 & 0 \end{bmatrix}$

27. **Proof** $(\mathbf{A} + \mathbf{B})^2 = (\mathbf{A} + \mathbf{B})(\mathbf{A} + \mathbf{B}) = \mathbf{A}(\mathbf{A} + \mathbf{B}) + \mathbf{B}(\mathbf{A} + \mathbf{B})$ distributive property
              $= \mathbf{A}^2 + \mathbf{AB} + \mathbf{BA} + \mathbf{B}^2$ distributive property

■ Find the Matrix

30.   Set $\begin{bmatrix} 1 & 2 \\ 4 & 1 \end{bmatrix}\begin{bmatrix} a & b \\ c & d \end{bmatrix} = \begin{bmatrix} 2 & 0 \\ 1 & 4 \end{bmatrix}$

$$a + 2c = 2 \qquad b + 2d = 0$$
$$4a + c = 1 \qquad 4b + d = 4$$

Thus $\quad c = 1, a = 0 \quad b = \dfrac{8}{7}, d = -\dfrac{4}{7}$ and $\begin{bmatrix} a & b \\ c & d \end{bmatrix} = \begin{bmatrix} 2 & \dfrac{8}{7} \\ 1 & -\dfrac{4}{7} \end{bmatrix}.$

■ Commuters

33. $\begin{bmatrix} 0 & 1 \\ 1 & 0 \end{bmatrix}\begin{bmatrix} a & b \\ c & d \end{bmatrix} = \begin{bmatrix} c & d \\ a & b \end{bmatrix}$

$\begin{bmatrix} a & b \\ c & d \end{bmatrix}\begin{bmatrix} 0 & 1 \\ 1 & 0 \end{bmatrix} = \begin{bmatrix} b & a \\ d & c \end{bmatrix} \qquad \begin{aligned} &\therefore b = c \\ &\quad a = d \end{aligned}$

Any matrix of the form $\begin{bmatrix} a & b \\ b & a \end{bmatrix} \quad a, b \in \mathbb{R}$ will commute with $\begin{bmatrix} 0 & 1 \\ 1 & 0 \end{bmatrix}.$

■ Reckoning

36.   Let a_{ij}, b_{ij} be the ijth elements of matrices **A** and **B** respectively, for $1 \le i \le m$, $1 \le j \le n$.

$\mathbf{A} + \mathbf{B} = \begin{bmatrix} a_{ij} \end{bmatrix} + \begin{bmatrix} b_{ij} \end{bmatrix} = \begin{bmatrix} a_{ij} + b_{ij} \end{bmatrix}$ from the commutative property of real numbers

$\qquad = \begin{bmatrix} b_{ij} + a_{ij} \end{bmatrix}$

$\qquad = \begin{bmatrix} b_{ij} \end{bmatrix} + \begin{bmatrix} a_{ij} \end{bmatrix} = \mathbf{B} + \mathbf{A}$

■ Properties of the Transpose

Rather than grinding out the proofs of Problems 39–42, we make the following observations:

39. $\left(\mathbf{A}^{\mathrm{T}} \right)^{\mathrm{T}} = \mathbf{A}$. Interchanging rows and columns of a matrix two times reproduce the original matrix.

42. $(\mathbf{AB})^{\mathrm{T}} = \mathbf{B}^{\mathrm{T}}\mathbf{A}^{\mathrm{T}}$. This identity is not so obvious. We verify the proof for 2×2 matrices to illustrate the method. The verification for 3×3 and higher-order matrices follows along exactly the same lines.

$$\mathbf{A} = \begin{bmatrix} a_{11} & a_{12} \\ a_{21} & a_{22} \end{bmatrix}$$

$$\mathbf{B} = \begin{bmatrix} b_{11} & b_{12} \\ b_{21} & b_{22} \end{bmatrix}$$

$$\mathbf{AB} = \begin{bmatrix} a_{11} & a_{12} \\ a_{21} & a_{22} \end{bmatrix}\begin{bmatrix} b_{11} & b_{12} \\ b_{21} & b_{22} \end{bmatrix} = \begin{bmatrix} a_{11}b_{11} + a_{12}b_{21} & a_{11}b_{12} + a_{12}b_{22} \\ a_{21}b_{11} + a_{22}b_{21} & a_{21}b_{12} + a_{22}b_{22} \end{bmatrix}$$

$$(\mathbf{AB})^{\mathrm{T}} = \begin{bmatrix} a_{11}b_{11} + a_{12}b_{21} & a_{21}b_{11} + a_{22}b_{21} \\ a_{11}b_{12} + a_{12}b_{22} & a_{21}b_{12} + a_{22}b_{22} \end{bmatrix}$$

$$\mathbf{B}^{\mathrm{T}}\mathbf{A}^{\mathrm{T}} = \begin{bmatrix} b_{11} & b_{21} \\ b_{12} & b_{22} \end{bmatrix}\begin{bmatrix} a_{11} & a_{21} \\ a_{12} & a_{22} \end{bmatrix} = \begin{bmatrix} a_{11}b_{11} + a_{12}b_{21} & a_{21}b_{11} + a_{22}b_{21} \\ a_{11}b_{12} + a_{12}b_{22} & a_{21}b_{12} + a_{22}b_{22} \end{bmatrix}$$

Hence, $(\mathbf{AB})^{\mathrm{T}} = \mathbf{B}^{\mathrm{T}}\mathbf{A}^{\mathrm{T}}$ for 2×2 matrices.

■   **Constructing Symmetry**

45.   We verify the statement that $\mathbf{A} + \mathbf{A}^{\mathrm{T}}$ is symmetric for any 2×2 matrix. The general proof follows along the same lines.

$$\mathbf{A} + \mathbf{A}^{\mathrm{T}} = \begin{bmatrix} a_{11} & a_{12} \\ a_{21} & a_{22} \end{bmatrix} + \begin{bmatrix} a_{11} & a_{21} \\ a_{12} & a_{22} \end{bmatrix} = \begin{bmatrix} 2a_{11} & a_{12} + a_{21} \\ a_{21} + a_{12} & 2a_{22} \end{bmatrix},$$

which is clearly symmetric.

■   **Trace of a Matrix**

48. $\mathrm{Tr}(c\mathbf{A}) = ca_{11} + \cdots + ca_{nn} = c(a_{11} + \cdots + a_{nn}) = c\mathrm{Tr}(\mathbf{A})$

■   **Matrices Can Be Complex**

51. $\mathbf{A} + 2\mathbf{B} = \begin{bmatrix} 3+i & 0 \\ 2+4i & 4-i \end{bmatrix}$ **54.** $\mathbf{A}^2 = \begin{bmatrix} 6i & 4+6i \\ 6-4i & -5-8i \end{bmatrix}$

57. $\mathbf{B}^{\mathrm{T}} = \begin{bmatrix} 1 & 2i \\ -i & 1+i \end{bmatrix}$

■   **Square Roots of Zero**

60.   If we assume

$$A = \begin{bmatrix} a & b \\ c & d \end{bmatrix}$$

is the square root of

$$\begin{bmatrix} 0 & 0 \\ 0 & 0 \end{bmatrix},$$

then we must have

$$A^2 = \begin{bmatrix} a & b \\ c & d \end{bmatrix}\begin{bmatrix} a & b \\ c & d \end{bmatrix} = \begin{bmatrix} a^2 + bc & ab + bd \\ ac + cd & bc + d^2 \end{bmatrix} = \begin{bmatrix} 0 & 0 \\ 0 & 0 \end{bmatrix},$$

which implies the four equations

$$a^2 + bc = 0$$
$$ab + bd = 0$$
$$ac + cd = 0$$
$$bc + d^2 = 0.$$

From the first and last equations, we have $a^2 = d^2$. We now consider two cases: first we assume $a = d$. From the middle two preceding equations we arrive at $b = 0$, $c = 0$, and hence $a = 0$, $d = 0$. The other condition, $a = -d$, gives no condition on b and c, so we seek a matrix of the form (we pick $a = 1$, $d = -1$ for simplicity)

$$\begin{bmatrix} 1 & b \\ c & -1 \end{bmatrix}\begin{bmatrix} 1 & b \\ c & -1 \end{bmatrix} = \begin{bmatrix} 1 + bc & 0 \\ 0 & 1 + bc \end{bmatrix}.$$

Hence, in order for the matrix to be the zero matrix, we must have $b = -\dfrac{1}{c}$, and hence

$$\begin{bmatrix} 1 & -\dfrac{1}{c} \\ c & -1 \end{bmatrix},$$

which gives

$$\begin{bmatrix} 1 & -\dfrac{1}{c} \\ c & -1 \end{bmatrix}\begin{bmatrix} 1 & -\dfrac{1}{c} \\ c & -1 \end{bmatrix} = \begin{bmatrix} 0 & 0 \\ 0 & 0 \end{bmatrix}.$$

■   **Taking Matrices Apart**

63.   (a) $A = \begin{bmatrix} A_1 | A_2 | A_3 \end{bmatrix} = \begin{bmatrix} 1 & 5 & 2 \\ -1 & 0 & 3 \\ 2 & 4 & 7 \end{bmatrix}$, $\bar{x} = \begin{bmatrix} 2 \\ 4 \\ 3 \end{bmatrix}$

where A_1, A_2, and A_3 are the three columns of the matrix A and $x_1 = 2$, $x_2 = 4$, $x_3 = 3$ are the elements of $\bar{x}$. We can write

$$A\bar{x} = \begin{bmatrix} 1 & 5 & 2 \\ -1 & 0 & 3 \\ 2 & 4 & 7 \end{bmatrix}\begin{bmatrix} 2 \\ 4 \\ 3 \end{bmatrix} = \begin{bmatrix} 1\times2 + 5\times4 + 2\times3 \\ -1\times2 + 0\times4 + 3\times3 \\ 2\times2 + 4\times4 + 7\times3 \end{bmatrix} = 2\begin{bmatrix} 1 \\ -1 \\ 2 \end{bmatrix} + 4\begin{bmatrix} 5 \\ 0 \\ 4 \end{bmatrix} + 3\begin{bmatrix} 2 \\ 3 \\ 7 \end{bmatrix}$$

$$= x_1 A_1 + x_2 A_2 + x_3 A_3.$$

(b) We verify the fact for a 3×3 matrix. The general $n \times n$ case follows along the same lines.

$$\mathbf{A}\vec{x} = \begin{bmatrix} a_{11} & a_{12} & a_{13} \\ a_{21} & a_{22} & a_{23} \\ a_{31} & a_{32} & a_{33} \end{bmatrix} \begin{bmatrix} x_1 \\ x_2 \\ x_3 \end{bmatrix} = \begin{bmatrix} a_{11}x_1 + a_{12}x_2 + a_{13}x_3 \\ a_{21}x_1 + a_{22}x_2 + a_{23}x_3 \\ a_{31}x_1 + a_{32}x_2 + a_{33}x_3 \end{bmatrix} = \begin{bmatrix} a_{11}x_1 \\ a_{21}x_1 \\ a_{31}x_1 \end{bmatrix} + \begin{bmatrix} a_{12}x_2 \\ a_{22}x_2 \\ a_{32}x_2 \end{bmatrix} + \begin{bmatrix} a_{13}x_3 \\ a_{23}x_3 \\ a_{33}x_3 \end{bmatrix}$$

$$= x_1 \begin{bmatrix} a_{11} \\ a_{21} \\ a_{31} \end{bmatrix} + x_2 \begin{bmatrix} a_{12} \\ a_{22} \\ a_{32} \end{bmatrix} + x_3 \begin{bmatrix} a_{13} \\ a_{23} \\ a_{33} \end{bmatrix} = x_1 \mathbf{A}_1 + x_2 \mathbf{A}_2 + x_3 \mathbf{A}_3$$

■ Upper Triangular Matrices

66. (a) Examples are

$$\begin{bmatrix} 1 & 2 \\ 0 & 3 \end{bmatrix}, \begin{bmatrix} 1 & 3 & 0 \\ 0 & 0 & 5 \\ 0 & 0 & 2 \end{bmatrix}, \begin{bmatrix} 2 & 7 & 9 & 0 \\ 0 & 3 & 8 & 1 \\ 0 & 0 & 4 & 2 \\ 0 & 0 & 0 & 6 \end{bmatrix}.$$

(b) By direct computation, it is easy to see that all the entries in the matrix product

$$\mathbf{AB} = \begin{bmatrix} a_{11} & a_{12} & a_{13} \\ 0 & a_{22} & a_{23} \\ 0 & 0 & a_{33} \end{bmatrix} \begin{bmatrix} b_{11} & b_{12} & b_{13} \\ 0 & b_{22} & b_{23} \\ 0 & 0 & b_{33} \end{bmatrix}$$

below the diagonal are zero.

(c) In the general case, if we multiply two upper-triangular matrices, it yields

$$\mathbf{AB} = \begin{bmatrix} a_{11} & a_{12} & a_{13} & \cdots & a_{1n} \\ 0 & a_{22} & a_{23} & \cdots & a_{2n} \\ 0 & 0 & a_{33} & \cdots & a_{3n} \\ \cdots & \cdots & \cdots & \cdots & \cdots \\ 0 & 0 & 0 & 0 & a_{nn} \end{bmatrix} \times \begin{bmatrix} b_{11} & b_{12} & b_{13} & \cdots & b_{1n} \\ 0 & b_{22} & b_{23} & \cdots & b_{2n} \\ 0 & 0 & b_{33} & \cdots & b_{3n} \\ \cdots & \cdots & \cdots & \cdots & \cdots \\ 0 & 0 & 0 & 0 & b_{nn} \end{bmatrix} = \begin{bmatrix} c_{11} & c_{12} & c_{13} & \cdots & c_{1n} \\ 0 & c_{22} & c_{23} & \cdots & c_{2n} \\ 0 & 0 & c_{33} & \cdots & c_{3n} \\ \cdots & \cdots & \cdots & \cdots & \cdots \\ 0 & 0 & 0 & 0 & c_{nn} \end{bmatrix}.$$

We won't bother to write the general expression for the elements c_{ij}; the important point is that the entries in the product matrix that lie below the main diagonal are clearly zero.

■ Orthogonality

69.
$$\begin{bmatrix} k \\ 2 \\ k \end{bmatrix} \cdot \begin{bmatrix} 1 \\ 0 \\ 4 \end{bmatrix} = k \cdot 1 + 2 \cdot 0 + k \cdot 4 = 0$$

$$5k = 0$$
$$k = 0$$

■   Orthogonality Subsets

72.   Set $\begin{bmatrix} a \\ b \\ c \end{bmatrix} \cdot \begin{bmatrix} 1 \\ 0 \\ 1 \end{bmatrix} = 0$

$$a \cdot 1 + b \cdot 0 + c \cdot 1 = 0$$
$$a + c = 0$$
$$c = -a$$

Orthogonal set $= \left\{ \begin{bmatrix} a \\ b \\ -a \end{bmatrix} : a, b \in \mathbb{R} \right\}$

75.   Set $\begin{bmatrix} a \\ b \\ c \end{bmatrix} \cdot \begin{bmatrix} 1 \\ 0 \\ 1 \end{bmatrix} = 0$ to get $c = -a$

Set $\begin{bmatrix} a \\ b \\ c \end{bmatrix} \cdot \begin{bmatrix} 2 \\ 1 \\ 0 \end{bmatrix} = 0$ to get $b = -2a$

Set $\begin{bmatrix} a \\ b \\ c \end{bmatrix} \cdot \begin{bmatrix} 0 \\ -1 \\ 2 \end{bmatrix} = a \cdot 0 + b(-1) + c(2)$

$$= -b + 2c = 0$$
$$= 2a - 2a = 0$$

$\left\{ \begin{bmatrix} a \\ -2a \\ -a \end{bmatrix} : a \in \mathbb{R} \right\}$ is the orthogonal set

■   Dot Products

78. $[2, 1, 2] \bullet [3, -1, 0] = 5$. Because the dot product is positive, this means the angle between the vectors is less than $90°$.

81. $[7, 5, 1, 5] \bullet [4, -3, 2, 3] = 30$, not orthogonal

■   Geometric Vector Operations

84. $\dfrac{1}{2}\mathbf{A} + \mathbf{B} = \dfrac{1}{2}[1, 2] + [-3, 1] = [-2.5, 2]$

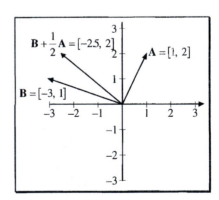

■   Triangles

87. $[2, -1, 2] \bullet [-1, 0, 1] = 0$ so in 3-space these vectors form a right angle, since dot product is zero.

■   **Properties of Scalar Products**

We let $\vec{a} = [a_1 \ldots a_n]$, $\vec{b} = [b_1 \ldots b_n]$, and $\vec{c} = [c_1 \ldots c_n]$ for simplicity.

90.   True.

$$(k\vec{a}) \bullet \vec{b} = [ka_1 \ \cdots \ ka_n] \bullet [b_1 \ \cdots \ b_n] = ka_1 b_1 + \cdots + ka_n b_n = a_1 k b_1 + \cdots + a_n k b_n$$

$$= [a_1 \ \cdots \ a_n] \bullet [kb_1 \ \cdots \ kb_n] = \vec{a} \bullet (k\vec{b})$$

■   **Tournament Play**

93.   The tournament graph had adjacency matrix

$$\mathbf{T} = \begin{bmatrix} 0 & 1 & 1 & 0 & 1 \\ 0 & 0 & 0 & 1 & 1 \\ 0 & 1 & 0 & 0 & 1 \\ 1 & 0 & 1 & 0 & 1 \\ 0 & 0 & 0 & 0 & 0 \end{bmatrix}.$$

Ranking players by the number of games won means summing the elements of each row of $\mathbf{T}$, which in this case gives two ties: 1 and 4, 2 and 3, 5. Players 1 and 4 have each won 3 games. Players 2 and 3 have each won 2 games. Player 5 has won none.

Second-order dominance can be determined from

$$\mathbf{T}^2 = \begin{bmatrix} 0 & 1 & 0 & 1 & 2 \\ 1 & 0 & 1 & 0 & 1 \\ 0 & 0 & 0 & 1 & 1 \\ 0 & 2 & 1 & 0 & 2 \\ 0 & 0 & 0 & 0 & 0 \end{bmatrix}$$

For example, $\mathbf{T}^2$ tells us that Player 1 can dominate Player 5 in two second-order ways (by beating either Player 2 or Player 4, both of whom beat Player 5). The sum

$$\mathbf{T} + \mathbf{T}^2 = \begin{bmatrix} 0 & 2 & 1 & 1 & 3 \\ 1 & 0 & 1 & 1 & 2 \\ 0 & 1 & 0 & 1 & 2 \\ 1 & 2 & 2 & 0 & 3 \\ 0 & 0 & 0 & 0 & 0 \end{bmatrix},$$

gives the number of ways one player has beaten another both directly and indirectly. Reranking players by sums of row elements of $\mathbf{T} + \mathbf{T}^2$ can sometimes break a tie: In this case it does so and ranks the players in order 4, 1, 2, 3, 5.

3.2   Systems of Linear Equations

■   **Matrix-Vector Form**

3.
$$\begin{bmatrix} 1 & 2 & 1 \\ 1 & -3 & 3 \\ 0 & 4 & -5 \end{bmatrix} \begin{bmatrix} r \\ s \\ t \end{bmatrix} = \begin{bmatrix} 1 \\ 1 \\ 3 \end{bmatrix}$$

Augmented matrix $= \begin{bmatrix} 1 & 2 & 1 & | & 1 \\ 1 & -3 & 3 & | & 1 \\ 0 & 4 & -5 & | & 3 \end{bmatrix}$

■   **Solutions in $\mathbb{R}^2$**

6.   (B)   9.   (A)

■   **Reduced Row Echelon Form**

12.   Not RREF (not all zeros above leading ones)

15.   RREF

18.   RREF

■   **Gauss-Jordan Elimination**

21. $\begin{bmatrix} 0 & 0 & 2 & 2 & -2 \\ 2 & 2 & 6 & 14 & 4 \end{bmatrix} \xrightarrow{R_1 \leftrightarrow R_2} \begin{bmatrix} 2 & 2 & 6 & 14 & 4 \\ 0 & 0 & 2 & 2 & -2 \end{bmatrix}$

$\xrightarrow{R_1^* = \frac{1}{2}R_1} \begin{bmatrix} 1 & 1 & 3 & 7 & 2 \\ 0 & 0 & 2 & 2 & -2 \end{bmatrix} \xrightarrow{R_2^* = \frac{1}{2}R_2} \begin{bmatrix} 1 & 1 & 3 & 7 & 2 \\ 0 & 0 & 1 & 1 & -1 \end{bmatrix}.$

The matrix is in row echelon form. To further reduce it to RREF we carry out the following elementary row operation.

$\xrightarrow{R_1^* = R_1 + (-3)R_2} \begin{bmatrix} 1 & 1 & 0 & 4 & 5 \\ 0 & 0 & 1 & 1 & -1 \end{bmatrix} \leftarrow$ RREF

The pivot columns of the original matrix are first and third.

$$\begin{bmatrix} \underline{\mathbf{0}} & 0 & \underline{\mathbf{2}} & 2 & -2 \\ \underline{\mathbf{2}} & 2 & \underline{\mathbf{6}} & 14 & 4 \end{bmatrix}.$$

■   **Solving Systems**

24. $\begin{bmatrix} 1 & 1 & | & 4 \\ 1 & -1 & | & 0 \end{bmatrix}$ $\xrightarrow{R_2^* = R_2 + (-1)R_1}$ $\begin{bmatrix} 1 & 1 & | & 4 \\ 0 & -2 & | & -4 \end{bmatrix}$ $\xrightarrow{R_2^* = -\frac{1}{2}R_2}$ $\begin{bmatrix} 1 & 1 & | & 4 \\ 0 & 1 & | & 2 \end{bmatrix}$

$\xrightarrow{R_1^* = R_1 + (-1)R_2}$ $\begin{bmatrix} 1 & 0 & | & 2 \\ 0 & 1 & | & 2 \end{bmatrix}$

unique solution; $x = 2$, $y = 2$

27. $\begin{bmatrix} 2 & 4 & -2 & | & 0 \\ 5 & 3 & 0 & | & 0 \end{bmatrix}$ $\xrightarrow{R_1^* = \frac{1}{2}R_1}$ $\begin{bmatrix} 1 & 2 & -1 & | & 0 \\ 5 & 3 & 0 & | & 0 \end{bmatrix}$ $\xrightarrow{R_2^* = R_2 + (-5)R_1}$ $\begin{bmatrix} 1 & 2 & -1 & | & 0 \\ 0 & -7 & 5 & | & 0 \end{bmatrix}$

$\xrightarrow{R_2^* = -\frac{1}{7}R_2}$ $\begin{bmatrix} 1 & 2 & -1 & | & 0 \\ 0 & 1 & -\frac{5}{7} & | & 0 \end{bmatrix}$ $\xrightarrow[\text{RREF}]{R_1^* = R_1 + (-2)R_2}$ $\begin{bmatrix} 1 & 0 & \frac{3}{7} & | & 0 \\ 0 & 1 & -\frac{5}{7} & | & 0 \end{bmatrix}$

nonunique solutions; $x = -\dfrac{3}{7}z$, $y = \dfrac{5}{7}z$, z is arbitrary

30. $\begin{bmatrix} 1 & 0 & 1 & | & 2 \\ 2 & -3 & 5 & | & 4 \\ 3 & 2 & -1 & | & 4 \end{bmatrix}$ $\begin{array}{c} R_2^* = R_2 + (-2)R_1 \\ \xrightarrow{\hspace{2cm}} \\ R_3^* = R_3 + (-3)R_1 \end{array}$ $\begin{bmatrix} 1 & 0 & 1 & | & 2 \\ 0 & -3 & 3 & | & 0 \\ 0 & 2 & -4 & | & -2 \end{bmatrix}$

$\xrightarrow{R_2^* = -\frac{1}{3}R_2}$ $\begin{bmatrix} 1 & 0 & 1 & | & 2 \\ 0 & 1 & -1 & | & 0 \\ 0 & 2 & -4 & | & -2 \end{bmatrix}$ $\xrightarrow{R_3^* = R_3 + (-2)R_2}$ $\begin{bmatrix} 1 & 0 & 1 & | & 2 \\ 0 & 1 & -1 & | & 0 \\ 0 & 0 & -2 & | & -2 \end{bmatrix}$

$\xrightarrow{R_3^* = -\frac{1}{2}R_3}$ $\begin{bmatrix} 1 & 0 & 1 & | & 2 \\ 0 & 1 & -1 & | & 0 \\ 0 & 0 & 1 & | & 1 \end{bmatrix}$ $\begin{array}{c} R_1^* = R_1 + (-1)R_3 \\ \xrightarrow{\hspace{2cm}} \\ R_2^* = R_2 + R_3 \end{array}$ $\text{RREF}\begin{bmatrix} 1 & 0 & 0 & | & 1 \\ 0 & 1 & 0 & | & 1 \\ 0 & 0 & 1 & | & 1 \end{bmatrix}$

unique solution; $x = y = z = 1$

33. $\begin{bmatrix} 1 & 1 & 2 & | & 1 \\ 2 & -1 & 1 & | & 2 \\ 4 & 1 & 5 & | & 4 \end{bmatrix}$ $\begin{array}{c} R_2^* = R_2 + (-2)R_1 \\ \xrightarrow{\hspace{2cm}} \\ R_3^* = R_3 + (-4)R_1 \end{array}$ $\begin{bmatrix} 1 & 1 & 2 & | & 1 \\ 0 & -3 & -3 & | & 0 \\ 0 & -3 & -3 & | & 0 \end{bmatrix}$

$\xrightarrow{R_2^* = -\frac{1}{3}R_2}$ $\begin{bmatrix} 1 & 1 & 2 & | & 1 \\ 0 & 1 & 1 & | & 0 \\ 0 & -3 & -3 & | & 0 \end{bmatrix}$ $\begin{array}{c} R_1^* = R_1 + (-1)R_2 \\ \xrightarrow{\hspace{2cm}} \\ R_3^* = R_3 + (3)R_2 \end{array}$ $\text{RREF}\begin{bmatrix} 1 & 0 & 1 & | & 1 \\ 0 & 1 & 1 & | & 0 \\ 0 & 0 & 0 & | & 0 \end{bmatrix}$

nonunique solutions; $x = 1 - z$, $y = -z$, z is arbitrary

36.

$$x \quad + 2x_3 - 4x_4 = 1$$
$$x_2 + \ x_3 - 3x_4 = 2$$

$\begin{bmatrix} 1 & 0 & 2 & -4 & | & 1 \\ 0 & 1 & 1 & -3 & | & 2 \end{bmatrix}$ is in RREF

$\therefore$ infinitely many solutions

$x_1 = -2r + 4s + 1$
$x_2 = -r + 3s + 2, \quad r, s \in \mathbb{R}$
$x_3 = r, \ x_4 = s$

■   **Using the Nonhomogenous Principle**

39.   In Problem 26, infinitely many solutions, $\begin{bmatrix} 1 \\ 1-z \\ z \end{bmatrix}$, and $\mathbb{W} = \left\{ \begin{bmatrix} 0 \\ -1 \\ 1 \end{bmatrix} : r \in \mathbb{R} \right\}$

hence $\vec{\mathbf{x}} = \begin{bmatrix} -1 \\ 1 \\ 0 \end{bmatrix} + r \begin{bmatrix} 0 \\ -1 \\ 1 \end{bmatrix}$ for any $r \in \mathbb{R}$

42.   In Problem 29, infinitely many solutions:

$x_1 = 4 - 3x_3$
$x_2 = -1 + 2x_3,$
x_3 arbitrary

$\mathbb{W} = \left\{ r \begin{bmatrix} -3 \\ 2 \\ 1 \end{bmatrix} : r \in \mathbb{R} \right\}$

$\vec{\mathbf{x}} = \begin{bmatrix} 4 \\ -1 \\ 0 \end{bmatrix} + r \begin{bmatrix} -3 \\ 2 \\ 1 \end{bmatrix}$ for any $r \in \mathbb{R}$

45.   In Problem 32, infinitely many solutions $x = -z, \ y = -z, \ z$ arbitrary, so

$\mathbb{W} = \left\{ r \begin{bmatrix} -1 \\ -1 \\ 1 \end{bmatrix} : r \in \mathbb{R} \right\}$

$\vec{\mathbf{x}} = \vec{\mathbf{0}} + r \begin{bmatrix} -1 \\ -1 \\ 1 \end{bmatrix}$ for any $r \in \mathbb{R}$

48.   In Problem 35, there is a unique solution: $x = \dfrac{24}{5}$, $y = \dfrac{4}{5}$, $z = -\dfrac{22}{5}$ so

$$\mathbb{W} = \{\vec{0}\} \quad \text{and} \quad \vec{x} = \begin{bmatrix} \dfrac{24}{5} \\[2mm] \dfrac{4}{5} \\[2mm] -\dfrac{22}{5} \end{bmatrix} + \vec{0}.$$

■ More Equations Than Variables

51.   Converting the augmented matrix to RREF yields

$$\left[\begin{array}{ccc|c} 3 & 5 & 0 & 1 \\ 3 & 7 & 3 & 8 \\ 0 & 5 & 0 & -5 \\ 0 & 2 & 3 & 7 \\ 1 & 4 & 1 & 1 \end{array}\right] \rightarrow \left[\begin{array}{ccc|c} 1 & 0 & 0 & 2 \\ 0 & 1 & 0 & -1 \\ 0 & 0 & 1 & 3 \\ 0 & 0 & 0 & 0 \\ 0 & 0 & 0 & 0 \end{array}\right]$$

consistent system; unique solution $x = 2$, $y = -1$, $z = 3$.

■ Homogeneous Systems

54.   The equations are

$$\begin{aligned} x \quad + \quad\quad 2z &= 0 \\ y \quad\quad &= 0 \end{aligned}$$

If we let $z = s$, we have $x = -2s$ and hence the solution is a line in $\mathbb{R}^3$ given by

$$\begin{bmatrix} x \\ y \\ z \end{bmatrix} = \begin{bmatrix} -2s \\ 0 \\ s \end{bmatrix} = s \begin{bmatrix} -2 \\ 0 \\ 1 \end{bmatrix}.$$

■ Making Systems Inconsistent

57. $\begin{bmatrix} 4 & 5 \\ 1 & 6 \\ 3 & -1 \end{bmatrix}$ Rank = 2

$$\left[\begin{array}{cc|c} 4 & 5 & a \\ 1 & 6 & b \\ 3 & -1 & c \end{array}\right] \xrightarrow{R_2 \leftrightarrow R_1} \left[\begin{array}{cc|c} 1 & 6 & b \\ 4 & 5 & a \\ 3 & -1 & c \end{array}\right] \xrightarrow[R_3^* = -3R_1 + R_3]{R_2^* = -4R_1 + R_2} \left[\begin{array}{cc|c} 1 & 0 & b \\ 0 & 5 & -5b + a \\ 0 & -1 & -3b + c \end{array}\right]$$

$$\xrightarrow[R_2 \leftrightarrow R_3]{R_3^* = -R_3} \left[\begin{array}{cc|c} 1 & 6 & b \\ 0 & 1 & 3b - c \\ 0 & 5 & -5b + a \end{array}\right] \xrightarrow{R_3^* = -5R_2 + R_3} \left[\begin{array}{cc|c} 1 & 6 & b \\ 0 & 1 & 3b - c \\ 0 & 0 & a - 20b + 5c \end{array}\right]$$

Thus the system is inconsistent for all vectors $\begin{bmatrix} a \\ b \\ c \end{bmatrix}$ for which $a - 20b + 5c \neq 0$.

60. $\begin{bmatrix} 1 & -1 & 1 \\ 1 & 1 & 0 \\ 1 & 2 & -1 \end{bmatrix} \xrightarrow[R_3^* = -R_1 + R_3]{R_2^* = -R_1 + R_2} \begin{bmatrix} 1 & -1 & 1 \\ 0 & 2 & -1 \\ 0 & 3 & -2 \end{bmatrix} \xrightarrow{R_2^* = \frac{1}{2}R_2} \begin{bmatrix} 1 & -1 & 1 \\ 0 & 1 & -\dfrac{1}{2} \\ 0 & 3 & -2 \end{bmatrix}$

$\xrightarrow[R_3^* = -3R_2 + R_3]{R_1^* = R_2 + R_1} \begin{bmatrix} 1 & 0 & \dfrac{1}{2} \\ 0 & 1 & -\dfrac{1}{2} \\ 0 & 0 & \dfrac{2}{3} \end{bmatrix} \xrightarrow{R_3^* = \frac{3}{2}R_2} \begin{bmatrix} 1 & 0 & \dfrac{1}{2} \\ 0 & 1 & -\dfrac{1}{8} \\ 0 & 0 & 1 \end{bmatrix}$

$\xrightarrow[R_2^* = -\frac{1}{8}R_3 + R_2]{R_1^* = -\frac{1}{2}R_3 + R_1} \begin{bmatrix} 1 & 0 & 0 \\ 0 & 1 & 0 \\ 0 & 0 & 1 \end{bmatrix} \quad \therefore \text{ rank } A = 3$

■ **Seeking Consistency**

63. $\begin{bmatrix} 1 & k & | & 0 \\ k & 1 & | & 2 \end{bmatrix} \xrightarrow{R_2^* = -kR_1 + R_2} \begin{bmatrix} 1 & k & | & 0 \\ 0 & -k^2 + 1 & | & 2 \end{bmatrix}$

so that $-k^2 + 1 \neq 0$. Thus $k \neq \pm 1$.

■ **Not Enough Equations**

66. (a) $\begin{bmatrix} 2 & 1 & 0 & 0 & | & 3 \\ 1 & -1 & 1 & 1 & | & 3 \\ 2 & -3 & 4 & 4 & | & 9 \end{bmatrix} \xrightarrow{R_1 \leftrightarrow R_2} \begin{bmatrix} 1 & -1 & 1 & 1 & | & 3 \\ 2 & 1 & 0 & 0 & | & 3 \\ 2 & -3 & 4 & 4 & | & 9 \end{bmatrix}$

$\xrightarrow[R_3^* = -2R_1 + R_3]{R_2^* = -2R_1 + R_2} \begin{bmatrix} 1 & -1 & 1 & 1 & | & 3 \\ 0 & 3 & 2 & 2 & | & -3 \\ 0 & -1 & 2 & 2 & | & 3 \end{bmatrix} \xrightarrow{R_2 \leftrightarrow R_3} \begin{bmatrix} 1 & -1 & 1 & 1 & | & 3 \\ 0 & 1 & -2 & -2 & | & 3 \\ 0 & 3 & 2 & 2 & | & -3 \end{bmatrix}$

$\xrightarrow{R_3^* = -3R_2 + R_3} \begin{bmatrix} 1 & 0 & -1 & -1 & | & 6 \\ 0 & 1 & -2 & -2 & | & 3 \\ 0 & 0 & 8 & 8 & | & -12 \end{bmatrix} \xrightarrow{R_3^* = \frac{1}{8}R_3} \begin{bmatrix} 1 & 0 & -1 & -1 & | & 6 \\ 0 & 1 & -2 & -2 & | & 3 \\ 0 & 0 & 1 & 1 & | & -\dfrac{3}{2} \end{bmatrix}$

This matrix is in row-echelon form and has 3 pivot colums Rank = 3
Consequently, there are infinitely many solutions because it represents a consistent system.

(b)
$$\begin{bmatrix} 2 & 1 & 0 & 0 & | & 3 \\ 1 & -1 & 1 & 1 & | & 3 \\ 1 & 2 & -1 & -1 & | & 6 \end{bmatrix} \xrightarrow{R_1 \leftrightarrow R_2} \begin{bmatrix} 1 & -1 & 1 & 1 & | & 3 \\ 2 & 1 & 0 & 0 & | & 3 \\ 1 & 2 & -1 & -1 & | & -6 \end{bmatrix}$$

$$\xrightarrow[\substack{R_3^* = -R_1 + R_2}]{R_2^* = -2R_1 + R_2} \begin{bmatrix} 1 & -1 & 1 & 1 & | & 3 \\ 0 & 3 & -2 & -2 & | & -3 \\ 0 & 3 & -2 & -2 & | & -9 \end{bmatrix} \longrightarrow \begin{bmatrix} 1 & -1 & 1 & 1 & | & 3 \\ 0 & 3 & -2 & -2 & | & -3 \\ 0 & 0 & 0 & 0 & | & -6 \end{bmatrix}$$

Clearly inconsistent, no solutions.

■ Equivalence of Systems

69.   Inverse of $R_i \leftrightarrow R_j$: The operation that puts the system back the way it was is $R_j \leftrightarrow R_i$. In other words, the operation $R_3 \leftrightarrow R_1$ will undo the operation $R_1 \leftrightarrow R_3$.

Inverse of $R_i = cR_i$: The operation that puts the system back the way it was is $R_i = \frac{1}{c}R_i$. In other words, the operation $R_1 = \frac{1}{3}R_1$ will undo the operation $R_1 = 3R_1$.

Inverse of $R_i = R_i + cR_j$: The operation that puts the system back is $R_i = R_i - cR_j$. In other words $R_i = R_i - cR_j$ will undo the operation $R_i = R_i + cR_j$. This is clear because if we add cR_j to row i and then subtract cR_j from row i, then row i will be unchanged. For example,

$$\begin{matrix} R_1 \\ R_2 \end{matrix}\begin{bmatrix} 1 & 2 & 3 \\ 2 & 1 & 1 \end{bmatrix}, \; R_1^* = R_1 + 2R_2, \; \begin{bmatrix} 5 & 4 & 5 \\ 2 & 1 & 1 \end{bmatrix}, \; R_1^* = R_1 + (-2)R_2, \; \begin{bmatrix} 1 & 2 & 3 \\ 2 & 1 & 1 \end{bmatrix}.$$

■ Tandem with a Twist

72.   (a)   We place the right-hand sides of the two systems in the last two columns of the augmented matrix

$$\begin{bmatrix} 1 & 1 & 0 & | & 3 & 5 \\ 0 & 2 & 1 & | & 2 & 4 \end{bmatrix}.$$

Reducing this matrix to RREF, yields

$$\begin{bmatrix} 1 & 0 & -\dfrac{1}{2} & | & 2 & 3 \\ 0 & 1 & \dfrac{1}{2} & | & 1 & 2 \end{bmatrix}.$$

Hence, the first system has solutions $x = 2 + \frac{1}{2}z$, $y = 1 - \frac{1}{2}z$, z arbitrary, and the second system has solutions $x = 3 + \frac{1}{2}z$, $y = 2 - \frac{1}{2}z$, z arbitrary.

(b) If you look carefully, you will see that the matrix equation

$$\begin{bmatrix} 1 & 1 & 0 \\ 0 & 2 & 1 \end{bmatrix} \begin{bmatrix} x_{11} & x_{12} \\ x_{21} & x_{22} \\ x_{31} & x_{32} \end{bmatrix} = \begin{bmatrix} 3 & 5 \\ 2 & 4 \end{bmatrix}$$

is equivalent to the two systems of equations

$$\begin{bmatrix} 1 & 1 & 0 \\ 0 & 2 & 1 \end{bmatrix} \begin{bmatrix} x_{11} \\ x_{21} \\ x_{31} \end{bmatrix} = \begin{bmatrix} 3 \\ 2 \end{bmatrix}$$

$$\begin{bmatrix} 1 & 1 & 0 \\ 0 & 2 & 1 \end{bmatrix} \begin{bmatrix} x_{12} \\ x_{22} \\ x_{32} \end{bmatrix} = \begin{bmatrix} 5 \\ 4 \end{bmatrix}.$$

We saw in part (a) that the solution of the system on the left was

$$x_{11} = 2 + \frac{1}{2} x_{31}, \ x_{21} = 1 - \frac{1}{2} x_{31}, \ x_{31} \text{ arbitrary,}$$

and the solution of the system on the right was

$$x_{12} = 3 + \frac{1}{2} x_{32}, \ x_{22} = 2 - \frac{1}{2} x_{32}, \ x_{32} \text{ arbitrary.}$$

Putting these solutions in the columns of our unknown matrix $\mathbf{X}$ and calling $x_{31} = \alpha$, $x_{32} = \beta$, we have

$$\mathbf{X} = \begin{bmatrix} x_{11} & x_{12} \\ x_{21} & x_{22} \\ x_{31} & x_{32} \end{bmatrix} = \begin{bmatrix} 2 + \dfrac{1}{2}\alpha & 3 + \dfrac{1}{2}\beta \\ 1 - \dfrac{1}{2}\alpha & 2 - \dfrac{1}{2}\beta \\ \alpha & \beta \end{bmatrix}.$$

■ Computerizing

75. The basic idea is to formalize a strategy like that used in Example 3. The augmented matrix for $\mathbf{Ax} = \mathbf{b}$ is

$$\begin{bmatrix} a_{11} & a_{12} & a_{13} & b_1 \\ a_{21} & a_{22} & a_{23} & b_2 \\ a_{31} & a_{32} & a_{33} & b_3 \end{bmatrix}.$$

A pseudocode might begin:

1. To get a one in first place in row 1, multiply every element of row 1 by $\dfrac{1}{a_{11}}$.

2. To get a zero in first place in row 2, replace row 2 by

$$\text{row } 2 - a_{21} (\text{row } 1).$$

$$\vdots$$

■   **More Circuit Analysis**

78.

$$I_1 - I_2 - I_3 - I_4 = 0$$
$$-I_1 + I_2 + I_3 + I_4 = 0$$

■   **Suggested Journal Entry I**

81.   Student Project

3.3   The Inverse of a Matrix

■   **Checking Inverses**

3.
$$\begin{bmatrix} 1 & 0 & 1 \\ 1 & 1 & -2 \\ 0 & 1 & 1 \end{bmatrix}\begin{bmatrix} 3/4 & 1/4 & -1/4 \\ -1/4 & 1/4 & 3/4 \\ 1/4 & -1/4 & 1/4 \end{bmatrix} = \begin{bmatrix} 3/4+0+1/4 & 1/4+0-1/4 & -1/4+0+1/4 \\ 3/4-1/4-2/4 & 1/4+1/4+2/4 & -1/4+3/4-2/4 \\ 0-1/4+1/4 & 0+1/4-1/4 & 0+3/4+1/4 \end{bmatrix}$$

$$= \begin{bmatrix} 1 & 0 & 0 \\ 0 & 1 & 0 \\ 0 & 0 & 1 \end{bmatrix}$$

■   **Matrix Inverses**

6.   We reduce $\begin{bmatrix} \mathbf{A} | \mathbf{I} \end{bmatrix}$ to RREF.

$$\begin{bmatrix} 1 & 3 & | & 1 & 0 \\ 2 & 5 & | & 0 & 1 \end{bmatrix} \xrightarrow{R_2^* = R_2 + (-2)R_1} \begin{bmatrix} 1 & 3 & | & 1 & 0 \\ 0 & -1 & | & -2 & 1 \end{bmatrix} \xrightarrow{R_2^* = (-1)R_2} \begin{bmatrix} 1 & 3 & | & 1 & 0 \\ 0 & 1 & | & 2 & -1 \end{bmatrix}$$

$$\xrightarrow{R_1^* = R_1 + (-3)R_2} \begin{bmatrix} 1 & 0 & | & -5 & 3 \\ 0 & 1 & | & 2 & -1 \end{bmatrix}.$$

Hence, $\mathbf{A}^{-1} = \begin{bmatrix} -5 & 3 \\ 2 & -1 \end{bmatrix}$.

9.   Dividing the first row by k gives

$$\begin{bmatrix} \mathbf{A} | \mathbf{I} \end{bmatrix} = \begin{bmatrix} k & 0 & 0 & | & 1 & 0 & 0 \\ 0 & 1 & 0 & | & 0 & 1 & 0 \\ 0 & 0 & 1 & | & 0 & 0 & 1 \end{bmatrix} \rightarrow \begin{bmatrix} \mathbf{I} | \mathbf{A}^{-1} \end{bmatrix} = \begin{bmatrix} 1 & 0 & 0 & | & \frac{1}{k} & 0 & 0 \\ 0 & 1 & 0 & | & 0 & 1 & 0 \\ 0 & 0 & 1 & | & 0 & 0 & 1 \end{bmatrix}$$

Hence $\mathbf{A}^{-1} = \begin{bmatrix} \frac{1}{k} & 0 & 0 \\ 0 & 1 & 0 \\ 0 & 0 & 1 \end{bmatrix}$.

12.

$$\left[\begin{array}{cccc|cccc} 1 & 0 & 1 & 1 & 1 & 0 & 0 & 0 \\ 0 & 0 & 1 & 0 & 0 & 1 & 0 & 0 \\ 1 & 1 & 1 & 0 & 0 & 0 & 1 & 0 \\ 1 & 0 & 0 & 2 & 0 & 0 & 0 & 1 \end{array}\right] \rightarrow \left[\begin{array}{cccc|cccc} 1 & 0 & 1 & 1 & 1 & 0 & 0 & 0 \\ 0 & 0 & 1 & 0 & 0 & 1 & 0 & 0 \\ 0 & 1 & 0 & -1 & -1 & 0 & 1 & 0 \\ 0 & 0 & -1 & 1 & -1 & 0 & 0 & 1 \end{array}\right] \rightarrow \left[\begin{array}{cccc|cccc} 1 & 0 & 1 & 1 & 1 & 0 & 0 & 0 \\ 0 & 1 & 0 & -1 & -1 & 0 & 1 & 0 \\ 0 & 0 & 1 & 0 & 0 & 1 & 0 & 0 \\ 0 & 0 & -1 & 1 & -1 & 0 & 0 & 1 \end{array}\right]$$

$$\rightarrow \left[\begin{array}{cccc|cccc} 1 & 0 & 0 & 1 & 1 & -1 & 0 & 0 \\ 0 & 1 & 0 & -1 & -1 & 0 & 1 & 0 \\ 0 & 0 & 1 & 0 & 0 & 1 & 0 & 0 \\ 0 & 0 & 0 & 1 & -1 & 1 & 0 & 1 \end{array}\right] \rightarrow \left[\begin{array}{cccc|cccc} 1 & 0 & 0 & 0 & 2 & -2 & 0 & -1 \\ 0 & 1 & 0 & 0 & -2 & 1 & 1 & 1 \\ 0 & 0 & 1 & 0 & 0 & 1 & 0 & 0 \\ 0 & 0 & 0 & 1 & -1 & 1 & 0 & 1 \end{array}\right] \text{ so } \mathbf{A}^{-1} = \left[\begin{array}{cccc} 2 & -2 & 0 & -1 \\ -2 & 1 & 1 & 1 \\ 0 & 1 & 0 & 0 \\ -1 & 1 & 0 & 1 \end{array}\right]$$

■ Inverse of the 2×2 Matrix

15. Verify $\mathbf{A}^{-1}\mathbf{A} = \mathbf{I} = \mathbf{A}\mathbf{A}^{-1}$. We have

$$\mathbf{A}^{-1}\mathbf{A} = \frac{1}{ad-bc}\begin{bmatrix} d & -b \\ -c & a \end{bmatrix}\begin{bmatrix} a & b \\ c & d \end{bmatrix} = \frac{1}{ad-bc}\begin{bmatrix} ad-bc & 0 \\ 0 & ad-bc \end{bmatrix} = \mathbf{I}$$

$$\mathbf{A}\mathbf{A}^{-1} = \begin{bmatrix} a & b \\ c & d \end{bmatrix}\frac{1}{ad-bc}\begin{bmatrix} d & -b \\ -c & a \end{bmatrix} = \frac{1}{ad-bc}\begin{bmatrix} ad-bc & 0 \\ 0 & ad-bc \end{bmatrix} = \mathbf{I}$$

Note that we must have $|\mathbf{A}| = ad-bc \neq 0$.

■ Finding Counterexamples

18. No.   Consider $\begin{bmatrix} 0 & 0 \\ 0 & 1 \end{bmatrix}\begin{bmatrix} 0 & 0 \\ 0 & 1 \end{bmatrix} = \begin{bmatrix} 0 & 0 \\ 0 & 1 \end{bmatrix}$

■ Solution by Invertible Matrix

21. Using the inverse found in Problem 7 yields

$$\begin{bmatrix} x \\ y \\ z \end{bmatrix} = \mathbf{A}^{-1}\vec{\mathbf{b}} = \begin{bmatrix} 1 & 0 & \dfrac{1}{3} \\ -2 & \dfrac{1}{2} & -\dfrac{5}{6} \\ 3 & -\dfrac{1}{2} & \dfrac{5}{6} \end{bmatrix}\begin{bmatrix} 5 \\ 2 \\ 0 \end{bmatrix} = \begin{bmatrix} 5 \\ -9 \\ 14 \end{bmatrix}.$$

■ Noninvertible 2×2 Matrices

24. If we reduce $\mathbf{A} = \begin{bmatrix} a & b \\ c & d \end{bmatrix}$ to RREF we get

$$\begin{bmatrix} 1 & \dfrac{b}{a} \\ 0 & \dfrac{ad-bc}{a} \end{bmatrix},$$

which says that the matrix is invertible when $\dfrac{ad-bc}{a} \neq 0$, or equivalently when $ad \neq bc$.

■   **Matrix Algebra with Inverses**

27.   Suppose $A(BA)^{-1} \vec{x} = \vec{b}$

$$\vec{x} = \left[\left(A(BA)^{-1} \right)^{-1} \right] \cdot \vec{b}$$
$$= (BA) A^{-1} \vec{b} = B \left(A A^{-1} \right) \vec{b}$$
$$= B\vec{b}$$

■   **Cancellation Works**

30.   Given that $AB = AC$ and A are invertible, we premultiply by A^{-1}, getting

$$A^{-1}AB = A^{-1}AC$$
$$IB = IC$$
$$B = C.$$

■   **Making Invertible Matrices**

33. $\begin{vmatrix} 1 & 0 & k \\ 0 & 1 & 0 \\ k & 0 & 1 \end{vmatrix} = 1 - k^2 \qquad k \neq \pm 1$

■   **Invertiblity of Triangular Matrices**

36.   Proof for ($\Rightarrow$): (Contrapositive) Let T be an upper triangular matrix with at least one diagonal element $= 0$, say a_{jj}. Then there is one column without a pivot. Therefore RREF (T) has a zero row.
Consequently T is not invertible.

Proof for ($\Leftarrow$): Let T be an upper $n \times n$ triangular matrix with no nonzero diagonal elements. Then every column is a pivot column so RREF(T) = I_n.
Therefore T is invertible.

■   **Inverse of a Transpose**

39.   To prove: If A is invertible, so is A^T and $(A^T)^{-1} = (A^{-1})^T$.
Proof: Let A be an invertible $n \times n$ matrix.
Then $(A^T)(A^{-1})^T = (A^{-1}A)^T = I_n^T = I_n$ because $(A^T)^T = A$ and $(AB)^T = B^T A^T$
$(A^{-1})^T A^T = (AA^{-1})^T = I_n^T = I_n$
Therefore $(A^T)^{-1} = (A^{-1})^T$

■   **Similar Matrices**

42.   Pick P as the identity matrix.

45.   Informal Discussion
$$\mathbf{B}^n = \underbrace{(\mathbf{P}^{-1}\mathbf{A}\mathbf{P})(\mathbf{P}^{-1}\mathbf{A}\mathbf{P}) \cdots (\mathbf{P}^{-1}\mathbf{A}\mathbf{P})}_{n \text{ factors}}$$

By generous application of the associative property of matrix multiplication we obtain.
$$\mathbf{B}^n = \mathbf{P}^{-1}\mathbf{A}(\mathbf{P}\mathbf{P}^{-1})\,\mathbf{A}(\mathbf{P}\mathbf{P}^{-1}) \cdots (\mathbf{P}\mathbf{P}^{-1})\mathbf{A}\mathbf{P}$$
$$= \mathbf{P}^{-1}\mathbf{A}^n\mathbf{P} \text{ by the facts that } \mathbf{P}\mathbf{P}^{-1} = \mathbf{I} \text{ and } \mathbf{A}\mathbf{I} = \mathbf{A}$$

Induction Proof

To Prove: $\mathbf{B}^n = \mathbf{P}^{-1}\mathbf{A}^n\mathbf{P}$ for all positive integers n

Proof: 1) $\mathbf{B}^1 = \mathbf{P}^{-1}\mathbf{A}\mathbf{P}$ by definition of $\mathbf{B}$

   2)   Assume for some k: $\mathbf{B}^k = \mathbf{P}^{-1}\mathbf{A}^k\mathbf{P}$

Now for $k + 1$:
$$\mathbf{B}^{k+1} = \mathbf{B}\mathbf{B}^k = (\mathbf{P}^{-1}\mathbf{A}\mathbf{P})(\mathbf{P}^{-1}\mathbf{A}^k\mathbf{P})$$
$$= (\mathbf{P}^{-1}\mathbf{A})(\mathbf{P}\mathbf{P}^{-1})(\mathbf{A}^k\mathbf{P})$$
$$= (\mathbf{P}^{-1}\mathbf{A})\mathbf{I}(\mathbf{A}^k\mathbf{P})$$
$$= \mathbf{P}^{-1}\mathbf{A}\mathbf{A}^k\mathbf{P} = \mathbf{P}^{-1}\mathbf{A}^{k+1}\mathbf{P}$$

So the case for $k \Rightarrow$ the case for $k + 1$

By Mathematical Induction, $\mathbf{B}^n = \mathbf{P}^{-1}\mathbf{A}^n\mathbf{P}$ for all n.

■   **Leontief Model**

48. $$\mathbf{T} = \begin{bmatrix} 0 & 0.1 \\ 0.2 & 0 \end{bmatrix}, \ \vec{\mathbf{d}} = \begin{bmatrix} 10 \\ 10 \end{bmatrix}$$

The basic equation is

   Total Output = External Demand + Internal Demand,

so we have
$$\begin{bmatrix} x_1 \\ x_2 \end{bmatrix} = \begin{bmatrix} 10 \\ 10 \end{bmatrix} + \begin{bmatrix} 0 & 0.1 \\ 0.2 & 0 \end{bmatrix} \begin{bmatrix} x_1 \\ x_2 \end{bmatrix}.$$

Solving these equations yields $x_1 = 11.2$, $x_2 = 12.2$.

■   **How Much Is Left Over?**

51.   The basic demand equation is

   Total Output = External Demand + Internal Demand,

so we have
$$\begin{bmatrix} 150 \\ 250 \end{bmatrix} = \begin{bmatrix} d_1 \\ d_2 \end{bmatrix} + \begin{bmatrix} 0.3 & 0.4 \\ 0.5 & 0.3 \end{bmatrix} \begin{bmatrix} 150 \\ 250 \end{bmatrix}.$$

Solving for d_1, d_2 yields $d_1 = 5$, $d_2 = 100$.

3.4   Determinants and Cramer's Rule

■ Calculating Determinants

3.   Expanding by cofactors down the third column we get

$$
\begin{vmatrix} 1 & 3 & 0 & -2 \\ 0 & 1 & -1 & 5 \\ -1 & -2 & 1 & 7 \\ 1 & 1 & 0 & -6 \end{vmatrix} = 1\begin{vmatrix} 1 & 3 & -2 \\ -1 & -2 & 7 \\ 1 & 1 & -6 \end{vmatrix} + \begin{vmatrix} 1 & 3 & -2 \\ 0 & 1 & 5 \\ 1 & 1 & -6 \end{vmatrix} = 6 + 6 = 12.
$$

6. $\begin{vmatrix} 0 & 0 & 1 \\ 0 & 2 & 1 \\ 3 & 1 & 1 \end{vmatrix} = 1\begin{vmatrix} 0 & 2 \\ 3 & 1 \end{vmatrix} = -6$

■ Find the Properties

9.   Factor out 3 from the second row of the matrix in the first determinant to get the matrix in the second determinant.

■ Basketweave for 3×3

12. $\begin{vmatrix} 0 & 7 & 9 \\ 2 & 1 & -1 \\ 5 & 6 & 2 \end{vmatrix} = 0 - 35 + 108 - 45 - 0 - 28 = 0$

■ Triangular Determinants

15.   We verify this for 4×4 matrices. Higher-order matrices follow along the same lines. Given the upper-triangular matrix

$$
|A| = \begin{vmatrix} a_{11} & a_{12} & a_{13} & a_{14} \\ 0 & a_{22} & a_{23} & a_{24} \\ 0 & 0 & a_{33} & a_{34} \\ 0 & 0 & 0 & a_{44} \end{vmatrix},
$$

we expand down the first column, getting

$$
\begin{vmatrix} a_{11} & a_{12} & a_{13} & a_{14} \\ 0 & a_{22} & a_{23} & a_{24} \\ 0 & 0 & a_{33} & a_{34} \\ 0 & 0 & 0 & a_{44} \end{vmatrix} = a_{11}\begin{vmatrix} a_{22} & a_{23} & a_{24} \\ 0 & a_{33} & a_{34} \\ 0 & 0 & a_{44} \end{vmatrix} = a_{11}a_{22}\begin{vmatrix} a_{33} & a_{34} \\ 0 & a_{44} \end{vmatrix} = a_{11}a_{22}a_{33}a_{44}.
$$

■   **Think Diagonal**

18.   The matrix is lower triangular, hence the determinant is the product of the diagonal elements.

$$\begin{vmatrix} 1 & 0 & 0 & 0 \\ -3 & 4 & 0 & 0 \\ 0 & 5 & -1 & 0 \\ 11 & 0 & -2 & 2 \end{vmatrix} = (1)(4)(-1)(2) = -8 \,.$$

■   **Invertibility**

21.   Not invertible if $\begin{vmatrix} 1 & k \\ k & -k \end{vmatrix} = 0$

$$-k + k^2 = 0$$
$$k(k-1) = 0$$

Invertible if $k \neq 0$ and $k \neq 1$

■   **Invertibility Test**

24.   The matrix has an inverse because its determinant is nonzero.

■   **Product Verification**

27. $\mathbf{A} = \begin{bmatrix} 1 & 2 \\ 3 & 4 \end{bmatrix}$

$\mathbf{B} = \begin{bmatrix} 1 & 0 \\ 1 & 1 \end{bmatrix}$

$\mathbf{AB} = \begin{bmatrix} 3 & 2 \\ 7 & 4 \end{bmatrix}$

$|\mathbf{A}| = \begin{bmatrix} 1 & 2 \\ 3 & 4 \end{bmatrix} = -2$

$|\mathbf{B}| = \begin{bmatrix} 1 & 0 \\ 1 & 1 \end{bmatrix} = 1$

$|\mathbf{AB}| = \begin{bmatrix} 3 & 2 \\ 7 & 4 \end{bmatrix} = -2$

Hence $|\mathbf{AB}| = |\mathbf{A}||\mathbf{B}|$.

■   **Do Determinants Commute?**

30. $$|\mathbf{AB}| = |\mathbf{A}||\mathbf{B}| = |\mathbf{B}||\mathbf{A}| = |\mathbf{BA}|,$$

because $|\mathbf{A}||\mathbf{B}|$ is a product of real or complex numbers.

■   **Determinants of Sums**

33.   An example is

$$\mathbf{A} = \begin{bmatrix} 1 & 0 \\ 0 & 1 \end{bmatrix}, \quad \mathbf{B} = \begin{bmatrix} -1 & 0 \\ 0 & -1 \end{bmatrix},$$

so

$$\mathbf{A} + \mathbf{B} = \begin{bmatrix} 0 & 0 \\ 0 & 0 \end{bmatrix},$$

which has the determinant

$$|\mathbf{A} + \mathbf{B}| = 0,$$

whereas

$$|\mathbf{A}| = |\mathbf{B}| = 1,$$

so $|\mathbf{A}| + |\mathbf{B}| = 2$. Hence,

$$|\mathbf{A} + \mathbf{B}| \neq |\mathbf{A}| + |\mathbf{B}|.$$

■   **Inversion by Determinants**

36.   Given the matrix

$$\mathbf{A} = \begin{bmatrix} 1 & 0 & 2 \\ 2 & 2 & 3 \\ 1 & 1 & 1 \end{bmatrix}$$

the matrix of minors can easily be computed and is

$$\mathbf{M} = \begin{bmatrix} -1 & -1 & 0 \\ -2 & -1 & 1 \\ -4 & -1 & 2 \end{bmatrix}.$$

The matrix of cofactors $\tilde{\mathbf{A}}$, which we get by multiplying the minors by $(-1)^{i+j}$, is given by

$$\tilde{\mathbf{A}} = (-1)^{i+j}\mathbf{M} = \begin{bmatrix} -1 & 1 & 0 \\ 2 & -1 & -1 \\ -4 & 1 & 2 \end{bmatrix}.$$

Taking the transpose of this matrix gives

$$\tilde{\mathbf{A}}^{\mathrm{T}} = \begin{bmatrix} -1 & 2 & -4 \\ 1 & -1 & 1 \\ 0 & -1 & 2 \end{bmatrix}.$$

Computing the determinant of $\mathbf{A}$, we get $|\mathbf{A}| = -1$. Hence, we have the inverse

$$\mathbf{A}^{-1} = \frac{1}{|\mathbf{A}|}\tilde{\mathbf{A}}^{\mathrm{T}} = \begin{bmatrix} 1 & -2 & 4 \\ -1 & 1 & -1 \\ 0 & 1 & -2 \end{bmatrix}.$$

■   Cramer's Rule

39.
$$x + 2y = 2$$
$$2x + 5y = 0$$

To solve this system we write it in matrix form as
$$\begin{bmatrix} 1 & 2 \\ 2 & 5 \end{bmatrix} \begin{bmatrix} x \\ y \end{bmatrix} = \begin{bmatrix} 2 \\ 0 \end{bmatrix}.$$

Using Cramer's rule, we compute the determinants
$$|A| = \begin{vmatrix} 1 & 2 \\ 2 & 5 \end{vmatrix} = 1, \quad |A_1| = \begin{vmatrix} 2 & 2 \\ 0 & 5 \end{vmatrix} = 10, \quad |A_2| = \begin{vmatrix} 1 & 2 \\ 2 & 0 \end{vmatrix} = -4.$$

Hence, the solution is
$$x = \frac{|A_1|}{|A|} = \frac{10}{1} = 10, \quad y = \frac{|A_2|}{|A|} = -\frac{4}{1} = -4.$$

42.
$$x_1 + 2x_2 - x_3 = 6$$
$$3x_1 + 8x_2 + 9x_3 = 10$$
$$2x_1 - x_2 + 2x_3 = -2$$

To solve this system, we write it in matrix form as
$$\begin{bmatrix} 1 & 2 & -1 \\ 3 & 8 & 9 \\ 2 & -1 & 2 \end{bmatrix} \begin{bmatrix} x_1 \\ x_2 \\ x_3 \end{bmatrix} = \begin{bmatrix} 6 \\ 10 \\ -2 \end{bmatrix}.$$

Using Cramer's rule, we compute the determinants

$$|A| = \begin{vmatrix} 1 & 2 & -1 \\ 3 & 8 & 9 \\ 2 & -1 & 2 \end{vmatrix} = 68, \quad |A_1| = \begin{vmatrix} 6 & 2 & -1 \\ 10 & 8 & 9 \\ -2 & -1 & 2 \end{vmatrix} = 68, \quad |A_2| = \begin{vmatrix} 1 & 6 & -1 \\ 3 & 10 & 9 \\ 2 & -2 & 2 \end{vmatrix} = 136, \quad |A_3| = \begin{vmatrix} 1 & 2 & 6 \\ 3 & 8 & 10 \\ 2 & -1 & -2 \end{vmatrix} = -68.$$

Hence, the solution is
$$x_1 = \frac{|A_1|}{|A|} = \frac{68}{68} = 1, \quad x_2 = \frac{|A_2|}{|A|} = \frac{136}{68} = 2, \quad x_3 = \frac{|A_3|}{|A|} = -\frac{68}{68} = -1.$$

■ **Alternative Derivation of Least Squares Equations**

45. (a) Equation (9) in the text

$$k + 1.7m = 1.1$$
$$k + 2.3m = 3.1$$
$$k + 3.1m = 2.3$$
$$k + 4.0m = 3.8$$

can be written in matrix form

$$\begin{bmatrix} 1 & 1.7 \\ 1 & 2.3 \\ 1 & 3.1 \\ 1 & 4.0 \end{bmatrix} \begin{bmatrix} k \\ m \end{bmatrix} = \begin{bmatrix} 1.1 \\ 3.1 \\ 2.3 \\ 3.8 \end{bmatrix}$$

which is the form of $\mathbf{A}\vec{\mathbf{x}} = \vec{\mathbf{b}}$.

(b) Given the matrix equation $\mathbf{A}\vec{\mathbf{x}} = \vec{\mathbf{b}}$, where

$$\mathbf{A} = \begin{bmatrix} 1 & x_1 \\ 1 & x_2 \\ 1 & x_3 \\ 1 & x_4 \end{bmatrix}, \quad \vec{\mathbf{x}} = \begin{bmatrix} k \\ m \end{bmatrix}, \quad \vec{\mathbf{b}} = \begin{bmatrix} y_1 \\ y_2 \\ y_3 \\ y_4 \end{bmatrix}$$

if we premultiply each side of the equation by $\mathbf{A}^{\mathrm{T}}$, we get $\mathbf{A}^{\mathrm{T}}\mathbf{A}\vec{\mathbf{x}} = \mathbf{A}^{\mathrm{T}}\vec{\mathbf{b}}$, or

$$\begin{bmatrix} 1 & 1 & 1 & 1 \\ x_1 & x_2 & x_3 & x_4 \end{bmatrix} \begin{bmatrix} 1 & x_1 \\ 1 & x_2 \\ 1 & x_3 \\ 1 & x_4 \end{bmatrix} \begin{bmatrix} k \\ m \end{bmatrix} = \begin{bmatrix} 1 & 1 & 1 & 1 \\ x_1 & x_2 & x_3 & x_4 \end{bmatrix} \begin{bmatrix} y_1 \\ y_2 \\ y_3 \\ y_4 \end{bmatrix}$$

or

$$\begin{bmatrix} 4 & \sum_{i=1}^{4} x_i \\ \sum_{i=1}^{4} x_i & \sum_{i=1}^{4} x_i^2 \end{bmatrix} \begin{bmatrix} k \\ m \end{bmatrix} = \begin{bmatrix} \sum_{i=1}^{4} y_i \\ \sum_{i=1}^{4} x_i y_i \end{bmatrix}.$$

Computer or Calculator

48. To find the least-square approximation of the form $y = k + mx$, we solve to a set of data points $\{(x_i, y_i) : i = 1, 2, ..., n\}$ to get the system

$$\begin{bmatrix} n & \sum_{i=1}^{n} x_i \\ \sum_{i=1}^{n} x_i & \sum_{i=1}^{n} x_i^2 \end{bmatrix} \begin{bmatrix} k \\ m \end{bmatrix} = \begin{bmatrix} \sum_{i=1}^{n} y_i \\ \sum_{i=1}^{n} x_i y_i \end{bmatrix}.$$

Using a spreadsheet to compute the elements of the coefficient matrix and the right-hand-side vector, we get

Spreadsheet to compute least squares

x	y	x^2	xy
0.91	1.35	0.8281	1.2285
1.07	1.96	1.1449	2.0972
2.56	3.13	6.5536	8.0128
4.11	5.72	16.8921	23.5092
5.34	7.08	28.5156	37.8072
6.25	8.14	39.0625	50.8750

sum x	sum y	sum x^2	sum xy
20.24	27.38	92.9968	123.5299

We must solve the system

$$\begin{bmatrix} 6.00 & 20.2400 \\ 20.24 & 92.9968 \end{bmatrix} \begin{bmatrix} k \\ m \end{bmatrix} = \begin{bmatrix} 27.3800 \\ 123.5299 \end{bmatrix}$$

getting $k = 0.309$, $m = 1.26$. Hence, the least-squares line is $y = 0.309 + 1.26x$.

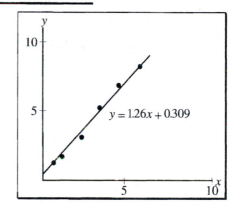

Suggested Journal Entry

51. Student Project

3.5   Vector Spaces and Subspaces

■   **They Do Not All Look Like Vectors**

3.   A typical vector is $[a,b,c,d]$, with negative $[-a, -b, -c, -d]$; the zero vector is $[0, 0, 0, 0]$.

6.   A typical vector is $\begin{bmatrix} a & b & c \\ d & e & f \\ g & h & i \end{bmatrix}$, with negative $\begin{bmatrix} -a & -b & -c \\ -d & -e & -f \\ -g & -h & -i \end{bmatrix}$; the zero vector is $\begin{bmatrix} 0 & 0 & 0 \\ 0 & 0 & 0 \\ 0 & 0 & 0 \end{bmatrix}$.

9.   A typical vector is a continuous and differentiable function, such as $f(t) = \sin t$, $g(t) = t^2$. The zero vector is $f_0(t) \equiv 0$ and the negative of $f(t)$ is $-f(t)$.

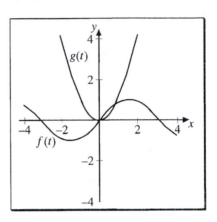

■   **Are They Vector Spaces?**

12.   First octant of space: No, the vectors have no negatives. For example, $[1, 3, 3]$ belongs to the set but $[-1, -3, -3]$ does not.

15.   Not a vector space since it is not closed under vector addition. See the example for Problem 14.

18.   Not a vector space; the set of all 2×2 invertible matrices is not closed under vector addition. For instance,

$$\begin{bmatrix} 1 & 0 \\ 0 & 1 \end{bmatrix} + \begin{bmatrix} -1 & 0 \\ 0 & -1 \end{bmatrix} = \begin{bmatrix} 0 & 0 \\ 0 & 0 \end{bmatrix}.$$

21.   Not a vector space; not closed under scalar multiplication; no additive inverse.

■   **A Familiar Vector Space**

24.   Yes a vector space. Straightforward verification of the 10 commandments of a vector space; that is, the sum of two vectors (real numbers in this case) is a vector (another real number), the product of a real number by a scalar (another real number) is a real number. The zero vector is the number 0. Every number has a negative. The distributivity and associatively properties are simply properties of the real numbers, and so on.

■ Another Solution Space

27.   Yes, the solution space of the linear homogeneous DE

$$(-y)'' + p(t)(-y') + q(t)(-y) = -\left[y'' + p(t)\, y' + q(t)\, y \right] = 0$$

is indeed a vector space; the linearity properties are sufficient to prove all the vector space properties.

■ Vector Space Properties

30.   *Unique Negative*: We show that if $\vec{v}$ is an arbitrary vector in some vector space, then there is only one vector $\vec{n}$ (which we call $-\vec{v}$) in that space that satisfies $\vec{v} + \vec{n} = \vec{0}$. Suppose another vector $\vec{n}*$ also satisfies $\vec{v} + \vec{n}* = \vec{0}$. Then

$$\vec{n} = \vec{n} + \vec{0} = \vec{n} + \left(\vec{v} + (-\vec{v}) \right) = (\vec{n} + \vec{v}) + (-\vec{v}) = \vec{0} + (-\vec{v}) = (\vec{v} + \vec{n}*) + (-\vec{v}) = \vec{v} + (-\vec{v}) + \vec{n}*$$

$$= \vec{0} + \vec{n}* = \vec{n}*.$$

■ A Vector Space Equation

33.   Let $\vec{v}$ be an arbitrary vector and c an arbitrary scalar. Set $c\vec{v} = \vec{0}$. Then either $c = 0$ or $\vec{v} = \vec{0}$. For $c \neq 0$,

$$\vec{v} = 1\vec{v} = \frac{1}{c}(c\vec{v}) = \frac{1}{c}(\vec{0}) = \vec{0},$$

which proves the result.

■ Nonstandard Definitions

36. $(x_1,\ y_1) + (x_2,\ y_2) \equiv (x_1 + x_2,\ y_1 + y_2)$ and $c(x,\ y) \equiv \left(\sqrt{c}\,x,\ \sqrt{c}\,y \right)$

Not a vector space, for example,

$$(c + d)\vec{x} \neq c\vec{x} + d\vec{x}.$$

For $c = 4$, $d = 9$ and vector $\vec{x} = (x_1,\ x_2)$, we have

$$(c + d)\vec{x} = 13(x_1,\ x_2) = \left(\sqrt{13}\,x_1,\ \sqrt{13}\,x_2 \right)$$

$$c\vec{x} + d\vec{x} = 4(x_1,\ x_2) + 9(x_1,\ x_2) = (2x_1,\ 2x_2) + (3x_1,\ 3x_2) = (5x_1,\ 5x_2).$$

■ Sifting Subsets for Subspaces

39. $\mathbb{W} = \left\{ (x_1,\ x_2,\ x_3) \,\middle|\, x_3 = 0 \right\}$ is a subspace of $\mathbb{R}^3$.

42. $\mathbb{W} = \left\{ f(t) \,\middle|\, f(0) = 0 \right\}$ is a subspace of $\mathscr{C}[0,\ 1]$.

45. $\mathbb{W} = \left\{ f(t) \,\middle|\, f'' + f = 0 \right\}$ is a subspace of $\mathscr{C}^2[0,\ 1]$.

48.   W is a subspace:

Nonempty: Note that $A\vec{0}=\vec{0}$ so $\vec{0}\in W$

Closure: Suppose $\vec{x},\vec{y}\in W$

$A\vec{x}=\vec{0}$ and $A\vec{y}=\vec{0}$,

so $A(a\vec{x}+b\vec{y})$

$$= A(a\vec{x})+A(b\vec{y})$$

$$= aA\vec{x}+bA\vec{y}=a\vec{0}+b\vec{0}=\vec{0}+\vec{0}=\vec{0}$$

■   **Are They Subspaces of $\mathbb{R}^n$?**

51.   No $[0, 0, 0, 0, 0] \notin \{[a, 0, b, 1, c]: a, b, c \in \mathbb{R}\}$
because the 4th coordinate $\neq 0$ for all $a, b, c \in \mathbb{R}$

■   **Differentiable Subspaces**

54. $\{f(t)\,|\,f'=1\}$. It is not a subspace, because it does not contain the zero vector and is not closed under vector addition. Hence, $f(t)=t$, $g(t)=t+2$ belongs to the subset but $(f+g)(t)$ does not belong. It is also not closed under scalar multiplication. For example $f(t)=t$ belongs to the subset, but $2f(t)=2t$ does not.

■   **Property Failures**

57.   The first quadrant (including the coordinate axes) is closed under vector addition, but not scalar multiplication.

■   **Solution Spaces of Homogenous Linear Algebraic Systems**

60. $x_1 - x_2 \quad\ \ + 4x_4 + 2x_5 - x_6 = 0$
$2x_1 - 2x_2 + x_3 + 2x_4 + 4x_5 - x_6 = 0$

The matrix of coefficients $A = \begin{vmatrix} 1 & -1 & 0 & 4 & 2 & -1 \\ 2 & -2 & 1 & 2 & 4 & -1 \end{vmatrix}$

has RREF $= \begin{vmatrix} 1 & -1 & 0 & 4 & 2 & -1 \\ 0 & 0 & 1 & -6 & 0 & 1 \end{vmatrix}$

$\quad\quad x_1 - x_2 \quad\ \ + 4x_4 + 2x_5 - x_6 = 0$
$\quad\quad\quad\quad x_3 - 6x_4 \quad\quad\ + x_6 = 0$

Let $x_2 = r$, $x_4 = s$, $x_5 = t$, $x_6 = u$ $\quad\quad x_1 = r - 4s - 2t + u$
$\quad\quad\quad\quad\quad\quad\quad\quad\quad\quad\quad\quad\quad\quad\quad x_3 = \quad\ 6s \quad\ - u$

$$\mathbb{S} = \left\{ r\begin{bmatrix} 1 \\ 1 \\ 0 \\ 0 \\ 0 \\ 0 \end{bmatrix} + s\begin{bmatrix} -4 \\ 0 \\ 6 \\ 1 \\ 0 \\ 0 \end{bmatrix} + t\begin{bmatrix} -2 \\ 0 \\ 0 \\ 0 \\ 1 \\ 0 \end{bmatrix} + u\begin{bmatrix} 1 \\ 0 \\ -1 \\ 0 \\ 0 \\ 1 \end{bmatrix} : r,s,t,u \in \mathbb{R} \right\}$$

■ Nonlinear Differential Equations

63. $y' = y^2$. Writing the equation in differential form, we have $y^{-2} dy = dt$. We get the general solution $y = \dfrac{1}{c - t}$. Hence, from $c = 0$ and 1, we have two solutions

$$y_1(t) = -\frac{1}{t}, \quad y_2(t) = \frac{1}{1-t}.$$

But, if we compute

$$y_1(t) + y_2(t) = -\frac{1}{t} + \frac{1}{1-t}$$

it would not be a solution of the DE. So the solution set of this nonlinear DE is not a vector space.

■ DE Solution Spaces

66. $y' + 2y = e^t$. Not a vector space, it doesn't contain the zero vector.

69. $y'' + (1 + \sin t) y = 0$. If y_1, y_2 satisfy the equation then

$$y_1'' + (1 + \sin t) y_1 = 0$$
$$y_2'' + (1 + \sin t) y_2 = 0.$$

By adding, we have

$$\left[y_1'' + (1 + \sin t) y_1 \right] + \left[y_1'' + (1 + \sin t) y_1 \right] = 0,$$

which from properties of the derivative is equivalent to

$$\left(c_1 y_1 + c_2 y_2 \right)'' + (1 + \sin t)\left(c_1 y_1 + c_2 y_2 \right) = 0,$$

which shows the set of solutions is a vector space. This is true for the solution set of *any* linear homogeneous DE.

■ Orthogonal Complements

72.   To prove: $\mathbb{V} \cap \mathbb{V}^{\perp} = \{\vec{0}\}$

$\vec{0} \in \mathbb{V}$ and $\vec{0} \in \mathbb{V}^{\perp}$ since $\vec{\mathbb{V}}$ is a subspace and $\vec{0} \cdot \vec{v} = 0$ for every $\vec{v} \in \mathbb{V}$, so $\vec{0} \in \mathbb{V}^{\perp}$

$\therefore \{\vec{0}\} \subset \mathbb{V} \cap \mathbb{V}^{\perp}$

Now suppose $\vec{w} \in \mathbb{V} \cap \mathbb{V}^{\perp}$ where $\vec{w} = [w_1, w_2, \ldots, w_n]$

Then $\vec{w} \cdot \vec{v} = \vec{0}$ for all $\vec{v} \in \mathbb{V}$
However $\vec{w} \in \mathbb{V}$ so $\vec{w} \cdot \vec{w} = 0$
$w_1^2 + w_2^2 + \ldots + w_n^2 = 0$
$\therefore w_1 = w_2 = \ldots = w_n = 0$

$\therefore \vec{w} = \vec{0}$

3.6   Basis and Dimension

■   **The Spin on Spans**

3. $V = \mathbb{R}^3$. Letting

$$[a, b, c] = c_1[1, 0, -1] + c_2[2, 0, 4] + c_3[-5, 0, 2] + c_4[0, 0, 1]$$

yields

$$a = c_1 + 2c_2 - 5c_3$$
$$b = 0$$
$$c = -c_1 + 4c_2 + 2c_3 + c_4.$$

These vectors do not span $\mathbb{R}^3$ because they cannot give any vector with $b \neq 0$.

6. $V = \mathbb{M}_{22}$. Letting

$$\begin{bmatrix} a & b \\ c & d \end{bmatrix} = c_1\begin{bmatrix} 1 & 1 \\ 0 & 0 \end{bmatrix} + c_2\begin{bmatrix} 0 & 0 \\ 1 & 1 \end{bmatrix} + c_3\begin{bmatrix} 1 & 0 \\ 1 & 0 \end{bmatrix} + c_4\begin{bmatrix} 0 & 1 \\ 0 & 1 \end{bmatrix}$$

we have the equations

$$c_1 + c_3 = a$$
$$c_1 + c_4 = b$$
$$c_2 + c_3 = c$$
$$c_2 + c_4 = d.$$

If we add the first and last equation, and then the second and third equations, we obtain the equations

$$c_1 + c_2 + c_3 + c_4 = a + d$$
$$c_1 + c_2 + c_3 + c_4 = b + c.$$

Hence, we have a solution if and only if $a + d = b + c$. This means we can solve for c_1, c_2, c_3, c_4 for only a subset of vectors in V. Hence, $\mathbb{W}$ does not span $\mathbb{R}^4$.

■   **Independence Day**

9. $V = \mathbb{R}^3$. Setting

$$c_1[1, 0, 0] + c_2[1, 1, 0] + c_3[1, 1, 1] = [0, 0, 0]$$

we get

$$c_1 + c_2 + c_3 = 0$$
$$c_2 + c_3 = 0$$
$$c_3 = 0$$

which implies $c_1 = c_2 = c_3 = 0$. Hence vectors in $\mathbb{W}$ are linearly independent.

12. $V = \mathbb{P}_1$. Setting

$$c_1 + c_2 t = 0,$$

we get $c_1 = 0$, $c_2 = 0$. Hence, the vectors in $\mathbb{W}$ are linearly independent.

15. $\mathbb{V} = \mathbb{P}_2$. Setting

$$c_1(1+t)+c_2(1-t)+c_3t^2 = 0$$

we get

$$
\begin{aligned}
c_1 + c_2 \quad &= 0 \\
c_1 - c_2 \quad &= 0 \\
c_3 &= 0
\end{aligned}
$$

which implies $c_1 = c_2 = c_3 = 0$. Hence, the vectors in $\mathbb{W}$ are linearly independent.

18. $\mathbb{V} = \mathbb{D}_{22}$. Setting

$$\begin{bmatrix} a & 0 \\ 0 & b \end{bmatrix} = c_1 \begin{bmatrix} 1 & 0 \\ 0 & 1 \end{bmatrix} + c_2 \begin{bmatrix} 1 & 0 \\ 0 & -1 \end{bmatrix}$$

we get $c_1 + c_2 = a$, $c_1 - c_2 = b$. We can solve these equations for c_1, c_2, and hence these vectors are linearly independent and span $\mathbb{D}_{22}$.

■   **Function Space Dependence**

21. $S = \{\sin t, \, \sin 2t, \, \sin 3t\}$. Letting

$$c_1 \sin t + c_2 \sin 2t + c_3 \sin 3t = 0$$

for all t. In particular if we choose three values of t, say $\dfrac{\pi}{6}$, $\dfrac{\pi}{4}$, $\dfrac{\pi}{2}$, we obtain three equations to solve for c_1, c_2, c_3, namely,

$$c_1\left(\frac{1}{2}\right) + c_2\left(\frac{\sqrt{3}}{2}\right) + c_3 = 0$$

$$c_1\left(\frac{\sqrt{2}}{2}\right) + c_2 + c_3\left(\frac{\sqrt{2}}{2}\right) = 0$$

$$c_1 - c_3 = 0.$$

We used Maple to compute the determinant of this coefficient matrix and found it to be $-\dfrac{3}{2} + \dfrac{1}{2}\sqrt{6}$. Hence, the system has a unique solution $c_1 = c_2 = c_3 = 0$. Thus, $\sin t$, $\sin 2t$, and $\sin 3t$ are linearly independent.

24. $S = \{e^t, e^{-t}, \cosh t\}$. Because

$$\cosh t = \frac{1}{2}\left(e^t + e^{-t}\right)$$

we have that $2\cosh t - e^t - e^{-t} = 0$ is a nontrivial linear combination that is identically zero for all t. Hence, the vectors are linearly dependent.

■   **Independence Testing**

27.   We will show that

$$c_1\begin{bmatrix} \sin t \\ \cos t \end{bmatrix} + c_2\begin{bmatrix} \cos t \\ -\sin t \end{bmatrix} = \begin{bmatrix} 0 \\ 0 \end{bmatrix}$$

for all t implies $c_1 = c_2 = 0$, and hence, the vectors are linearly independent. If it is true for all t, then it must be true for $t = 0$, which gives the two equations $c_2 = 0$, $c_1 = 0$. This proves the vectors are linearly independent.

Another approach is to say that the vectors are linearly independent because clearly there is no constant k such that one vector is k times the other vector for all t.

■   **Twins?**

30.   We have $\text{span}\{\cos t + \sin t, \cos t - \sin t\} = \{c_1(\cos t + \sin t) + c_2(\cos t - \sin t)\}$

$$= \{(c_1 + c_2)\cos t + (c_1 - c_2)\sin t\}$$

$$= \{C_1\cos t + C_2\sin t\}$$

$$= \text{span}\{\sin t, \cos t\}.$$

■   **Zero Wronskian Does Not Imply Linear Dependence**

33.   (a) $f(t) = t^2$ $g(t) = \begin{cases} t^2 & t \ge 0 \\ -t^2 & t < 0 \end{cases}$

$f'(t) = 2t$ $g'(t) = \begin{cases} 2t & t \ge 0 \\ -2t & t < 0 \end{cases}$

For $t \ge 0$ $W = \begin{vmatrix} t^2 & t^2 \\ 2t & 2t \end{vmatrix} = 0$

For $t < 0$ $W = \begin{vmatrix} t^2 & -t^2 \\ 2t & -2t \end{vmatrix} = -2t^3 + 2t^3 = 0$

$\therefore W = 0$ on $(-\infty, \infty)$

(b) f and g are linearly independent because $f(t) \ne kg(t)$ on $(-\infty, \infty)$ for every $k \in \mathbb{R}$.

■ Revisiting Linear Independence

36. The Wronskian is

$$W = \begin{vmatrix} e^t & 5e^{-t} & e^{3t} \\ e^t & -5e^{-t} & 3e^{3t} \\ e^t & 5e^{-t} & 9e^{3t} \end{vmatrix} = e^t \begin{vmatrix} -5e^{-t} & 3e^{3t} \\ 5e^{-t} & 9e^{3t} \end{vmatrix} - e^t \begin{vmatrix} 5e^{-t} & e^{3t} \\ 5e^{-t} & 9e^{3t} \end{vmatrix} + e^t \begin{vmatrix} 5e^{-t} & e^{3t} \\ -5e^{-t} & 3e^{3t} \end{vmatrix}$$

$$= e^{3t}\left[(-45-15)-(45-5)+(15+5)\right] = -80e^{3t} \neq 0$$

Hence, the vectors are linearly independent.

■ Independence Checking

39.

$$W = \begin{vmatrix} 2+t & 2-t & t^2 \\ 1 & -1 & 2t \\ 0 & 0 & 2 \end{vmatrix} = 2 \begin{vmatrix} 2+t & 2-t \\ 1 & -1 \end{vmatrix} = 2(-2-t-2+t)$$

$$= -8 \neq 0$$

$\therefore \{2+t,\ 2-t,\ t^2\}$ is linearly independent on $\mathbb{R}$

42.

$$W = \begin{vmatrix} e^t \cos t & e^t \sin t \\ e^t(-\sin t)+e^t \cos t & e^t \cos t + e^t \sin t \end{vmatrix}$$

$$= e^{2t}\cos^2 t + e^{2t}\cos t \sin t + e^{2t}\sin^2 t - e^t \sin t \cos t$$

$$= e^{2t}(\cos^2 t + \sin^2 t) = e^{2t} \neq 0 \text{ for all } t$$

$\{e^t \cos t, e^t \sin t\}$ is linearly independent on $\mathbb{R}$

■ Getting on Base in $\mathbb{R}^2$

45. $\{(-1, -1),\ (1, 1)\}$ is not a basis because $[-1, -1] = -[1, 1]$, hence they are linearly dependent.

48. $\{[0, 0],\ [1, 1],\ [2, 2],\ [-1, -1]\}$ is not a basis because the vectors are linearly dependent.

■ The Base for the Space

51. $\mathbb{V} = \mathbb{R}^3$: S is not a basis because four vectors are linearly dependent in $\mathbb{R}^3$.

54. $\mathbb{V} = \mathbb{P}_4$: We assume that

$$c_1 t^4 + c_2 (t+3) + c_3 \left(t^3 + 4\right) + c_4 (t-1) + c_5 \left(t^2 - 5t + 1\right) = 0$$

and compare coefficients. We find a homogeneous system of equations that has only the zero solution $c_1 = c_2 = c_3 = c_4 = c_5 = 0$. Hence, the vectors are linearly independent. To show the vectors span $\mathbb{P}_4$, we set the above linear combination equal to an arbitrary vector $at^4 + bt^3 + ct^2 + dt + e$, and compare coefficients to arrive at a system of equations, which can besolved for c_1, c_2, c_3, c_4, and c_5 in terms of a, b, c, d, e. Hence, the vectors span $\mathbb{P}_4$ so that they are a basis for $\mathbb{P}_4$.

■   **Sizing Them Up**

57. $W = \left\{ \left[x_1,\ x_2,\ x_3 \right] \middle| x_1 + x_2 + x_3 = 0 \right\}$

Letting $x_2 = \alpha$, $x_3 = \beta$, we can write $x_1 = -\alpha - \beta$. Any vector in W can be written as

$$\begin{bmatrix} x_1 \\ x_2 \\ x_3 \end{bmatrix} = \begin{bmatrix} -\alpha - \beta \\ \alpha \\ \beta \end{bmatrix} = \alpha \begin{bmatrix} -1 \\ 1 \\ 0 \end{bmatrix} + \beta \begin{bmatrix} -1 \\ 0 \\ 1 \end{bmatrix}$$

where α and β are arbitrary real numbers. Hence, The dimension of W is 2; a basis is

$$\left\{ \left[-1,\ 1,\ 0 \right], \left[-1,\ 0,\ 1 \right] \right\}.$$

■   **Polynomial Dimensions**

60. $\left\{ t,\ t-1,\ t^2 + 1 \right\}$. We write

$$at^2 + bt + c = c_1 t + c_2 (t - 1) + c_3 (t^2 + 1)$$

yielding the equations

$$
\begin{aligned}
t^2: & & & & c_3 & = a \\
t: & & c_1 & + c_2 & & = b \\
1: & & & - c_2 & + c_3 & = c.
\end{aligned}
$$

Because we can solve this system for c_1, c_2, c_3 in terms of a, b, c getting

$$c_1 = -a + c + b$$
$$c_2 = a - c$$
$$c_3 = a$$

the subspace spans the entire three-dimensional vector space $\mathbb{P}_2$.

■   **Solution Spaces for Linear Algebraic Systems**

63.   The matrix of coefficients for the system in Problem 61, Section 3.5

has RREF $\begin{bmatrix} 1 & 0 & 3 & 0 \\ 0 & 1 & 1 & 0 \\ 0 & 0 & 0 & 1 \\ 0 & 0 & 0 & 0 \end{bmatrix}$, so $\begin{aligned} x_1 + 3x_3 &= 0 \\ x_2 + x_3 &= 0. \\ x_4 &= 0 \end{aligned}$

Let $r = x_3$; then

$$W = \left\{ \begin{bmatrix} 3 \\ -1 \\ 1 \\ 0 \end{bmatrix} : r \in \mathbb{R} \right\} \text{ so a basis is } \left\{ \begin{bmatrix} 3 \\ -1 \\ 1 \\ 0 \end{bmatrix} \right\}.$$

Dim $W = 1$.

■   **DE Solution Spaces**

66. $y' - 2y = 0$ This is a first order linear DE with solution (by either method of Section 2.2)

$$y = Ce^{2t}$$

(a)   The solution space $\mathbb{S} = \{Ce^{2t} : C \in \mathbb{R}\} \subseteq \mathscr{C}^n(\mathbb{R})$

(b)   A basis $B = \{e^{2t}\}$,   dim $\mathbb{S} = 1$.

69. $y' + y^2 = 0$ y^2 is not a linear function so
$y' + y^2 = 0$ is not a linear differential equation.
By separation of variables

$$y' = -y^2$$

$$\int \frac{dy}{y^2} = -\int dt$$

$$\frac{y^{-1}}{-1} = -t + c$$

$$\frac{1}{y} = t - c$$

$$y = \frac{1}{t - c}$$

But these solutions do not form a vector space

Let $k \in \mathbb{R}$, $k \neq 1$; then

$$\frac{k}{t - c}$$ is not a solution of the ODE.

Hence $\left\{ \dfrac{1}{t - c} : c \in \mathbb{R} \right\}$ is not a vector space.

■   **Bases for Subspaces of $\mathbb{R}^n$**

72. $\mathbb{W} = \{(a, a - b, 2a + 3b) : a, b \in \mathbb{R}\}$

$$= \left\{ a \begin{bmatrix} 1 \\ 1 \\ 2 \end{bmatrix} + b \begin{bmatrix} 0 \\ -1 \\ 3 \end{bmatrix} : a, b \in \mathbb{R} \right\}$$

so $\left\{ \begin{bmatrix} 1 \\ 1 \\ 2 \end{bmatrix}, \begin{bmatrix} 0 \\ -1 \\ 3 \end{bmatrix} \right\}$ is a basis for $\mathbb{W}$.

Dim $\mathbb{W} = 2$.

■   **Basis for Zero Trace Matrices**

75.   Letting

$$c_1\begin{bmatrix} 1 & 0 \\ 0 & -1 \end{bmatrix} + c_2\begin{bmatrix} 0 & 1 \\ 0 & 0 \end{bmatrix} + c_3\begin{bmatrix} 0 & 0 \\ 1 & 0 \end{bmatrix} = \begin{bmatrix} a & b \\ c & d \end{bmatrix}$$

we find $a = c_1$, $b = c_2$, $c = c_3$, $d = -c_1$. Given $a = b = c = d = 0$ implies $c_1 = c_2 = c_3 = c_4 = 0$, which shows the vectors (matrices) are linearly independent. It also shows they span the set of 2×2 matrices with trace zero because if $a + d = 0$, we can solve for $c_1 = a = -d$, $c_2 = b$, $c_3 = c$. In other words we can write any zero trace 2×2 matrix as follows as a linear combination of the three given vectors (matrices):

$$\begin{bmatrix} a & b \\ c & -a \end{bmatrix} = a\begin{bmatrix} 1 & 0 \\ 0 & -1 \end{bmatrix} + b\begin{bmatrix} 0 & 1 \\ 0 & 0 \end{bmatrix} + c\begin{bmatrix} 0 & 0 \\ 1 & 0 \end{bmatrix}.$$

Hence, the vectors (matrices) form a basis for the 2×2 zero trace matrices.

■   **Making New Basis From Old:**

78. $B_1 = \{\vec{i}, \vec{j}, \vec{k}\}$ (Many correct answers)

A typical answer is $B_2 = \{\vec{i}, \vec{i} + \vec{j}, \vec{i} + \vec{k}\}$

To show linear independence:

Set $c_1\vec{i} + c_2(\vec{i} + \vec{j}) + c_3(\vec{i} + \vec{k}) = \vec{0}$

$$c_1 + c_2 + c_3 = 0$$
$$c_1 + c_2 = 0$$
$$c_1 + c_3 = 0$$

$$\begin{vmatrix} 1 & 1 & 1 \\ 1 & 1 & 0 \\ 1 & 0 & 1 \end{vmatrix} = -1 \neq 0$$

$\therefore B_2$ is a basis since dim $\mathbb{R}^3 = 3$

■   **Basis for $\mathbb{P}_2$**

81.   We first show the vectors span $\mathbb{P}_2$ by selecting an arbitrary vector from $\mathbb{P}_2$ and show it can be written as a linear combination of the three given vectors. We set

$$at^2 + bt + c = c_1\left(t^2 + t + 1\right) + c_2\left(t + 1\right) + c_3$$

and try to solve for c_1, c_2, c_3 in terms of a, b, c. Setting the coefficients of t^2, t, and 1 equal to each other yields

$$
\begin{aligned}
t^2: \quad & c_1 && && = a \\
t: \quad & c_1 & + c_2 && && = b \\
1: \quad & c_1 & + c_2 & + c_3 && = c,
\end{aligned}
$$

giving the solution $c_1 = a$, $c_2 = -a + b$, $c_3 = -b + c$.

continued on the next page

Hence, the set spans $\mathbb{P}_2$. We also know that the vectors

$$\left\{t^2+t+1,\ t+1,\ 1\right\}$$

are independent because setting

$$c_1\left(t^2+t+1\right)+c_2\left(t+1\right)+c_3=0$$

we get

$$
\begin{aligned}
c_1 &&&&&= 0\\
c_1 &+& c_2 &&&= 0\\
c_1 &+& c_2 &+& c_3 &= 0
\end{aligned}
$$

which has only the solution $c_1=c_2=c_3=0$. Hence, the vectors are a basis for $\mathbb{P}_2$, for example,

$$3t^2+2t+1=3\left(t^2+t+1\right)-1(t+1)-1(1).$$

■ **Convergent Sequence Space**

84. $\mathbb{V}$ is a vector space since the addition and scalar multiplication operations follow the rules for $\mathbb{R}$

and the operations

$$\{a_n\}+\{b_n\}=\{a_n+b_n\}\ \text{and}\ c\{a_n\}=\{ca_n\}$$

are the precise requirements for closure under vector addition and scalar multiplication.

Zero element is $\{0\}$ where $a_n=0$ for all n

Additive Inverse for $\{a_n\}$ is $\{-a_n\}$

Let $\mathbb{W}=\left\{\{2a_n\}:\{a_n\}\in\mathbb{V}\right\}$ Clearly $\{0\}=\mathbb{W}$ and $\{2a_n\}+\{2b_n\}=\{2a_n+2b_n\}=2\{a_n+b_n\}$

Also $k\{2a_n\}=\{2ka_n\}$ for every $k\in\mathbb{R}$

$\therefore\ \mathbb{W}$ is a subspace

$\dim\mathbb{W}=\infty$

A basis is $\left\{\{1,0,0,0,\ldots\},\{0,1,0,0,\ldots\},\ \text{and so forth}\right\}$.

■ **More Cosets**

87.   The coset through the point $(1, -2, 1)$ is given by the points

$$\left\{(1,\ -2,\ 1)+t(1,\ 3,\ 2)\right\};$$

t is an arbitrary number. This describes a line through $(1, -2, 1)$ parallel to the line $t(1,\ 3,\ 2)$.

■ **Suggested Journal Entry I**

90.   Student Project

CHAPTER 4 — Higher-Order Linear Differential Equations

4.1 The Harmonic Oscillator

■ The Undamped Oscillator

3. $\ddot{x} + 9x = 0$, $x(0) = 1$, $\dot{x}(0) = 1$

The general solution of the harmonic oscillator equation $\ddot{x} + 9x = 0$ is given by

$$x(t) = c_1 \cos 3t + c_2 \sin 3t$$
$$\dot{x}(t) = -3c_1 \sin 3t + 3c_2 \cos 3t.$$

Substituting the initial conditions $x(0) = 1$, $\dot{x}(0) = 1$, gives

$$x(0) = c_1 = 1$$
$$\dot{x}(0) = 3c_2 = 1$$

so $c_1 = 1$, $c_2 = \dfrac{1}{3}$. Hence, the IVP has the solution

$$x(t) = \cos 3t + \frac{1}{3}\sin 3t.$$

In polar form, this would be

$$x(t) = \frac{\sqrt{10}}{3}\cos(3t - \delta)$$

where $\delta = \tan^{-1}\dfrac{1}{3}$. This would be in the first quadrant.

6. $\ddot{x} + 16x = 0$, $x(0) = 0$, $\dot{x}(0) = 4$

The general solution of the harmonic oscillator equation $\ddot{x} + 16x = 0$ is given by

$$x(t) = c_1 \cos 4t + c_2 \sin 4t$$
$$\dot{x}(t) = -4c_1 \sin 4t + 4c_2 \cos 4t.$$

Substituting the initial conditions $x(0) = 0$, $\dot{x}(0) = 4$, we get

$$x(0) = c_1 = 0$$
$$\dot{x}(0) = 4c_2 = 4$$

so $c_1 = 0$, $c_2 = 1$. The IVP has the solution

$$x(t) = \sin 4t .$$

■   **Graphing by Calculator**

9. $y = \cos t + \sin t$

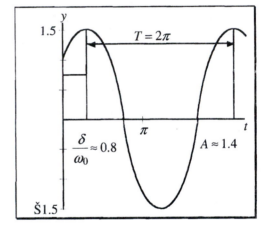

The equation tells us $T = 2\pi$ and because $T = \dfrac{2\pi}{\omega_0}$, $\omega_0 = 1$. We then measure the delay $\dfrac{\delta}{\omega_0} \approx 0.8$ which we can compute as the phase angle $\delta \approx 0.8(1) = 0.8$. The amplitude A can be measured directly giving $A \approx 1.4$. Hence,

$$\cos t + \sin t \approx 1.4 \cos(t - 0.8) .$$

Compare with the algebraic form in Problem 15.

12. $y = \cos 3t + 5\sin 3t$

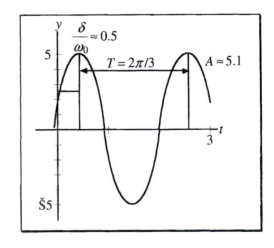

The equation tells us the period is $T = \dfrac{2\pi}{3}$ and because $T = \dfrac{2\pi}{\omega_0}$, $\omega_0 = 3$. We then measure the delay $\dfrac{\delta}{\omega_0} \approx 0.5$, which we can compute as the phase angle $\delta \approx 0.5(3) = 1.5$. The amplitude A can be measured directly giving $A \approx 5.1$. Hence,

$$\cos 3t + 5\sin 3t \approx 5.1 \cos(3t - 1.5) .$$

■ **Single-Wave Forms of Simple Harmonic Motion**

15. $\cos t + \sin t$

By Equation (4) $c_1 = 1$, $c_2 = 1$, and $\omega_0 = 1$. By Equation (5)

$$A = \sqrt{2}, \ \delta = \frac{\pi}{4}$$

yielding

$$\cos t + \sin t = \sqrt{2} \cos\left(t - \frac{\pi}{4}\right).$$

(Compare with solution to Problem 9.)

18. $-\cos t - \sin t$

By Equation (5) $c_1 = -1$, $c_2 = -1$, and $\omega_0 = 1$. By Equation (6)

$$A = \sqrt{2}, \ \delta = \frac{5\pi}{4}$$

yielding

$$-\cos t - \sin t = \sqrt{2} \cos\left(t - \frac{5\pi}{4}\right).$$

Because c_1 and c_2 are negative, the phase angle is in the 3rd quadrant.

■ **Component Form of Simple Harmonic Motion**

Using $\cos(A + B) = \cos A \cos B - \sin A \sin B$, we write:

21. $3\cos\left(t - \frac{\pi}{4}\right) = 3\left\{\cos t \cos\left(-\frac{\pi}{4}\right) - \sin t \sin\left(-\frac{\pi}{4}\right)\right\} = 3\left\{\frac{\sqrt{2}}{2}\cos t + \frac{\sqrt{2}}{2}\sin t\right\} = \frac{3\sqrt{2}}{2}\left\{\cos t + \sin t\right\}$

■ **Interpreting Oscillator Solutions**

24. $\ddot{x} + x = 0$, $x(0) = 1$, $\dot{x}(0) = 1$

Because $\omega_0 = 1$ radians per second, we know the natural frequency is $\frac{1}{2\pi}$ Hz (cycles per second), and the period is 2π. Using the initial conditions, we find the solution (see Problem 2)

$$x(t) = \sqrt{2} \cos\left(t - \frac{\pi}{4}\right),$$

which tells us the amplitude is $\sqrt{2}$ and the phase angle is $\delta = \frac{\pi}{4}$ radians.

27. $\ddot{x} + 16x = 0$, $x(0) = -1$, $\dot{x}(0) = 0$

Because $\omega_0 = 4$ radians per second, we know the natural frequency is $\dfrac{2}{\pi}$ Hz (cycles per second),

and the period is $\dfrac{\pi}{2}$. Using the initial conditions, we find the solution (see Problem 5)

$$x(t) = \cos(4t - \pi),$$

which tells us the amplitude is 1 and the phase angle is $\delta = \pi$ radians.

30. $4\ddot{x} + \pi^2 x = 0$, $x(0) = 1$, $\dot{x}(0) = \pi$

$4r^2 + \pi^2 = 0$

$r = \pm \dfrac{\pi}{2}i$ $x = c_1 \cos\dfrac{\pi}{2}t + c_2 \sin\dfrac{\pi}{2}t$ $1 = c_1$

$\dot{x} = -\dfrac{\pi}{2}c_1 \sin\dfrac{\pi}{2}t + \dfrac{\pi}{2}c_2 \cos\dfrac{\pi}{2}t$ $\pi = \dfrac{\pi}{2}c_2$

$x = \cos\dfrac{\pi}{2}t + 2\sin\dfrac{\pi}{2}t$ $c_2 = 2$

Amplitude: $A = \sqrt{1 + 2^2} = \sqrt{5}$

$x = \sqrt{5}\cos\left(\dfrac{\pi}{2}t - 1.11\right)$

■   **Phase Portraits**

For comparison of phase portraits, the main observation is that the elliptical shape depends on ω_0, which is $\sqrt{k}$ in all of these problems because $\ddot{x} + kx = 0$.

If $\omega_0 = 1$, trajectories are circular. As ω_0 increases above 1, ellipses become taller and thinner. As ω_0 decreases from 1 to 0, ellipses become shorter and wider.

The aspect ratio of $\dfrac{x_{max}}{\dot{x}_{max}} = \dfrac{1}{\omega}$.

Other observations include:

- All these phase portraits show closed elliptical trajectories that circulate clockwise.
- The trajectory of Problem 33 has a greater radius than that of Problem 32 because the initial condition is further from the origin.
- The trajectories in Problems 36 and 37 are on the same ellipse with different starting points that give different solution equations.

33. $\ddot{x} + x = 0$ $\vec{x}_0 = \begin{bmatrix} 1 \\ 1 \end{bmatrix}$

From Problem 2 $x(t) = \cos(t) + \sin(t)$,

so $\dot{x}(t) = -\sin(t) + \cos(t)$.

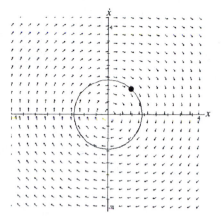

36. $\ddot{x} + 16x = 0$ $\vec{x}_0 = \begin{bmatrix} -1 \\ 0 \end{bmatrix}$

From Problem 5, $x(t) = -\cos 4t$,

so $x'(t) = 4 \sin 4t$.

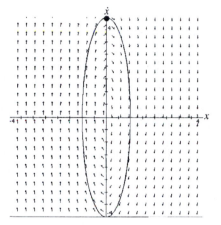

39. $4\ddot{x} + \pi^2 x = 0$ $\vec{x}_0 = \begin{bmatrix} 1 \\ \pi \end{bmatrix}$

From Problem 8, $x(t) = \cos\frac{\pi}{2}t + 2\sin\frac{\pi}{2}t$,

so $\dot{x}(t) = -\frac{\pi}{2}\sin\frac{\pi}{2}t + \pi\cos\frac{\pi}{2}t$.

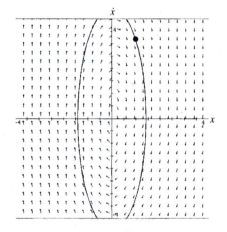

■   **Matching Problems**

42.   D

■   **Detective Work**

45.   (a)   The curve $y = 1.4\cos\left(t - \dfrac{8\pi}{5}\right)$ is a sinusoidal curve with period 2π, amplitude $A \approx 1.4$,

and phase angle $\delta \approx \dfrac{8\pi}{5}$.

(b)   From this graph we estimate $\omega_0 = 1$, $A \approx 2.3$, and $\delta \approx \dfrac{\pi}{4}$. Thus, we have

$$x(t) = A\cos(\omega_0 t - \delta_0) = 2.3\cos\left(t - \frac{\pi}{4}\right) = 2.3\left[\cos t\left(\cos - \frac{\pi}{4}\right) - \sin t\left(-\sin\frac{\pi}{4}\right)\right]$$

$$= 2.3\left\{\frac{\sqrt{2}}{2}\cos t + \frac{\sqrt{2}}{2}\sin t\right\} \approx 1.6(\cos t + \sin t).$$

■   **Initial-Value Problems**

48.   (a)   The weight is 16 lbs, so the mass is roughly $\dfrac{16}{32} = \dfrac{1}{2}$ slugs. (See Table 4.1.1 in text.) This

mass stretches the spring $\dfrac{1}{2}$ foot, hence $k = \dfrac{16}{\frac{1}{2}} = 32$ lb/ft. This yields the equation

$\dfrac{1}{2}(\ddot{x}) + 32x = 0$, or

$$\ddot{x} + 64x = 0.$$

The initial conditions are that the mass is pulled down 4 inches ($\dfrac{1}{3}$ foot) from equilib-

rium and then given an upward velocity of 4 ft/sec. This gives the initial conditions of

$x(0) = \dfrac{1}{3}$ ft, $\dot{x}(0) = -4$ ft/sec, using the engineering convention that for x, *down* is

positive.

(b)   We have the same equation $\ddot{x} + 64x = 0$, but the initial conditions are $x(0) = -\dfrac{1}{6}$ ft,

$\dot{x}(0) = 1$ ft/sec.

■   **Testing Your Intuition**

51. $\ddot{x} + x + x^3 = 0$

Here we have a vibrating spring with no friction, but a nonlinear restoring force $F = -x - x^3$ that

is stronger than a purely linear force $-x$. For small displacement x the nonlinear F will not be

much different (for small x, x^3 is very small), but for larger x, the force F will be much stronger

than in a linear spring; as F increases, the frequency of the vibration increases. This equation is

called Duffing's (strong) equation, and the associated springs are called strong springs.

54. $\ddot{x} + \dfrac{1}{t}\dot{x} + x = 0$

This equation can be interpreted as describing the motion of a vibrating mass that has infinite friction $\dfrac{1}{t}\dot{x}$ at $t = 0$, but friction immediately begins to diminish and approaches zero as t becomes very large. You may simulate in your mind the motion of such a system. Do you think for large t that the oscillation might behave much like simple harmonic motion? (See 4.3 Problem 68.)

■   **LR-Circuit**

57.   (a)   Without having a capacitor to store energy, we do not expect the current in the circuit to oscillate. If there had been a constant voltage V_0 on in the past, we would expect the current to be (by Ohm's law) $I = \dfrac{V_0}{R}$. If we then shut off the voltage, we would expect the current to die off in the presence of a resistance.

(b)   If a current I passes through a resistor with resistance R, then the voltage drop is RI; the voltage drop across an inductor of inductance L is $L\dot{I}$. We obtain the IVP:

$$L\dot{I} + RI = 0, \; I(0) = \frac{V_0}{R}$$

(c)   The solution of the IVP is

$$I(t) = \frac{V_0}{R}e^{-(R/L)t}.$$

(d)   If $R = 40$ ohms, $L = 5$ henries, $V_0 = 10$ volts, then $I(t) = \dfrac{1}{4}e^{-8t}$ ohms.

■   **Changing into Systems**

60. $4\ddot{x} - 2\dot{x} + 3x = 17 - \cos t$

$$\dot{x} = y$$
$$\dot{y} = \frac{1}{4}(-3x + 2y + 17 - \cos t)$$

63. $t^2\ddot{x} + 4t\dot{x} + x = t\sin 2t$ $t > 0$

$$\ddot{x} + \frac{4}{t}\dot{x} + \frac{1}{t^2}x = \frac{\sin 2t}{t}$$
$$\dot{x} = y$$
$$\dot{y} = -\frac{x}{t^2} - \frac{4}{t}y + \frac{\sin 2t}{t}$$

■   Another Harmonic Motion

66.   For simple harmonic motion the circular frequency ω_0 is

$$\omega_0 = \sqrt{\frac{kR^2}{mR^2 + I}},$$

so the natural frequency f_0 is

$$f_0 = \frac{1}{2\pi} \sqrt{\frac{kR^2}{mR^2 + I}}.$$

■   Factoring Out Friction

69.   (a)   Letting $x(t) = e^{(-b/2m)t} X(t)$, we have

$$\dot{x}(t) = \frac{-b}{2m} e^{(-b/2m)t} X(t) + e^{(-b/2m)t} \dot{X}(t)$$

$$\ddot{x}(t) = \frac{b^2}{4m^2} e^{(-b/2m)t} X(t) + \frac{-b}{m} e^{(-b/2m)t} \dot{X}(t) + e^{-(b/2m)t} \ddot{X}(t).$$

Substituting this into the original equation (1) and dividing through by $e^{-(b/2m)t}$, we arrive at

$$m\left[\ddot{X} - \frac{b}{m}\dot{X} + \frac{b^2}{4m^2}X \right] + b\left[-\frac{b}{2m}X + \dot{X} \right] + k[X] = 0.$$

Rearranging terms gives

$$m\ddot{X} + [-b + b]\dot{X} + \left[\frac{b^2}{4m} - \frac{b^2}{2m} + k \right]X = 0$$

or

$$m\ddot{X} + \left(k - \frac{b^2}{4m} \right)X = 0.$$

(b)   If we assume $k - \dfrac{b^2}{4m} > 0$, then divide by m and let

$$\omega_0 = \frac{1}{2m}\sqrt{4mk - b^2}\ )$$

we find the solution of this DE in X is

$$X(t) = c_1 \cos \omega_0 t + c_2 \sin \omega_0 t = A\cos(\omega_0 t - \delta).$$

Thus, we have

$$x(t) = e^{-(b/2m)t} X(t) = Ae^{-(b/2m)t} \cos(\omega_0 t - \delta).$$

4.2   Real Characteristic Roots

■   **Real Characteristic Roots**

3. $y'' - 9y = 0$

The characteristic equation is $r^2 - 9 = 0$, which has roots 3, –3. Thus, the general solution is

$$y(t) = c_1 e^{3t} + c_2 e^{-3t}.$$

6. $y'' - y' - 2y = 0$

The characteristic equation is $r^2 - r - 2 = 0$, which factors into $(r - 2)(r + 1) = 0$, and hence has roots 2, –1. Thus, the general solution is

$$y(t) = c_1 e^{2t} + c_2 e^{-t}.$$

9. $2y'' - 3y' + y = 0$

The characteristic equation is $2r^2 - 3r + 1 = 0$, which factors into $(2r - 1)(r - 1) = 0$, and hence has roots $\dfrac{1}{2}$, 1. Thus, the general solution is

$$y(t) = c_1 e^{t/2} + c_2 e^t.$$

12. $y'' - y' - 6y = 0$

The characteristic equation is $r^2 - r - 6 = 0$, which factors into $(r + 2)(r - 3) = 0$, and hence has roots –2, 3. Thus, the general solution is

$$y(t) = c_1 e^{-2t} + c_2 e^{3t}.$$

■   **Initial Values Specified**

15. $y'' - 25y = 0$, $y(0) = 1$, $y'(0) = 0$

The characteristic equation of the differential equation is $r^2 - 25 = 0$, which factors into $(r - 5)(r + 5) = 0$, and thus has roots 5, –5. Hence,

$$y(t) = c_1 e^{5t} + c_2 e^{-5t}.$$

Substituting in the initial conditions $y(0) = 1$ gives $c_1 + c_2 = 1$. Substituting in $\dot{y}(0) = 0$ gives $5c_1 - 5c_2 = 0$. Solving for c_1, c_2 gives $c_1 = c_2 = \dfrac{1}{2}$. Thus the general solution is

$$y(t) = \frac{1}{2} e^{5t} + \frac{1}{2} e^{-5t}.$$

18. $y'' - 9y = 0$, $y(0) = -1$, $y'(0) = 0$

The characteristic equation is $r^2 - 9 = 0$, which factors into $(r-3)(r+3) = 0$, and hence has

roots are 3, –3. Thus, the general solution is

$$y(t) = c_1 e^{3t} + c_2 e^{-3t}.$$

Substituting into $y(0) = -1$, $y'(0) = 0$ yields $c_1 = c_2 = -\dfrac{1}{2}$, so

$$y(t) = -\frac{1}{2} e^{3t} - \frac{1}{2} e^{-3t}.$$

21. $y'' - y' = 0$ $y(0) = 2,$ $y(0) = -1$

$r^2 - r = 0$ (Characteristic equation)

$r(r - 1) = 0$ $r = 0, 1$

$y = c_1 + c_2 e^t \Rightarrow 2 = c_1 + c_2$

$y' = c_2 e^t \Rightarrow -1 = c_2, \quad c_1 = 3$

$y = 3 - e^t$

■   **Bases and Solution Spaces**

24. $y'' - 10y' + 25y = 0$

$r^2 - 10r + 25 = 0$ (Characteristic equation)

$(r - 5)^2 = 0 \Rightarrow r = 5, 5$

Basis: $\{e^{5t}, te^{5t}\}$

Solution Space: $\{y \mid y = c_1 e^{5t} + c_2 te^{5t}; c_1, c_2 \in \mathbb{R}\}$

■   **Other Bases**

27. $y'' - 4y = 0$

$r^2 - 4 = 0$ (Characteristic equation)

$r = \pm 2 \; \therefore \; \{e^{2t}, e^{-2t}\}$ is a basis

To show $\{\cosh 2t, \sinh 2t\}$ is a basis, we need only show that $\cosh 2t$ and $\sinh 2t$ are linearly

independent solutions:

$$W = \begin{vmatrix} \cosh 2t & \sinh 2t \\ 2\sinh 2t & 2\cosh 2t \end{vmatrix} = 4\cosh^2 2t - 4\sinh^2 2t$$

$$\cosh^2 2t = \left(\frac{e^{2t} + e^{-2t}}{2}\right)^2 = \left(\frac{e^{4t} + 1 + e^{-4t}}{2}\right)$$

$$\sinh^2 2t = \left(\frac{e^{2t} - e^{-2t}}{2}\right)^2 = \left(\frac{e^{4t} - 1 - e^{-4t}}{2}\right)$$

so $\cosh^2 2t - \sinh^2 2t = 1$ and $W = 4 \neq 0$.

$\therefore$ $\cosh 2t$, $\sinh 2t$ are linearly independent.

Substitute $y = \cosh 2t$, $y' = 2\sinh 2t$, $y'' = 4\cosh 2t$

Then $y'' - 4y = 4\cosh 2t - 4\cosh 2t = 0$ $\qquad \therefore y = \cosh 2t$ is a solution.

In similar fashion, we can show that $y = \sinh 2t$ is also a solution.

To show that $\{e^{2t}, \cosh 2t\}$ is a basis, we use the facts that e^{2t} and $\cosh 2t$ are solutions.

Then:

$$W = \begin{vmatrix} e^{2t} & \cosh 2t \\ 2e^{2t} & 2\sinh 2t \end{vmatrix} = \begin{vmatrix} e^{2t} & \dfrac{e^{2t} + e^{-2t}}{2} \\ 2e^{2t} & e^{2t} - e^{-2t} \end{vmatrix}$$

$$= (e^{4t} - 1) - (e^{4t} + 1) = -2 \neq 0$$

$\therefore$ e^{2t} and $\cosh 2t$ are linearly independent

■ **The Wronskian Test**

30. $\qquad W = \begin{vmatrix} te^{5t} & e^{5t} & 2e^{5t} - 1 \\ (5t+1)e^{5t} & 5e^{5t} & 10e^{5t} \\ (25t+10)e^{5t} & 25e^{5t} & 50e^{5t} \end{vmatrix} = e^{5t} \begin{vmatrix} t & 1 & 2 - e^{-5t} \\ 5t+1 & 5 & 10 \\ 25t+10 & 25 & 50 \end{vmatrix}$

$$= e^{5t} \left[t \begin{vmatrix} 5 & 10 \\ 25 & 50 \end{vmatrix} - 1 \begin{vmatrix} 5t+1 & 10 \\ 25t+10 & 50 \end{vmatrix} + (2 - e^{-5t}) \begin{vmatrix} 5t+1 & 5 \\ 25t+10 & 25 \end{vmatrix} \right]$$

$$= 25e^{5t} \neq 0$$

Yes, $\{te^{5t}, e^{5t}, 2e^{5t} - 1\}$ is a basis for the solution space for $y''' - 10y'' + 25y' = 0$.

■   Relating Graphs

For Problem 33, $\ddot{x} + 5\dot{x} + 6x = 0$ has (from Example 1) solutions

$$x(t) = c_1 e^{-2t} + c_2 e^{-3t} \qquad (1)$$

$$\dot{x}(t) = -2c_1 e^{-2t} - 3c_2 e^{-3t} \qquad (2)$$

33.   (a), (b)

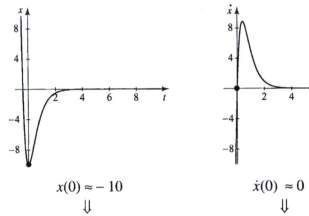 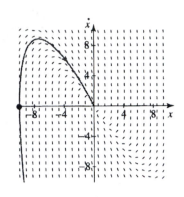

$$x(0) \approx -10 \qquad\qquad\qquad \dot{x}(0) \approx 0$$

$$\Downarrow \qquad\qquad\qquad\qquad \Downarrow$$

$$\underbrace{c_1 + c_2 = -10 \qquad\qquad -2c_1 - 3c_2 = 0}$$

$$c_1 = -30, \; c_2 = 20$$

(c)   From (1) in box, $x(t) = -30e^{-2t} + 20e^{-3t}$.

For $t > 0$, each term diminishes as t increases; the result remains negative, below the t-axis.

For $t < 0$, each exponential increases as t decreases; the negative term cancels the positive term when $30e^{-2t} = 20e^{-3t}$ or $e^{-t} = 1.5$,

that is, when $t = -\ln 1.5 \approx -.405$ which looks about right on the tx-graph.

(d)   From (2), $\dot{x}(t) = 60e^{-2t} - 60e^{-3t} = 60e^{-2t}(1 - e^{-t})$ which is always positive for $t > 0$, decreasing as t increases.

$\dot{x}(t)$ reaches a maximum when $\ddot{x}(t) = -120e^{-2t} + 180e^{-3t} = 0$

$$-2 + 3e^{-t} = 0$$

$$e^{-t} = \frac{2}{3},$$

so $t = -\ln\dfrac{2}{3} \approx 0.406$, which looks about right on the $t\dot{x}$ -graph.

> For Problems 36–39, $\ddot{x} - \dot{x} - 6x = 0$ has (from Problem 12) solutions
>
> $$x(t) = c_1 e^{-2t} + c_2 e^{3t} \qquad (1)$$
>
> $$\dot{x}(t) = -2c_1 e^{-2t} + 3c_2 e^{3t} \qquad (2)$$

36.   (a)

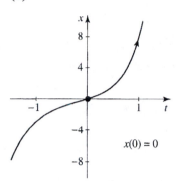   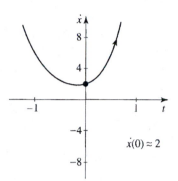   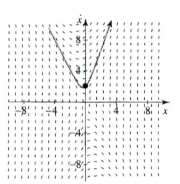

$x(0) = 0$ $\dot{x}(0) \approx 2$

(b)   From (1) $c_1 + c_2 = 0$

From (2) $-2c_1 + 3c_2 = 2$ $\Bigg\}$ $c_1 = -\dfrac{2}{5};\ \ c_2 = \dfrac{2}{5}$

$$x(t) = -\frac{2}{5}e^{-2t} + \frac{2}{5}e^{3t}$$

$$\dot{x}(t) = \frac{4}{5}e^{-2t} + \frac{6}{5}e^{3t}$$

(c)   For $t > 0$, $e^{-2t} < e^{3t}$, so $x(t)$ is always positive, and as t increases, so does $x(t)$.

This result agrees with the tx-graph.

For $t < 0$, $e^{3t} < e^{-2t}$ so $x(t)$ is always negative, and as t becomes *more* negative, $x(t)$ becomes more negative.

(d) $\dot{x}(t)$ is always positive.

For $t > 0$, $e^{-2t} < e^{3t}$, so the second term dominates as t increases, and $\dot{x}(t)$ increases as well. These facts are in agreement with the $t\dot{x}$ -graph.

39.   (a)

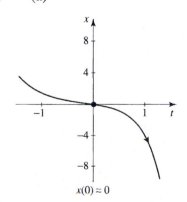   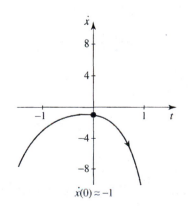   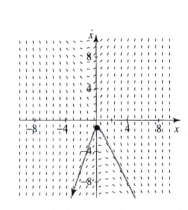

$x(0) \approx 0$ $\dot{x}(0) \approx -1$

(b)   From (1), $c_1 + c_2 = 0$

From (2), $-2c_1 + 3c_2 = -1$ $c_1 = \dfrac{1}{5};\ \ c_2 = -\dfrac{1}{5}$

$$x(t) = \frac{1}{5}e^{-2t} - \frac{1}{5}e^{3t}$$

$$\dot{x}(t) = -\frac{2}{5}e^{-2t} - \frac{3}{5}e^{3t}$$

(c)   For $t > 0$ the second term dominates, so $x(t)$ is negative, ever more so as t increases.

For $t < 0$ the first term dominates, so $x(t)$ is positive, ever more so as t becomes more negative.

These facts agree with the tx graph.

(d) $\dot{x}(t)$ is always negative. The maximum value will occur when $t = 0$, as shown on the $t\dot{x}$ - graph.

■   **Phase Portraits**

Careful inspection shows:

42.   (A)

■   **Second Solution**

45.   Substituting $y = v(t)e^{-bt/2a}$ into

$$ay'' + by' + cy = 0$$

gives

$$y' = v'e^{-bt/2a} - \frac{b}{2a}ve^{-bt/2a}$$

$$y'' = v''e^{-bt/2a} - \frac{b}{a}v'e^{-bt/2a} + \frac{b^2}{4a^2}ve^{-bt/2a}.$$

Substituting v, v', v'' into the differential equation gives the new equation (after dividing by $e^{-bt/2a}$)

$$a\left(v'' - \frac{b}{a}v' + \frac{b^2}{4a^2}v\right) + b\left(v' - \frac{b}{2a}v\right) + cv = 0.$$

Simplifying gives

$$av'' - \left(\frac{b^2}{4a} - c\right)v = 0.$$

Because we have assumed $b^2 = 4ac$, we have the equation $v'' = 0$, which was the condition to be proved.

■ Negative Roots

48.   We have $r = -b \pm \sqrt{b^2 - 4mk}$, so in the overdamped case where $b^2 - 4mk > 0$, these characteristic roots are real. Because m and k are both nonnegative, $b^2 - 4mk < b^2$ causing

$$r_1 = -b + \sqrt{b^2 - 4mk} \quad \text{to be a negative sum of negative and positive terms}$$

and

$$r_2 = -b - \sqrt{b^2 - 4mk} \quad \text{to be a negative sum of two negative terms.}$$

■ An Overdamped Spring

51.   (a)   The solution of an overdamped equation has the form

$$x(t) = c_1 e^{r_1 t} + c_2 e^{r_2 t}.$$

Suppose that

$$c_1 e^{r_1 t_1} + c_2 e^{r_2 t_1} = 0$$

for some t_1. Because $e^{r_2 t_1}$ is never zero, we can divide by $e^{r_2 t_1}$ to get $c_1 e^{(r_1 - r_2)t_1} + c_2 = 0$. Solving for t_1 gives

$$t_1 = \frac{1}{r_1 - r_2} \ln \frac{-c_2}{c_1}.$$

This unique number is the only value for which the curve may pass through 0. If the argument of the logarithm is negative or if the value of t_1 is negative, then the solution does not cross the equilibrium point.

(b)   By a similar argument, we can show that the derivative $\dot{x}(t)$ also has one zero.

■ Linking Graphs

After inspection, we have labeled the yt and $y't$ graphs as follows.

54.

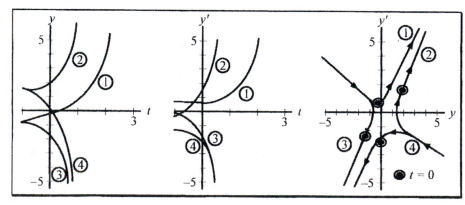

■   **Surge Functions**

57.   For $m\ddot{x} + b\dot{x} + kx = 0$, let $m = 1$, find b, k and initial conditions for the solution $x = Ate^{-rt}$

$$\ddot{x} + b\dot{x} + kx = 0$$

$$r^2 + br + k = 0 \quad \text{(characteristic equation)}$$

$$r = \frac{-b \pm \sqrt{b^2 - 4 \cdot 1 \cdot k}}{2}$$

$b^2 - 4k = 0$ to obtain repeated roots, $r = -\dfrac{b}{2}, -\dfrac{b}{2}$

$$x = c_1 e^{-rt} + c_2 t e^{-rt}$$

$$\dot{x} = -rc_1 e^{-rt} + c_2(t(-r)e^{-rt} + e^{-rt}) \qquad \therefore c_1 = 0 = x(0)$$
$$c_2 = A = \dot{x}(0)$$

$$\dot{x}(0) = c_2$$

$$\therefore b = -2r, \ 4k = b^2$$

and from above we know $4k = 4r^2$ so that $k = \pm r$, for $k > 0$.

Results: r and A are given, and

$$b = -2r$$

$$k = r$$

$$x(0) = 0$$

$$x'(0) = A$$

■   **The Euler-Cauchy Equation** $at^2 y'' + bty' + cy = 0$

60.   Let $y(t) = t^r$, so

$$y' = rt^{r-1}$$
$$y'' = r(r-1)t^{r-2}.$$

Hence

$$at^2 y'' + bty' + cy = ar(r-1)t^r + brt^r + ct^r = 0.$$

Dividing by t^r yields the characteristic equation

$$ar(r-1) + br + c = 0,$$

which can be written as

$$ar^2 + (b-a)r + c = 0.$$

If r_1 and r_2 are two distinct roots of this equation, we have solutions

$$y_1(t) = t^{r_1}$$
$$y_2(t) = t^{r_2}.$$

Because these two functions are clearly linearly independent (one not a constant multiple of the other) for $r_1 \neq r_2$, we have

$$y(t) = c_1 t^{r_1} + c_2 t^{r_2}$$

for $t > 0$.

■   **The Euler-Cauchy Equation with Distinct Roots**

For Problem 63, see Problem 60 for the form of the characteristic equation for the Euler-Cauchy DE.

63. $t^2 y'' + 4ty' + 2y = 0$

In this case $a = 1$, $b = 4$, $c = 2$, so the characteristic equation is

$$r(r-1) + 4r + 2 = r^2 + 3r + 2 = (r+1)(r+2) = 0.$$

Hence, we have roots $r_1 = -1$, $r_2 = -2$, and thus

$$y(t) = c_1 t^{-1} + c_2 t^{-2}.$$

■   **Solutions for Repeated Euler-Cauchy Roots**

For Problems 66 and 67 use the result of Problem 60, $y(t) = c_1 t^r + c_2 t^r \ln t$.

66. $t^2 y'' + 5ty' + 4y = 0$

In this case, $a = 1$, $b = 5$, and $c = 4$, so our characteristic equation for r is $r^2 + 4r + 4 = 0$, with a double root at -2. The general solution is

$$y(t) = c_1 t^{-2} + c_2 t^{-2} \ln t$$

for $t > 0$.

69. $4t^2 y'' + 8ty' + y = 0$ Euler-Cauchy method: $y = t^m, t > 0$

$$4m(m-1) + 8m + 1 = 0$$
$$4m^2 + 4m + 1 = 0$$
$$(2m+1)^2 = 0 \qquad m = -\frac{1}{2}$$

$$y(t) = c_1 t^{-1/2} + c_2 t^{-1/2} \ln t$$

■   **Computer: Phase-Plane Trajectories**

72. $y(t) = e^t + e^{-t}$

(a)   The roots of the characteristic equation are 1 and –1, so the characteristic equation is

$$(r-1)(r+1) = r^2 - 1 = 0.$$

$y(t)$ satisfies the differential equation

$$y'' - y = 0.$$

(b)   The derivative is

$$y'(t) = e^t - e^{-t}.$$

The IC for the given trajectory in yy' space is

$$\left(y(0),\ y'(0)\right) = (2,\ 0).$$

(c)   We plot this and a few other trajectories of this DE in yy' space.

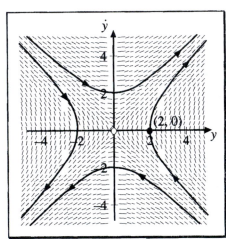

DE trajectories in yy' space

■   **Reduction of Order**

75.   (a)   Let $y_2 = vy_1$ and

$$y_2' = v'y_1 + vy_1'$$
$$y_2'' = v''y_1 + 2v'y_1' + vy_1''.$$

Then

$$y_2'' + p(x)y_2' + q(x)y_2 = v''y_1 + 2v'y_1' + pv'y_1 + \left(vy_1'' + pvy_1' + qvy_1\right) = 0.$$

Because $y_1'' + py_1' + qy_1 = 0$, cancel the terms involving v, and arrive at the new equation

$$y_1 v'' + \left(2y_1' + p(x)y_1\right)v' = 0$$

(b)   Setting $v' = w$ and using the fact that $y_1' dx = dy_1'$, we obtain

$$y_1 w' + \left(2y_1' + p(x) y_1 \right) w = 0$$

$$w' + \left(\frac{2y_1' + p(x) y_1}{y_1} \right) w = 0$$

$$\frac{dw}{w} = \left(\frac{-2y_1' - p(x) y_1}{y_1} \right) dx$$

$$\ln |w| = \int \left(\frac{-2y_1'}{y} - p(x) \right) dx$$

$$\ln |w| = \int \frac{-2}{y_1} dy_1 - \int p(x) dx$$

$$\ln |w| = \ln y_1^{-2} - \int p(x) dx$$

$$w = \pm \frac{e^{-\int p(x) dx}}{y_1^2} = v'$$

$$v = \pm \int \frac{e^{-\int p(x) dx}}{y_1^2}$$

By convention, the positive sign is chosen.

(c)   If v is a constant function on I, then $v' \equiv 0$ and $w \equiv 0$ because $v' = w$. The condition $w \equiv 0$ contradicts our work in part (b) as $\ln |w|$ where $w = 0$ is undefined. Because v is not constant on I, $\{ y_1, y_2 \}$ is a linearly independent set of I.

■   **Reduction of Order: Second Solution**

78. $t^2 y'' - ty' + y = 0$, $y_1 = t$

We won't use the formula this time. We simply redo the steps in Problem 75. We seek a second solution of the form $y_2 = vy_1 = tv$. Differentiating, we have

$$y_2' = tv' + v$$
$$y_2'' = tv'' + 2v'.$$

Substituting into the equation we obtain

$$t^2 y_2'' - ty_2' + y_2 = t^3 v'' + t^2 v' = 0.$$

Letting $w = v'$ and dividing by t^3 yields

$$w' + \frac{1}{t} w = 0.$$

We can solve by integrating the factor method, getting $w = c_1 t^{-1}$. Integrating we find

$$v = c_1 \ln |t| + c_2,$$

so

$$y_2 = tv = c_1 t \ln|t| + c_2 t .$$

Letting $c_1 = 1$, $c_2 = 0$, we get a second linearly independent solution

$$y_2 = t \ln|t| .$$

■ Classical Equation

81. $(1-t^2) y'' - ty' + y = 0$, $y_1(t) = t$ (Chebyshev's Equation)

Letting $y_2 = vy_1 = vt$, we have

$$y_2' = tv' + v , \quad y_2'' = tv'' + 2v' ,$$

hence we have the equation

$$(1-t^2) y_2'' - ty_2' + y_2 = t(1-t^2)v'' + (2-3t^2)v' = 0 .$$

Dividing by $t(1-t^2)$, and letting $w = v'$,

$$w' + \frac{2-3t^2}{t(1-t^2)} w = 0 .$$

Using partial fractions yields

$$\int \frac{2-3t^2}{t(1-t^2)} dt = 2\ln|t| + \frac{1}{2}\ln|t-1| + \frac{1}{2}\ln|1+t| ,$$

so our integrating factor is $t^2 \sqrt{1-t^2}$ and

$$w = c_1 \frac{1}{t^2 \sqrt{1-t^2}} .$$

Letting $c_1 = 1$ and multiplying by t yields a final answer of

$$y_2(t) = tv = t \int \frac{1}{t^2 \sqrt{1-t^2}} dt .$$

This is a perfect example of a formula that does not tell us much about how the solutions behave. Check out the IDE tool Chebyshev's Equation to see the value of graphical solutions.

■ Suggested Journal Entry

84.   Student Project

4.3   Complex Characteristic Roots

■ Solutions in General

3. $y'' - 4y' + 5y = 0$

The characteristic equation is $r^2 - 4r + 5 = 0$, which has roots $2 \pm i$. The general solution is

$$y(t) = e^{2t} \left(c_1 \cos t + c_2 \sin t \right).$$

6. $y'' - 4y' + 7y = 0$

The characteristic equation is $r^2 - 4r + 7 = 0$, which has roots $2 \pm i\sqrt{3}$. The general solution is

$$y(t) = e^{2t} \left(c_1 \cos \sqrt{3}t + c_2 \sin \sqrt{3}t \right).$$

9. $y'' - y' + y = 0$

The characteristic equation is $r^2 - r + 1 = 0$, which has roots $\dfrac{1}{2} \pm i\dfrac{\sqrt{3}}{2}$. The general solution is

$$y(t) = e^{t/2} \left(c_1 \cos \frac{\sqrt{3}}{2}t + c_2 \sin \frac{\sqrt{3}}{2}t \right).$$

■ Initial-Value Problems

12. $y'' - 4y' + 13y = 0$, $y(0) = 1$, $y'(0) = 0$

The characteristic equation is $r^2 - 4r + 13 = 0$, which has roots $2 \pm 3i$. The general solution is

$$y(t) = e^{2t} \left(c_1 \cos 3t + c_2 \sin 3t \right).$$

Substituting this into the initial conditions yields $y(0) = c_1 = 1$, $y'(0) = 2c_1 + 3c_2 = 0$, resulting in $c_1 = 1$, $c_2 = -\dfrac{2}{3}$. Hence, the solution of the initial-value problem is

$$y(t) = e^{2t} \left(\cos 3t - \frac{2}{3} \sin 3t \right).$$

15. $y'' - 4y' + 7y = 0$, $y(0) = 0$, $y'(0) = -1$

From Problem 6,

$$y(t) = e^{2t} \left\{ c_1 \cos\left(\sqrt{3}t\right) + c_2 \sin\left(\sqrt{3}t\right) \right\}.$$

Subsituting this into the initial conditions yields $y(0) = 0$, $y'(0) = -1$, resulting in

$$c_1 = 0, \ c_2 = -\frac{1}{3}\sqrt{3}.$$

Hence, the solution of the initial-value problem is

$$y(t) = -\frac{1}{3}\sqrt{3}e^{2t} \sin\left(\sqrt{3}t\right).$$

■   **Working Backwards**

18. $(r - 4)(r - (1 - i))(r - (1 + i)) = r^3 - 6r^2 + 10r - 8 = 0$

$$y''' - 6y'' + 10y' - 8y = 0$$

■   **Matching Problems**

21. $y'' - y' = 0 \Rightarrow r = 0, 1$

$\qquad y(t) = c_1 + c_2 e^t$ Graph D

24. $y'' - 5y' + 6y = 0 \Rightarrow r = 2, 3$

$\qquad y(t) = c_1 e^{2t} + c_2 e^{3t}$ Graph C

27. $y'' + 4y' + 4y = 0 \Rightarrow r = -2, -2$

$\qquad y(t) = (c_1 + c_2 t)e^{-2t}$ Graph E

■   **Long-Term Behavior of Solutions**

30. $r_1 < 0$, $r_2 < 0$. When $r_1 \neq r_2$, the solution is

$$y(t) = c_1 e^{r_1 t} + c_2 e^{r_2 t}$$

and goes to 0 as $t \to \infty$. When $r = r_1 = r_2 < 0$, the solution has the form

$$y(t) = c_1 e^{rt} + c_2 t e^{rt}.$$

In this case using l'Hôpital's rule we prove the second term te^{rt} goes to zero as $t \to \infty$ when $r < 0$.

33. $r_1 = 0$, $r_2 = 0$. The solution

$$y(t) = c_1 + c_2 t$$

approaches ∞ as $t \to \infty$ when $c_2 > 0$ and $-\infty$ when $c_2 < 0$.

■ Linear Independence

36.   Suppose

$$c_1 e^{\alpha t} \cos \beta t + c_2 e^{\alpha t} \sin \beta t = 0$$

on an arbitrary interval. Dividing both sides by $e^{\alpha t}$, then differentiating the new equation and dividing by β, yields

$$c_1 \cos \beta t + c_2 \sin \beta t = 0$$
$$c_2 \cos \beta t - c_1 \sin \beta t = 0.$$

Hence, $c_1 = 0$, $c_2 = 0$ and we have proven linear independence of the given functions.

■ Higher-Order DEs

39. $$\frac{d^5 y}{dt^5} - \frac{4d^4 y}{dt^4} + \frac{4d^3 y}{dt^3} = 0$$

The characteristic equation is

$$r^5 - 4r^4 + 4r^3 = r^3 \left(r^2 - 4r + 4 \right) = r^3 \left(r - 2 \right)^2 = 0,$$

which has roots, 0, 0, 0, 2, 2. Hence,

$$y(t) = c_1 + c_2 t + c_3 t^2 + c_4 e^{2t} + c_5 t e^{2t}.$$

42. $y''' - 4y'' + 5y' - 2y = 0$

$r^3 - 4r^2 + 5r - 2 = 0$ (characteristic equation)

$f(1) = 1 - 4 + 5 - 2 = 0$

$\therefore r = 1$ is a root

By long division, we obtain

$(r-1)(r^2 - 3r + 2) = 0$

$(r-1)(r-2)(r-1) = 0$ $r = 1, 1, 2$

$y(t) = c_1 e^t + c_2 t e^t + c_3 e^{2t}$

■ Linking Graphs

45.

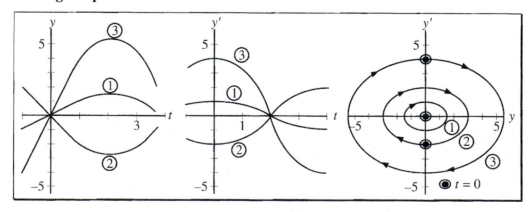

■ Changing the Spring

48.   (a)   The solutions of

$$\ddot{x} + \dot{x} + kx = 0, \ x(0) = 4, \ \dot{x}(0) = 0$$

are shown for $k = \dfrac{1}{4}$, $\dfrac{1}{2}$, 1, 2, 4. For

larger k we have more oscillations.

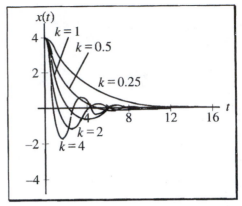

(b)   For larger k, since there are more oscillations, the phase-plane trajectory spirals
further around the origin.

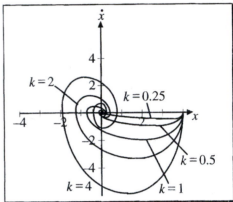

■ Oscillating Euler-Cauchy

51.   We used the substitution $y = t^r$ and obtained for $r_1 = \alpha + i\beta$ and $r_2 = \alpha - i\beta$ the solution

$$y(t) = k_1 t^{\alpha + i\beta} + k_2 t^{\alpha - i\beta} = k_1 e^{(\alpha + i\beta)\ln t} + k_2 e^{(\alpha - i\beta)\ln t} = k_1 e^{\alpha \ln t + i\beta \ln t} + k_2 e^{\alpha \ln t - i\beta \ln t}$$

$$= e^{\alpha \ln t}\left(c_1 \cos\left(\beta \ln t\right) + c_2 \sin\left(\beta \ln t\right)\right) = t^{\alpha}\left(c_1 \cos\left(\beta \ln t\right) + c_2 \sin\left(\beta \ln t\right)\right).$$

This is the same process as that used at the start of Case 3 in the text utilizing the Euler's Formula (4).

54. $t^2 y'' + 17ty' + 16y = 0$ Euler-Cauchy: $y = t^m$, $t > 0$

$m(m - 1) + 17m + 16 = 0$ (characteristic equation)

$m^2 + 16m + 16 = 0$

$$m = \frac{-16 \pm \sqrt{(16)^2 - 4(16)}}{2} = -8 \pm 4\sqrt{3}$$

$$y(t) = t^{-8}\left(c_1 \cos(4\sqrt{3}\ln t) + c_2 \sin(4\sqrt{3}\ln t)\right)$$

■   Third-Order Euler-Cauchy Problems

57. $t^3 y''' + 3t^2 y'' + 5ty = 0$ Let $y = t^m$, $t > 0$

$m(m-1)(m-2) + 3m(m-1) + 5m = 0$ (characertistic equation)

$m^3 - 3m^2 + 2m + 3m^2 - 3m + 5 = 0$

$m^3 + 4m = 0$ $m = 0, \pm 2i$

$y(t) = c_1 + c_2 \cos (2 \ln t) + c_3 \sin (2 \ln t)$

■   Finding the Damped Oscillation

60.   The initial conditions

$$x(0) = 1, \ \dot{x}(0) = 1$$

give the constants $c_1 = 1$, $c_2 = 2$. Hence, we have

$$x(t) = e^{-t}\left(\cos t + 2\sin t\right).$$

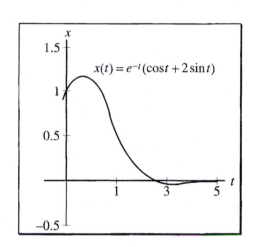

■   Damped Mass-Spring System

63.   The IVP is

$$\ddot{x} + b\dot{x} + 64x = 0, \ x(0) = 1, \ \dot{x}(0) = 0.$$

(a) $b = 10$: (underdamped), $x(t) = e^{-5t}\left(\cos \sqrt{39}t + \dfrac{5}{\sqrt{39}}\sin \sqrt{39}t\right)$

(b) $b = 16$: (critically damped), $x(t) = (1 + 8t)e^{-8t}$

(c) $b = 20$: (overdamped), $x(t) = \dfrac{1}{3}\left(4e^{-4t} - e^{-16t}\right)$

■   Computer Lab: Damped Free Vibrations

66.   IDE Lab

■   **Effects of Nonconstant Coefficients**

69. $t\ddot{x} + x = 0$

(a)   If you divide by t, you will see that this equation is the same as the equation in Problem 67.

72. $\ddot{x} + \dfrac{1}{t}\dot{x} + tx = 0$

(a)   For this ODE damping is initially large, but vanishes as time increases; the restoring force on the other hand is initially small but increases with time. How will these effects combine?

(b)   We plotted the solution with IC $x(0.1) = 2$, $\dot{x}(0.1) = 0$ in the tx and $x\dot{x}$ planes.

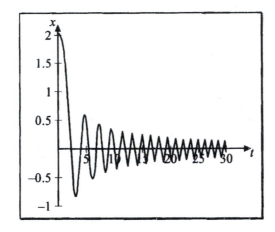
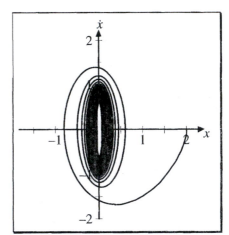

(c)   As we expected, the tx graph shows initially large damping, which rapidly decreases the amplitude of the solution, and increasing frequency, due to the effect of the increasing spring "constant", which shortens the period. The center of the $x\dot{x}$ graph will continue to fill in, very slowly, if you give it a much longer time interval.

■   **Boundary-Value Problems**

75. $y'' + y = 0$, $y(0) = 0$, $y\left(\dfrac{\pi}{2}\right) = 1$

$y(t) = c_1 \cos t + c_2 \sin t$

$y(0) = 0 = c_1$

$y\left(\dfrac{\pi}{2}\right) = 1 = c_2$, so $y(t) = \sin t$ is the solution.

■   **Exact Second-Order Differential Equations**

78. $y'' + \dfrac{1}{t} y' - \dfrac{1}{t^2} y = 0$ is the same as $y'' + \left[\dfrac{1}{t} y \right]' = 0$.

Integrating we obtain the linear equation $y' + \dfrac{1}{t} y = c_1$,

for which $\mu = e^{\int \frac{1}{t} dt} = e^{\ln t} = t$ so we have $ty' + y = c_1 t$.

Thus, $\dfrac{d}{dt}(ty) = c_1 t$, so $ty = \dfrac{c_1}{2} t^2 + c_2$ and $y(t) = \dfrac{c_1}{2} t + \dfrac{c_2}{t}$.

Substituting back into the original equation we find $c_1 = 0$, so $y(t) = \dfrac{c}{t}$ is the general solution.

■   **Suggested Journal Entry**

81.   Student Project

4.4   Undetermined Coefficients

■   **Inspection First**

3. $y'' = 2 \Rightarrow y_p(t) = t^2$

6. $y'' - y = -2\cos t \Rightarrow y_p(t) = \cos t$

■   **Educated Prediction**

The homogeneous equation $y'' + 2y' + 5y = 0$ has characteristic equation $r^2 + 2r + 5 = 0$, which has complex roots $-1 \pm 2i$. Hence,

$$y_h(t) = c_1 e^{-t} \sin 2t + c_2 e^{-t} \cos 2t,$$

so for the right-hand sides $f(t)$, we try the following:

9. $f(t) = 2t^3 - 3t \Rightarrow y_p(t) = At^3 + Bt^2 + Ct + D$

12. $f(t) = 2e^{-t} \sin t \Rightarrow y_p(t) = e^{-t}(A\cos t + B\sin t)$

■   **Guess Again**

The homogeneous equation $y'' - 6y' + 9y = 0$ has characteristic equation $r^2 - 6r + 9 = 0$, which has a double root 3, 3. Hence,

$$y_h(t) = c_1 e^{3t} + c_2 t e^{3t}.$$

We try particular solutions of the form:

15. $f(t) = e^{-t} + \sin t \Rightarrow y_p(t) = Ae^{-t} + B\sin t + C\cos t$

■   **Determining the Undetermined**

18. $y' + y = 1$. The homogeneous solution is $y_h(t) = ce^{-t}$ where c is any constant. By simple inspection we observe that $y_p(t) = 1$ is a solution of the nonhomogeneous equation. Hence, the general solution is

$$y(t) = ce^{-t} + 1.$$

21. $y'' + 4y' = 1$. The characteristic equation is $r^2 + 4r = 0$, which has roots 0, –4. Hence, the homogeneous solution is

$$y_h(t) = c_1 + c_2 e^{-4t}.$$

The constant on the right-hand side of the differential equation indicates we seek a particular solution of the form $y_p(t) = A$, except that the homogeneous solution has a constant solution;

thus we seek a solution of the form $y_p(t) = At$. Substituting this expression into the differential equation yields $4A = 1$, or $A = \dfrac{1}{4}$. Hence, we have a particular solution

$$y_p(t) = \frac{1}{4}t,$$

so the general solution is

$$y(t) = c_1 + c_2 e^{-4t} + \frac{1}{4}t.$$

24. $y'' + y' - 2y = 3 - 6t$. The characteristic equation is $r^2 + r - 2 = 0$, which has roots -2 and 1. Hence, the homogeneous solution

$$y_h(t) = c_1 e^{-2t} + c_2 e^t.$$

The linear polynomial on the right-hand side of the equation indicates we seek a particular solution of the form

$$y_p(t) = At + B.$$

(Note that we don't have any matches with the homogeneous solution.) Substituting this expression into the differential equation yields the equation

$$y'' + y' - 2y = A - 2At - 2B = 3 - 6t$$

so $A = 3$, $B = 0$. Hence, we have the general solution

$$y(t) = c_1 e^{-2t} + c_2 e^t + 3t.$$

27. $y'' + y' = 6\sin 2t$. The characteristic equation is $r^2 + r = 0$, which has roots 0 and -1. Hence, the homogeneous solution is

$$y_h(t) = c_1 + c_2 e^{-t}.$$

The sine term on the right-hand side of the differential equation indicates we seek a particular solution of the form

$$y_p(t) = A\cos 2t + B\sin 2t.$$

Substituting into the differential equation yields

$$y'' + y' = (-4A + 2B)\cos 2t + (-4B - 2A)\sin 2t = 6\sin 2t.$$

Comparing coefficients yields the equations

$$-4A + 2B = 0$$
$$-4B - 2A = 6,$$

which has the solution $A = -\dfrac{3}{5}$, $B = -\dfrac{6}{5}$. Hence, we have

$$y_p(t) = -\frac{3}{5}\cos 2t - \frac{6}{5}\sin 2t,$$

and the general solution is

$$y(t) = c_1 + c_2 e^{-t} - \frac{3}{5}\cos 2t - \frac{6}{5}\sin 2t.$$

30. $y'' - y = t\sin t$. The characteristic equation is $r^2 - 1 = 0$, which has roots ± 1. Hence, the homogeneous solution is

$$y_h(t) = c_1 e^t + c_2 e^{-t}.$$

The term on the right-hand side of the differential equation indicates we seek a particular solution

$$y_p(t) = (At + B)\cos t + (Ct + D)\sin t.$$

Differentiating this expression two times and substituting it into the differential equation yields the algebraic equation

$$y'' - y = -2Ct\sin t - 2At\cos t + (-2A - 2D)\sin t + (2C - 2B)\cos t = t\sin t.$$

Comparing terms in $\sin t$, $\cos t$, $t\sin t$, $t\cos t$, we get equations that yield

$$A = 0, \ B = -\frac{1}{2}, \ C = -\frac{1}{2}, \ D = 0.$$

Hence,

$$y(t) = c_1 e^t + c_2 e^{-t} - \frac{1}{2}(t\sin t + \cos t).$$

33. $y'' - 4y' + 4y = te^{2t}$. The characteristic equation of the differential equation is $r^2 - 4r + 4 = 0$, which has a double root of 2. Hence, the homogeneous solution is

$$y_h(t) = c_1 e^{2t} + c_2 t e^{2t}.$$

The term on the right-hand side of the differential equation indicates we seek a particular solution of the form

$$y_p(t) = Ate^{2t} + Be^{2t},$$

but both terms are linearly dependent with terms in the homogeneous solution, so we choose

$$y_p(t) = At^3 e^{2t} + Bt^2 e^{2t}.$$

Differentiating and substituting this expression into the differential equation yields the algebraic equation

$$y'' - 4y' + 4y = e^{2t}\left(0 + 0 + 6At + 2B\right) = te^{2t}.$$

Comparing coefficients, we get $A = \dfrac{1}{6}$, $B = 0$. Hence, the general solution is

$$y(t) = c_1 e^{2t} + c_2 t e^{2t} + \frac{1}{6} t^3 e^{2t}.$$

36. $y'' + 3y' = \sin t + 2\cos t$. The characteristic equation is $r^2 + 3r = 0$, which has roots 0, –3. Hence, the homogeneous solution is

$$y_h(t) = c_1 + c_2 e^{-3t}.$$

The sine and cosine terms on the right-hand side of the equation indicate we seek a particular solution of the form

$$y_p(t) = A\cos t + B\sin t.$$

Substituting this into the equation yields

$$y'' + 3y' = (-A + 3B)\cos t + (-B - 3A)\sin t = \sin t + 2\cos t.$$

Comparing terms, we arrive at $(-A + 3B) = 2$, $(-B - 3A) = 1$, yielding $A = -\dfrac{1}{2}$, $B = \dfrac{1}{2}$. From this, that the general solution is

$$y(t) = c_1 + c_2 e^{-3t} + \frac{1}{2}\left(\sin t - \cos t\right).$$

39. $y^{(4)} - y = 10$

(1)   Find y_h:

$$r^4 - 1 = 0 \qquad (r^2 + 1)(r^2 - 1) = (r^2 + 1)(r - 1)(r + 1) \qquad r = \pm i, \pm 1$$

$$y_h = c_1 \cos t + c_2 \sin t + c_3 e^t + c_4 e^{-t}$$

(2)   Find y_p:

$$y_p = A, \text{ so that } \quad y_p' = y_p'' = y_p''' = y_p^{(4)} = 0$$

$$\Rightarrow y_p^{(4)} - y_p = 0 - A = 10 \Rightarrow A = -10 \Rightarrow y_p = -10$$

(3) $y(t) = y_h + y_p = c_1 \cos t + c_2 \sin t + c_3 e^t + c_4 e^{-t} - 10$

■   **Initial-Value Problems**

42. $y'' + 4y' + 4y = te^{-t}, y(0) = -1, y'(0) = 1$

(1)   Find y_h:

$$r^2 + 4r + 4 = 0 \Rightarrow (r+2)^2 = 0 \Rightarrow r = -2, -2$$

$$\therefore y_h = c_1 e^{-2t} + c_2 t e^{-2t} = (c_1 + c_2 t)e^{-2t}$$

(2)   Find y_p:

$$y_p = e^{-t}(At + B) \Rightarrow y'_p = -e^{-t}(At + B) + Ae^{-t}$$

$$\Rightarrow y''_p = e^{-t}(At + B) - Ae^{-t} - Ae^{-t} = e^{-t}(At + B) - 2Ae^{-t}$$

So $y''_p + 4y'_p + 4y_p = e^{-t}(At + B) - 2Ae^{-t} + 4(-e^{-t}(At + B) + Ae^{-t}) + 4e^{-t}(At + B)$

$$= e^{-t}(At + 2A + B)$$

This gives $A = 1$, $2A + B = 0$ and so $A = 1$ and $B = -2$.

Therefore $y_p = e^{-t}(t - 2)$.

(3) $y = y_h + y_p = c_1 e^{-2t} + c_2 t e^{-2t} + e^{-t}(t - 2)$

$$y' = -2c_1 e^{-2t} + c_2 e^{-2t} - 2c_2 t e^{-2t} - e^{-t}(t - 2) + e^{-t}$$

$$\left.\begin{array}{l} y(0) = -1 \Rightarrow c_1 - 2 = -1 \\ y'(0) = 1 \Rightarrow -2c_1 + c_2 + 2 + 1 = 1 \end{array}\right\} \Rightarrow \begin{array}{l} c_1 = 1 \\ c_2 = 0 \end{array}$$

$$\therefore y(t) = e^{-2t} + e^{-t}(t - 2)$$

45. $4y'' + y = \cos 2t, y(0) = 1, y'(0) = 0$

(1)   Find y_h: $4r^2 + 1 = 0 \Rightarrow r^2 = -\dfrac{1}{4} \Rightarrow r = \pm\dfrac{1}{2}i \Rightarrow y_h = c_1 \cos\left(\dfrac{1}{2}t\right) + c_2 \sin\left(\dfrac{1}{2}t\right)$

(2)   Find y_p: $y_p = A\cos 2t + B\sin 2t$, $y'_p = -2A\sin 2t + 2B\cos 2t$, $y''_p = -4A\cos 2t - 4B\sin 2t$

$$\Rightarrow 4y''_p + y_p = -16A\cos 2t - 16B\sin 2t + A\cos 2t + B\sin 2t$$

$$= -15A\cos 2t - 15B\sin 2t = \cos 2t$$

coefficient of $\cos 2t$: $-15A = 1$, coefficient of $\sin 2t$: $-15B = 0$

$$\Rightarrow A = -\frac{1}{15}, B = 0$$

$$\therefore y_p = -\frac{1}{15}\cos 2t$$

(3) $y = y_h + y_p = c_1 \cos\left(\dfrac{1}{2}t\right) + c_2 \sin\left(\dfrac{1}{2}t\right) - \dfrac{1}{15}\cos 2t$

$y' = -\dfrac{1}{2}c_1 \sin\left(\dfrac{1}{2}t\right) + \dfrac{1}{2}c_2 \cos\left(\dfrac{1}{2}t\right) + \dfrac{2}{15}\sin 2t$

$y(0) = 1 \Rightarrow c_1 - \dfrac{1}{15} = 1 \Rightarrow c_1 = \dfrac{16}{15}$

$y'(0) = 0 \Rightarrow c_2 = 0$

$\therefore\ y(t) = \dfrac{16}{15}\cos\left(\dfrac{1}{2}t\right) - \dfrac{1}{15}\cos 2t$

48. $y'' - 4y' + 3y = e^{-t} + t,\ y(0) = 0,\ y'(0) = 0$

(1)   Find y_h:

$r^2 - 4r + 3 = 0 \Rightarrow (r-1)(r-3) = 0$ so that $y_h = c_1 e^t + c_2 e^{3t}$

(2)   Find y_p:

$y_p = Ae^{-t} + Bt + C,\ y_p' = -Ae^{-t} + B,\ y_p'' = Ae^{-t}$

Thus $y_p'' - 4y_p' + 3y_p = Ae^{-t} - 4(-Ae^{-t} + B) + 3(Ae^{-t} + Bt + C)$

$\qquad\qquad = Ae^{-t} - 4(-Ae^{-t} + B) + 3(Ae^{-t} + Bt + C)$

$\qquad\qquad = 8Ae^{-t} + 3Bt - 4B + 3C = e^{-t} + t$

$\qquad\qquad \Rightarrow 8A = 1,\ 3B = 1,\ -4B + 3C = 0$

$\qquad\qquad \Rightarrow A = \dfrac{1}{8},\ B = \dfrac{1}{3},\ C = \dfrac{4}{3}B = \dfrac{4}{9}.$

Thus, $y_p = \dfrac{1}{8}e^{-t} + \dfrac{1}{3}t + \dfrac{4}{9}.$

(3)   Find y:

$y = y_h + y_p = c_1 e^t + c_2 e^{3t} + \dfrac{1}{8}e^{-t} + \dfrac{1}{3}t + \dfrac{4}{9}.$

$y' = c_1 e^t + 3c_2 e^{3t} - \dfrac{1}{8}e^{-t} + \dfrac{1}{3}.$

$y(0) = 0 \Rightarrow c_1 + c_2 + \dfrac{1}{8} + \dfrac{4}{9} = 0$

$y'(0) = 0 \Rightarrow c_1 + 3c_2 - \dfrac{1}{8} + \dfrac{1}{3} = 0$

$\Rightarrow$

$c_1 + c_2 = -\dfrac{41}{72}$

$c_1 + 3c_2 = -\dfrac{5}{24}$

$\Rightarrow$

$c_1 = -\dfrac{3}{4}$

$c_2 = \dfrac{13}{72}$

Therefore, $y(t) = -\dfrac{3}{4}e^t + \dfrac{13}{72}e^{3t} + \dfrac{1}{8}e^{-t} + \dfrac{1}{3}t + \dfrac{4}{9}.$

51. $y^{(4)} - y = e^{2t}$, $y(0) = y'(0) = y''(0) = y'''(0) = 0$

    (a)   Find y_h:

$$r^4 - 1 = 0 \Rightarrow (r^2 + 1)(r^2 - 1) \Rightarrow r = \pm i, \pm 1$$

$$y_h = c_1 \cos t + c_2 \sin t + c_3 e^t + c_4 e^{-t}$$

    (2)   Find y_p:

$$y_p = Ae^{2t}, \; y_p' = 2Ae^{2t}, \; y_p'' = 4Ae^{2t}, \; y_p''' = 8Ae^{2t}, \; y_p^{(4)} = 16Ae^{2t}$$

Thus $y_p^{(4)} - y_p = 16Ae^{2t} - Ae^{2t} = e^{2t} \Rightarrow 15Ae^{2t} = e^{2t} \Rightarrow A = \dfrac{1}{15}$ so that $y_p = \dfrac{1}{15}e^{2t}$.

    (3) $y = y_h + y_p = c_1 \cos t + c_2 \sin t + c_3 e^t + c_4 e^{-t} + \dfrac{1}{15}e^{2t}$

$$y' = -c_1 \sin t + c_2 \cos t + c_3 e^t - c_4 e^{-t} + \dfrac{2}{15}e^{2t}$$

$$y'' = -c_1 \cos t - c_2 \sin t + c_3 e^t + c_4 e^{-t} + \dfrac{4}{15}e^{2t}$$

$$y''' = c_1 \sin t - c_2 \cos t + c_3 e^t - c_4 e^{-t} + \dfrac{8}{15}e^{2t}$$

$$y(0) = 0 \Rightarrow c_1 + c_3 + c_4 + \dfrac{1}{15} = 0$$

$$y'(0) = 0 \Rightarrow c_2 + c_3 - c_4 + \dfrac{2}{15} = 0$$

$$y''(0) = 0 \Rightarrow -c_1 + c_3 + c_4 + \dfrac{4}{15} = 0$$

$$y'''(0) = 0 \Rightarrow -c_2 + c_3 - c_4 + \dfrac{8}{15} = 0$$

From these 4 equations in 4 unknowns, we obtain (by the methods of Chapter 3),

$$c_1 = \dfrac{1}{10}, \; c_2 = \dfrac{1}{5}, \; c_3 = -\dfrac{1}{4} \text{ and } c_4 = \dfrac{1}{12}$$

$$\therefore \; y(t) = \dfrac{1}{10}\cos t + \dfrac{1}{5}\sin t - \dfrac{1}{4}e^t + \dfrac{1}{12}e^{-t} + \dfrac{1}{15}e^{2t}$$

■   **Trial Solutions**

54. $y''' - y'' = t^2 + e^t$

Find y_h:

$r^3 - r^2 = 0 \Rightarrow r^2(r-1) = 0 \Rightarrow r = 0, 0, 1$

$\therefore\ y_h = c_1 + c_2t + c_3e^t$

Find y_p:

$y_{p_1} = t^2(At^2 + Bt + C),\ y_{p_2} = t(De^t)$

$\therefore\ y_p(t) = y_{p_1} + y_{p_2} = At^4 + Bt^3 + Ct^2 + Dte^t$

■   **Judicious Superposition**

57.   (a)   The characteristic equation is $r^2 - r - 6 = 0$ has roots $r = 3, -2$, so the general solution is

$$y_h(t) = c_1e^{3t} + c_2e^{-2t}.$$

(b)   (i)   Substituting $y_p(t) = Ae^t$ yields

$$Ae^t - Ae^t - 6Ae^t = e^t,$$

which yields $A = -\dfrac{1}{6}$. Hence, $y_p(t) = -\dfrac{1}{6}e^t$.

(ii)   Substituting $y_p(t) = Ae^{-t}$ yields

$$Ae^{-t} + Ae^{-t} - 6Ae^{-t} = e^{-t}$$

or $A = -\dfrac{1}{4}$. Hence, $y_p(t) = -\dfrac{1}{4}e^{-t}$.

continued on the next page

(c)   Calling $L(y) = y'' - y' - 6y$ we found in part (b) that

$$L\left(-\frac{1}{6}e^{t}\right) = e^{t}, \text{ and } L\left(-\frac{1}{4}e^{-t}\right) = e^{-t}.$$

Multiplying each equation by $\frac{1}{2}$ and using basic properties of derivatives yields

$$L\left(-\frac{1}{12}e^{t}\right) = \frac{1}{2}e^{t}, \text{ and } L\left(-\frac{1}{8}e^{t}\right) = \frac{1}{2}e^{-t}$$

and

$$L\left(-\frac{1}{12}e^{t} - \frac{1}{8}e^{-t}\right) = \frac{1}{2}\left(e^{t} + e^{-t}\right) = \cosh t.$$

Hence, a solution of $y'' - y' - 6y = \cosh t$ is

$$y_{p}(t) = -\frac{1}{12}e^{t} - \frac{1}{8}e^{-t}.$$

■ Discontinuous Forcing Functions

60. $y'' + 16y = \begin{cases} \cos t & 0 \le t \le \pi \\ 0 & t > \pi \end{cases} \qquad y(0) = 1, y'(0) = 0$

Part 1: $y_1'' + 16y_1 = \cos t \qquad\qquad 0 \le t \le \pi \qquad y_1(0) = 1, \ y_1'(0) = 0$

Find $(y_1)_h$: $(y_1)_h = c_1 \cos 4t + c_2 \sin 4t$

Find $(y_1)_p$: $(y_1)_p = A \cos t + B \sin t$
$(y_1)_p'' = -A\cos t - B\sin t$ $\left.\phantom{\begin{matrix}a\\b\end{matrix}}\right\} \Rightarrow B = 0, A = \dfrac{1}{15}$

Therefore, $y_1 = c_1 \cos 4t + c_2 \sin 4t + \dfrac{1}{15}\cos t$

$$y_1' = -4c_1 \sin 4t + 4c_2 \cos 4t - \frac{1}{15}\sin t$$

$\left. \begin{matrix} y_1(0) = 1 = c_1 + \dfrac{1}{15} \\[2mm] y_1'(0) = 0 = 4c_2 \end{matrix} \right\} \Rightarrow \begin{matrix} c_1 = -\dfrac{14}{15} \\[2mm] c_2 = 0 \end{matrix}$

$\therefore \ y_1(t) = -\dfrac{14}{15}\cos 4t + \dfrac{1}{15}\cos t$

Part 2: $y_2'' + 16y_2 = 0$ $t > \pi$

$y_2 = c_1 \cos 4t + c_2 \sin 4t$

$y_2' = -4c_1 \sin 4t + 4c_2 \cos 4t$

$y_1(\pi) = -\dfrac{14}{15} + \dfrac{1}{15} = -\dfrac{13}{15} = y_2(\pi) = -c_1$ $\qquad c_1 = -\dfrac{13}{15}$

$y_1'(\pi) = 4 = y_2'(\pi) = 4c_2$ $\qquad\qquad\qquad\qquad c_2 = 1$

Thus $y_2(t) = -\dfrac{13}{15} \cos 4t + \sin 4t$

and $y(t) = \begin{cases} -\dfrac{14}{15} \cos 4t + \dfrac{1}{15} \cos t & 0 \le t \le \pi \\[2mm] -\dfrac{13}{15} \cos 4t + \sin 4t & t > \pi \end{cases}$

■   **Solutions of Differential Equations Using Complex Functions**

63. $y'' + 25y = 20\sin 5t$ $\qquad$ We will use $y'' + 25y = 20e^{i5t}$

The homogeneous solution is $y_h = c_1 \cos 5t + c_2 \sin 5t$

For the particular solution we note that e^{5it} is included in y_h, so we must use an extra factor of t in y_p. We want $Im(y_p)$ where $y_p = Ate^{5it}$, so $y_p' = A(t5ie^{5it} + e^{5it})$,

and $y_p'' = A5i(t5ie^{5it} + e^{5it}) + Ai5e^{5it} = A(10ie^{5it} - 25te^{5it})$.

By substitution, we obtain

$A(10ie^{-5i} - 25te^{5it}) + 25Ate^{5it} = 20e^{5it}$ so that $10Ai = 20$. Thus $A = -2i$ and $y_p = -2ite^{5it}$.

$Im(y_p) = Im\left(-2it(\cos 5t + i\sin 5t)\right) = -2t \cos 5t$

$y = c_1 \cos 5t + c_2 \sin 5t - 2t \cos 5t$

4.5   Variation of Parameters

■   **Straight Stuff**

3. $y'' - 2y' + y = \dfrac{1}{t}e^t$, $(t > 0)$

The two linear independent solutions y_1 and y_2 of the homogeneous equation are

$$y_1(t) = e^t \text{ and } y_2(t) = te^t.$$

Using the method of variation of parameters, we seek the particular solution

$$y_p(t) = v_1(t)e^t + v_2(t)te^t.$$

In order for $y_p(t)$ to satisfy the differential equation, v_1 and v_2 must satisfy

$$y_1 v_1' + y_2 v_2' = e^t v_1' t e^t v_2' = 0$$

$$y_1 v_1' + y_2 v_2' = e^t v_1' e^t (t+1)v_2' = \frac{1}{t}e^t.$$

Solving algebraically for v_1' and v_2' we obtain $v_1'(t) = -1$ and $v_2' = \dfrac{1}{t}$.

Integrating gives the values $v_1(t) = -t$ and $v_2 = \ln t$.

Substituting these values into y_p yields the particular solution

$$y_p(t) = -te^t + te^t \ln t.$$

Hence, the general solution is $y(t) = c_1 e^t + c_2 te^t + te^t \ln t$.

6. $y'' - 2y' + 2y = e^t \sin t$

The homogeneous solutions are $y_1(t) = e^t \cos t$ and $y_2(t) = e^t \sin t$.

To find a particular solution of the form $y_p(t) = v_2 e^t \cos t + v_2 e^t \sin t$,

we solve the equations $e^t \cos t v_1' + e^t \sin t v_2' = 0$

$$\left(e^t \cos t - e^t \sin t\right)v_1' + \left(e^t \sin t + e^t \cos t\right)v_2' = e^t \sin t$$

for v_1' and v_2'. This yields $v_1' = -\sin^2 t$ and $v_2' = \sin t \cos t$.

Integrating yields the functions $v_1(t) = \dfrac{1}{2}(-t + \cos t \sin t)$ and $v_2(t) = \dfrac{1}{2}\sin^2 t$.

Hence, a particular solution

$$y_p(t) = y_1v_1 + y_2v_2 = \frac{1}{2}e^t\cos t\left(-t + \cos t\sin t\right) + \frac{1}{2}e^t\sin t\left(\sin^2 t\right) = \frac{1}{2}e^t\left(\sin t - t\cos t\right)$$

and the general solution is

$$y(t) = c_1e^t\cos t + c_2e^t\sin t - \frac{1}{2}te^t\cos t.$$

9. $y'' + 4y = \tan 2t$

$$y_h = c_1\underbrace{\cos 2t}_{y_1} + c_2\underbrace{\sin 2t}_{y_2}$$

$$v_1 = \int\frac{(-\tan 2t)(\sin 2t)}{2}\,dt = -\frac{1}{2}\int\frac{\sin^2 2t}{\cos 2t}\,dt = -\frac{1}{2}\int\frac{1 - \cos^2 2t}{\cos 2t}\,dt = -\frac{1}{2}\int\sec 2t - \cos 2t\,dt$$

$$= -\frac{1}{4}\left(\ln|\sec 2t + \tan 2t| - \sin 2t\right)$$

$$v_2 = \int\frac{\tan 2t\cos 2t}{2}\,dt = -\frac{1}{4}\cos 2t$$

So $y_p = y_1v_1 + y_2v_2 = -\frac{1}{4}\cos 2t\left(\ln|\sec 2t + \tan 2t| - \sin 2t\right) - \frac{1}{4}\sin 2t\cos 2t$.

General solution: $y(t) = c_1\cos 2t + c_2\sin 2t + y_p.$

12. $y'' - y = \dfrac{e^t}{t}$

$$y_h = c_1\underbrace{e^t}_{y_1} + c_2\underbrace{e^{-t}}_{y_2}$$

$$v_1 = \frac{1}{2}\int\left(\frac{e^t}{t}\right)(e^{-t})\,dt = \frac{1}{2}\int\frac{1}{t}\,dt = \frac{1}{2}\ln|t|$$

$$v_2 = -\frac{1}{2}\int\left(\frac{e^t}{t}\right)e^t\,dt = -\frac{1}{2}\int\frac{e^{2t}}{t}\,dt = -\frac{1}{2}\int_{t_0}^t\frac{e^{2s}}{s}\,ds$$

So $y_p = v_1y_1 + v_2y_2 = \dfrac{1}{2}e^t\ln|t| - \dfrac{1}{2}e^{-t}\displaystyle\int_{t_0}^t\frac{e^{2s}}{s}\,ds$

General solution: $y(t) = c_1e^t + c_2e^{-t} + y_p$

■   **Variable Coefficients**

15. $(1-t)y'' + ty' - y = 2(t-1)^2 e^{-t}$, $y_1(t) = t$, $y_2(t) = e^t$

We begin by dividing the equation by $(1-t)$, to get the proper form for variation of parameters

$$y'' + \frac{t}{1-t}y' - \frac{1}{1-t}y = 2(t-1)e^{-t}$$

Susbtitution verifies that y_1 and y_2 form a fundamental set of solutions to the associated homogeneous equation, so $y_h = c_1 t + c_2 e^t$

We seek a particular solution $y_p(y) = v_1 t + v_2 e^t$,

where v_1 and v_2 satisfy the conditions $\qquad t v_1' + e^t v_2' = 0$

$$v_1' + e^t v_2' = -2(t-1)e^{-t}$$

Solving algebraically for v_1' and v_2' yields $v_1' = 2e^{-t}$ and $v_2' = -2te^{-2t}$. Integrating

yields $\qquad\qquad\qquad v_1 = -2e^{-t}$ and $v_2 = e^{-2t}\left(t + \frac{1}{2}\right)$.

Thus, $\qquad\qquad\qquad y_p(t) = -2te^{-t} + te^{-t} + \frac{1}{2}e^{-t} = e^{-t}\left(\frac{1}{2} - t\right)$.

Hence, the general solution of this equation is $y(t) = c_1 t + c_2 e^t + e^{-t}\left(\frac{1}{2} - t\right)$.

■   **Third-Order DEs**

18. $y''' - 2y'' - y' + 2y = e^t$

The characteristic equation $(\lambda - 1)(\lambda + 1)(\lambda - 2) = 0$ and has roots 1, –1, and 2. The fundamental set is $y_1 = e^t$, $y_2 = e^{-t}$, and $y_3 = e^{2t}$. Hence,

$$y_h = c_1 e^t + c_2 e^{-t} + c_3 e^{2t}.$$

By variation of parameters, we seek $y_p(t) = v_1 e^t + v_2 e^{-t} + v_3 e^{2t}$, as in Problem 17. Hence the system to solve is

$$e^t v_1' + e^{-t} v_2' + e^{2t} v_3' = 0$$
$$e^t v_1' - e^{-t} v_2' + 2e^{2t} v_3' = 0$$
$$e^t v_1' + e^{-t} v_2' + 4e^{2t} v_3' = e^t.$$

Using Cramer's rule and computing the determinants yields:

$$W = \begin{bmatrix} e^t & e^{-t} & e^{2t} \\ e^t & -e^{-t} & 2e^{2t} \\ e^t & e^{-t} & 4e^{2t} \end{bmatrix} = -6e^{2t};$$

$$v_1' = \frac{\begin{bmatrix} 0 & e^{-t} & e^{2t} \\ 0 & -e^{-t} & 2e^{2t} \\ e^t & e^{-t} & 4e^{2t} \end{bmatrix}}{W} = \frac{3e^{2t}}{-6e^{2t}} = -\frac{1}{2}$$

$$v_2' = \frac{\begin{bmatrix} e^t & 0 & e^{2t} \\ e^t & 0 & 2e^{2t} \\ e^t & e^t & 4e^{2t} \end{bmatrix}}{W} = \frac{-e^{4t}}{-6e^{2t}} = \frac{1}{6}e^{2t}$$

$$v_3' = \frac{\begin{bmatrix} e^t & e^{-t} & 0 \\ e^t & -e^{-t} & 0 \\ e^t & e^{-t} & e^t \end{bmatrix}}{W} = \frac{-2e^t}{-6e^{2t}} = \frac{1}{3}e^{-t}$$

Hence we obtain $\quad v_1' = -\dfrac{1}{2} \quad v_2' = \dfrac{1}{6}e^{2t} \quad v_3' = \dfrac{1}{3}e^{-t}$.

Hence, $\quad\quad\quad\quad v_1 = -\dfrac{t}{2} \quad v_2 = \dfrac{1}{12}e^{2t} \quad v_3 = -\dfrac{1}{3}e^{-t}$.

We get a particular solution of $\quad y_p(t) = -\dfrac{1}{2}te^t + \dfrac{1}{12}e^t - \dfrac{1}{3}e^t$ and the general solution is

$$y(t) = c_1 e^t + c_2 e^{-t} + c_3 e^{2t} - \frac{1}{2}te^t - \frac{1}{4}e^t.$$

■   **Method Choice**

21. $y''' - y' = f(t)$

We first find the homogeneous solution. The characteristic equation $\lambda(\lambda^2 - 1) = 0$ has roots $0, \pm 1$, so the homogeneous solution is $y_h = c_1 + c_2 e^t + c_3 e^{-t}$.

(a) $y''' - y' = 2e^{-t}$. Because e^{-t} is in y_h, we must try $y_p = a\, te^{-t}$. The method of undetermined coefficients is straightforward and gives $a = 1$, so $y_p = te^{-t}$ and the general solution can be written $y(t) = c_1 + c_2 e^t + c_3 e^{-t} + te^{-t}$.

(b) $y''' - y' = \sin^2 t$. We *can* use undetermined coefficients on $\sin^2 t$, with the trigonometric identity $\sin^2 t = 1/2(1 - \cos 2t)$; however we choose to use variation of parameters to seek a particular solution of the form $y_p(t) = v_1 + v_2 e^t + v_3 e^{-t}$, with the derivatives of v_1, v_2, and v_3 determined from the equations

$$1v_1' + e^t v_2' + e^{-t} v_3' = 0$$
$$0v_1' + e^t v_2' - e^{-t} v_3' = 0$$
$$0v_1' + e^t v_2' + e^{-t} v_3' = \sin^2 t.$$

Using Cramer's rule (as outlined in Problem 18), we obtain

$$v_1' = -\sin^2 t \qquad v_2' = \frac{1}{2} e^{-t} \sin^2 t \qquad v_3' = \frac{1}{2} e^t \sin^2 t.$$

The antiderivative of v_1' is easy to find; the other two must be left as integrals

$$v_1 = \frac{1}{2}(\sin t \cos t - t) \qquad v_2 = \int \frac{1}{2} e^{-t} \sin^2 t\, dt \qquad v_3 = \int \frac{1}{2} e^t \sin^2 t\, dt.$$

Hence, the general solution is

$$y(t) = c_1 + c_2 e^t + c_3 e^{-t} + \frac{1}{2}(\sin t \cos t - t) + e^t \int \frac{1}{2} e^{-t} \sin^2 t\, dt + e^{-t} \int \frac{1}{2} e^t \sin^2 t\, dt.$$

(c) $y''' - y' = \tan t$. As in Part (b), we must use variation of parameters to find y_p, with

$$1v_1' + e^t v_2' + e^{-t} v_3' = 0$$
$$0v_1' + e^t v_2' - e^{-t} v_3' = 0$$
$$0v_1' + e^t v_2' + e^{-t} v_3' = \tan t.$$

Using Cramer's rule (as outlined in Problem 18), to solve these equations we find

$$v_1' = \tan t \qquad v_2' = \frac{1}{2} e^{-t} \tan t \qquad v_3' = \frac{1}{2} e^t \tan t.$$

The antiderivative of v_1' is easy to find; the other two must be left as integrals

$$v_1 = \ln|\cos t| \qquad v_2 = \int \frac{1}{2} e^{-t} \tan t\, dt \qquad v_3 = \int \frac{1}{2} e^t \tan t\, dt.$$

Hence, the general solution is

$$y(t) = c_1 + c_2 e^t + c_3 e^{-t} + \frac{1}{2} \ln|\cos t| + \frac{1}{2} e^t \int e^{-t} \tan t\, dt + \frac{1}{2} e^{-t} \int e^t \tan t\, dt.$$

Parts (b) and (c) demonstrate the power of graphical methods because the algebraic expressions for $y(t)$ are pretty meaningless. It is easier and more informative to use DE software to approximate solutions of this equation in ty space than it is to pursue the analytical formula for the solution. The figures show curves for several initial conditions to show the variety that can occur. For any IVP there would be only one solution.

For a *graphical* approach, we revisit Problem 21 using a 3D graphic DE solver with the following equations for $y''' - y' = f(t)$:

$$y' = x \qquad\qquad x' = y$$
$$y'' = x' = z \qquad \text{relisted as} \qquad y' = x$$
$$y''' = z' = x + f(t) \qquad\qquad z' = x + f(t)$$

(b) $f(t) = \sin^2 t.$

The expression for $y(t)$ on the previous page can be further evaluated using the identity

$$\sin^2 t = \frac{1}{2}(1 - \cos 2t),$$

but solution behavior is more easily seen on a graph of $y(t)$.

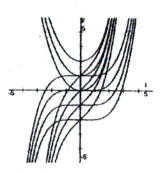

(c) $f(t) = \tan t$

The expression for $y(t)$ on the previous page is even more complicated than that for part (b); again, solution behavior is more readily understood with a graph of $y(t)$.

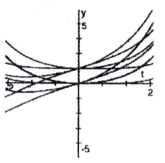

■   **Green's Follow-Up**

24.   From the Leibniz Rule in multivariable calculus we have the following result:

For a continuous function $g(t,s)$,

$$\frac{d}{dt}\int_0^t g(t, s)\,ds = \lim_{r \to t}\left[g(r,r) + \int_0^r \frac{\partial g}{\partial t}(t,s)\,ds \right].$$

In Problem 22, the solution of the equation $y'' + y = f(t)$ is

$$y(t) = \int_0^t \sin(t - s)f(s)\,ds$$

Differentiating yields

$$y' = \sin(t - t)f(t) + \int_0^t \cos(t - s)f(s)\,ds = \int_0^t \cos(t - s)f(s)\,ds$$

and

$$y'' = \cos(t - t)f(t) - \int_0^t \sin(t - s)f(s)\,ds = f(t) - \int_0^t \sin(t - s)f(s)\,ds.$$

Hence,

$$y'' + y = f(t) - \int_0^t \sin(t - s)f(s)\,ds + \int_0^t \sin(t - s)f(s)\,ds = f(t).$$

4.6   Forced Oscillations

■   **Mass-Spring Problems**

3. $2x'' + 3x = 4\cos 8t$

Find x_h: $2r^2 + 3 = 0 \Rightarrow r^2 - \dfrac{3}{2} \Rightarrow r = \pm\sqrt{\dfrac{3}{2}}i$

$$\Rightarrow x_h = c_1 \cos\sqrt{\dfrac{3}{2}}t + c_2 \sin\sqrt{\dfrac{3}{2}}t$$

Find x_p: $x_p = A\cos 8t, \quad x_p' = -8A\sin 8t, \quad x_p'' = -64A\cos 8t$

$$2x_p'' + 3x_p = 2(-64A\cos 8t)$$
$$\underline{+3(A\cos 8t)}$$
$$-125A\cos 8t = 4\cos 8t$$

$$\Rightarrow -125A = 4$$

$$\Rightarrow A = -\dfrac{4}{125}$$

$$\Rightarrow x_p = -\dfrac{4}{125}\cos 8t$$

$$x(t) = x_h + x_p = c_1 \cos\sqrt{\dfrac{3}{2}}t + c_2 \sin\sqrt{\dfrac{3}{2}}t - \dfrac{4}{125}\cos 8t$$

$$x_{ss} = -\dfrac{4}{125}\cos 8t = \dfrac{4}{125}\cos(8t - \pi)$$

Amplitude $C = \dfrac{4}{125}$; phase shift $\dfrac{\delta}{\beta} = \dfrac{\pi}{8}$ radians

6. $x'' + 4x' + 5x = 2\cos 2t$

Find x_h: $r^2 + 4r + 5 = 0 \Rightarrow r = -2 \pm i$

$x_h = e^{-2t}(c_1 \cos t + c_2 \sin t)$

Find x_p: $x_p = A\cos 2t + B\sin 2t, \quad x_p' = -2A\sin 2t + 2B\cos 2t, \quad x_p'' = -4A\cos 2t - 4B\sin 2t$

$$x_p'' + 4x_p' + 5x_p = \quad -4A\cos 2t - 4B\sin 2t)$$
$$+ 4(2B\cos 2t - 2A\sin 2t)$$
$$\underline{+ 5(A\cos 2t + B\sin 2t)}$$
$$(A + 8B)\cos 2t + (-8A + B)\sin 2t = 2\cos 2t$$

$$\Rightarrow A + 8B = 2, \;\; -8A + B = 0$$

$$\Rightarrow A = \frac{2}{65}, \;\; B = \frac{16}{65}$$

$$x_p = \frac{2}{65}\cos 2t \; + \; \frac{16}{65}\sin 2t$$

$$x(t) = x_h + x_p = e^{-2t}(c_1 \cos t + c_2 \sin t) + \frac{2}{65}\cos 2t + \frac{16}{65}\sin 2t$$

$$x_{ss} = \frac{2}{65}\cos 2t + \frac{16}{65}\sin 2t \; = \; \frac{2}{\sqrt{65}}\cos(2t - 1.4)$$

Amplitude $C = \dfrac{2}{\sqrt{65}}$; phase shift $\dfrac{\delta}{\beta} \approx 0.73$ radians

■   **Mass-Spring Again**

9.   (a)   The mass is $m = 100$ kg; gravitational force (weight) acting on the spring is $mg = 100(9.8) = 980$ newtons. Because the weight stretches the spring by 20 cm $= 0.2$ m, we have

$$k = \frac{980}{0.20} = 4900 \text{ nt/m}.$$

(b)   The initial-value problem for this mass is

$$\ddot{x} + 49x = 0, \; x(0) = 0.40, \; \dot{x}(0) = 0.$$

Solving we write the transient solution in polar form

$$x(t) = C\cos(\omega_0 t - \delta) = C\cos(7t - \delta)$$

where the circular frequency is $\omega_0 = 7$ radians per second. Using the initial conditions gives

$$x(0) = C\cos\delta = 0.4$$
$$\dot{x}(0) = -7C\sin\delta = 0$$

or $\delta = 0$, $C = 0.4$. Hence,

$$x(t) = 0.4\cos(7t).$$

(c)   Amplitude: $C = 0.4$ meter; period: $T = \dfrac{2\pi}{7}$ seconds.

(d)   If $b = 500$, then

$$b^2 - 4mk = 250,000 - 4(100)(4900) < 0.$$

The system is underdamped.

(e) $100\ddot{x} + 500\dot{x} + 4900x = 0$ has characteristic equation

$$r^2 + 5r + 49 = 0,$$

which has roots $x_{1,2} = -\dfrac{5}{2} \pm i\dfrac{1}{2}\sqrt{171}$. Hence, the general solution is

$$x(t) = e^{-5t/2}\left[c_1 \cos\left(\frac{\sqrt{171}}{2}t\right) + c_2 \sin\left(\frac{\sqrt{171}}{2}t\right)\right].$$

Using the initial conditions $x(0) = 0.4$, $\dot{x}(0) = 0$ gives

$$x(0) = c_1 = 0.4$$

$$\dot{x}(0) = c_2\frac{\sqrt{171}}{2} - \frac{5}{2}c_1 = 0$$

which implies $c_1 = 0.4$, $c_2 = \dfrac{2\sqrt{171}}{171}$. Hence, the solution is

$$x(t) = e^{-5t/2}\left(0.4\cos\frac{\sqrt{171}}{2}t + \frac{2\sqrt{171}}{171}\sin\frac{\sqrt{171}}{2}t\right).$$

■ Damped Forced Motion I

12. $\ddot{x} + 8\dot{x} + 36x = 72\cos 6t$

The given characteristic equation has roots $-4 \pm 2i\sqrt{5}$, hence in the long run the homogeneous equation solution always decays to zero. We are only interested in a particular solution, and in this case that solution is

$$x = A\cos 6t + B\sin 6t.$$

Differentiating and substituting into the differential equation gives

$$A = 0,\ B = \frac{3}{2}.$$

Hence, the steady-state solution is given by

$$x_{ss}(t) = \frac{3}{2}\sin 6t.$$

The graph of the steady-state solution is shown.

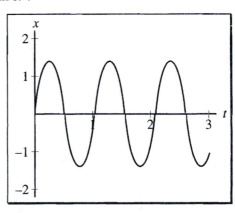

Charge and Current

15. $\ddot{Q} + 12\dot{Q} + 100Q = 12\cos 10t, \ \ Q(0) = 0, \ Q'(0) = 0$

(a)   Find Q_h: $r^2 + 12r + 100 = 0 \Rightarrow r = -6 \pm 8i \Rightarrow Q_h = e^{-6t}(c_1 \cos 8t + c_2 \sin 8t)$.

Find Q_p: Q_p has the form $A \cos 10t + B \sin 10t$.

Substitution in $4\ddot{Q}_p + 12\dot{Q}_p + 100Q_p = 12\cos 10t$ leads to $A = 0, \ B = \dfrac{1}{10}$

and so $Q_p = \dfrac{1}{10}\sin 10t$.

Thus $Q = Q_h + Q_p = e^{-6t}(c_1 \cos 8t + c_2 \sin 8t) + \dfrac{1}{10}\sin 10t$.

The initial conditions $Q(0) = 0$ and $\dot{Q}(0) = 0$ give us $c_1 = 0 \ c_2 = -\dfrac{1}{8}$

and the solution of the IVP is $Q(t) = -\dfrac{1}{8}e^{-6t}\sin 8t + \dfrac{1}{10}\sin 10t$.

(b) $I(t) = \dot{Q}(t) = e^{-6t}\left(\dfrac{3}{4}\sin 8t - \cos 8t\right) + \cos 10t$.

Beats

18.   The identity

$$\cos(A+B) - \cos(A-B) = -2\sin A \sin B$$

may be used here. In this case, if $A = 2t$, $B = t$, and we have $A + B = 3t$, $A - B = t$. Hence,

$$\cos 3t - \cos t = -2\sin 2t \sin t.$$

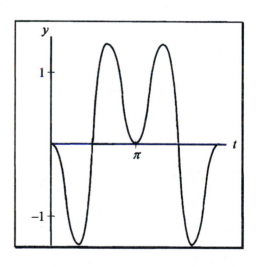

■   **Steady State**

Note: We must be careful in finding the phase angle using the formula

$$\delta = \tan^{-1}\frac{B}{A}$$

because we don't know in which quadrant δ lies using $\delta = \tan^{-1}\frac{B}{A}$. The value of δ you get might be π units different from the correct value. Unless you know by some other means in which quadrant δ lies, it is best to use the two equations

$$C\cos\delta = A, \ C\sin\delta = B.$$

A good rule of thumb is to think of the AB plane; when both A, B are positive, δ will be in the first quadrant (i.e., δ between 0 and $\frac{\pi}{2}$), but when both A and B are negative, δ will be in the third quadrant, and so on.

21. $\ddot{x} + 2\dot{x} + 2x = 2\cos t$

The roots of the characteristic equation are $-1 \pm i$. We use for the particular solution:

$$x_p(t) = A\cos t + B\sin t = C\cos(t - \delta).$$

Another approach to x_p is to note that $F_0 = 2$, $\omega_0 = \sqrt{2}$, $\omega_f = 1$, $m = 1$, $b = 2$ and simply substitute these numbers into the text solution to find

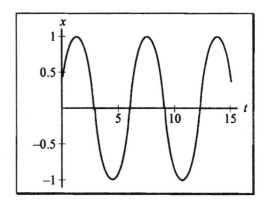

$$x_{ss}(t) = \frac{2}{\sqrt{5}}\cos(t - \delta)$$

where the phase angle is

$$\delta = \tan^{-1}(2) \approx 1.1 \text{ radians.}$$

■   **Resonance**

24.   If resonance exists, the input frequency ω_f is the same as the natural frequency $\omega_0 = 2\sqrt{3}$ (see Problem 23). Hence, we have the initial-value problem

$$\ddot{x} + 12x = 16\cos(2\sqrt{3}t), \ x(0) = \dot{x}(0) = 0.$$

This equation has homogeneous solution

$$x_h(t) = c_1\cos(2\sqrt{3}t) + c_2\sin(2\sqrt{3}t).$$

To find a particular solution we seek a function of the form

$$x_p = At\cos\left(2\sqrt{3}t\right) + Bt\sin\left(2\sqrt{3}t\right).$$

Differentiating and substituting into the differential equation yields

$$A = 0, \quad B = \frac{16}{4\sqrt{3}} = 4\frac{\sqrt{3}}{3},$$

so the general solution is

$$x(t) = c_1\cos\left(2\sqrt{3}t\right) + c_2\sin\left(2\sqrt{3}t\right) + \frac{4\sqrt{3}}{3}t\sin\left(2\sqrt{3}t\right).$$

Substituting this into $x(0) = \dot{x}(0) = 0$ yields $c_1 = 0$, $c_2 = 0$. Hence, the solution to the IVP is

$$x(t) = \frac{4\sqrt{3}}{3}t\sin\left(2\sqrt{3}t\right).$$

■   **Phase Portrait Recognition**

27. $\ddot{x} + 0.3\dot{x} + x = \cos t$

(C) We have damping but we also have a sinusoidal forcing term. Hence, the homogeneous solution goes to zero and particular solutions consist of sines and cosines, which give rise to circles in the phase plane. Therefore, starting from the origin $x(0) = \dot{x}(0) = 0$ we get a curve that approaches a circle from the inside.

30. $\ddot{x} + 0.3\dot{x} + x = 0$

(B) The system is unforced but damped, and hence trajectories must approach $x(0) = \dot{x}(0) = 0$.

■   **Electrical Version**

33.   (a) $Q_h = 4\cos 4t - 5\sin 4t$

(b)   The amplitude of the transient solution is $A = \sqrt{4^2 + 5^2} = \sqrt{41}$

(c) $Q_s = Q_p = 6t\cos 4t$ (d) $Q_p = 6t\cos 4t$

(e) $4 = \sqrt{\dfrac{1}{LC}} = \sqrt{\dfrac{1}{C}}, \quad C = \dfrac{1}{16}$

(f)   The charge on the capacitor will oscillate with ever-increasing amplitude due to pure resonance.

■   **Extrema of the Amplitude Response**

36.   We write

$$A(\omega_f) = \frac{F_0}{\sqrt{\left(k - m\omega_f^2\right)^2 + b^2\omega_f^2}} = \frac{F_0/m}{\sqrt{\left[\left(\dfrac{k}{m}\right) - \omega_f^2\right]^2 + \left(\dfrac{b}{m}\right)^2 \omega_f^2}}.$$

Differentiating A with respect to ω_f, we find

$$A'(\omega_f) = \frac{-\left(\dfrac{2\omega_f}{m}\right)\left[\omega_f^2 - \left(\dfrac{k}{m} - \dfrac{b^2}{2m^2}\right)\right]}{\left[\left(\dfrac{k}{m} - \omega_f^2\right)^2 + \left(\dfrac{b}{m}\right)^2 \omega_f^2\right]^{3/2}} \times F_0$$

from which it follows that $A'(\omega_f) = 0$ if and only if $\omega_f = 0$ or

$$\omega_f = \sqrt{\frac{k}{m} - \frac{b^2}{2m^2}}\,.$$

When $b^2 > 2mk$, ω_f is not real. Hence $A'(\omega_f) = 0$ only when $\omega_f = 0$. In this case $A(\omega_f)$ damps to zero as ω_f goes from 0 to ∞. It is clear then that the maximum of $A(\omega_f)$ occurs when $\omega_f = 0$ and has the value

$$A(0) = \frac{1}{k}\,.$$

When $b^2 < 2mk$, then ω_f is real and positive. It is easy using the sign of the derivative to see that the maximum of $A(\omega_f)$ occurs at

$$\omega_f = \sqrt{\frac{k}{m} - \frac{b^2}{2m^2}}\,.$$

Evaluating the amplitude response at this value of ω_f yields the expression

$$A_{\max} = \frac{F_0}{b\sqrt{\dfrac{k}{m} - \dfrac{b^2}{4m^2}}}\,.$$

4.7   Conservation and Conversion

- **General Formula for Total Energy in an *LC*-Circuit**

3. $L\ddot{Q} + \dfrac{1}{C}Q = 0$, $Q(0) = Q_0$, $I(0) = I_0$

The total energy of this LC system is the constant value

$$E(t) = \frac{1}{2}L\dot{Q}^2 + \frac{1}{2C}Q^2 = \frac{1}{2}LI_0^2 + \frac{1}{2C}Q_0^2.$$

- **Questions of Energy**

6. $\ddot{x} - x + x^3 = 0$

(a) $\text{KE} = \dfrac{1}{2}\dot{x}^2$, $V = \displaystyle\int\left(-x + x^3\right)dx = -\frac{1}{2}x^2 + \frac{1}{4}x^4$

$$E(x,\ \dot{x}) = \text{KE} + V = \frac{1}{2}\dot{x}^2 - \frac{1}{2}x^2 + \frac{1}{4}x^4$$

(b)   To find the equilibrium points, we seek the solutions of

$$\frac{\partial E}{\partial x} = -x + x^3 = 0, \quad \frac{\partial E}{\partial \dot{x}} = \dot{x} = 0.$$

Solving these equations, we find three equilibrium points at $(-1,\ 0)$, $(0,\ 0)$ and $(1,\ 0)$.

Because $\dot{x} = 0$ for all these points, we determine which points are stable (local maxima) by simply drawing the graph of $V(x)$ (shown in part (c)).

(c)   Graph of the potential energy $V(x)$ is shown. Note that $V(x)$ has local minima at $x = \pm 1$ and a local maxima at $x = 0$. Hence,

$$(-1,\ 0) \text{ and } (1,\ 0)$$

are stable points, and

$$(0,\ 0)$$

is an unstable point.

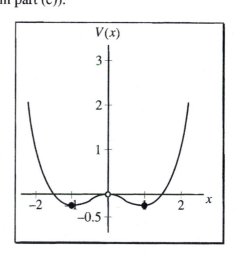

Potential energy of $\ddot{x} - x + x^3 = 0$

9. $\ddot{x} + x^2 = 0$

(a) $\text{KE} = \dfrac{1}{2}\dot{x}^2$, $V = \displaystyle\int x^2 dx = \dfrac{1}{3}x^3$

$E(x,\,\dot{x}) = \text{KE} + V = \dfrac{1}{2}\dot{x}^2 + \dfrac{1}{3}x^3$

(b)   To find the equilibrium points, we seek the solutions of the equations

$$\frac{\partial E}{\partial x} = x^2 = 0, \ \frac{\partial E}{\partial \dot{x}} = \dot{x} = 0.$$

Solving these equations, we find one equilibrium point at

$$(0,\,0).$$

To determine if the point is stable, we note that the potential energy

$$V(x) = \frac{1}{3}x^3$$

does not have a local maxima or minima at $(0,0)$, so $(0,0)$ is an unstable (or semistable) equilibrium point.

(c)   The graph of $V(x)$ is a simple cubic.

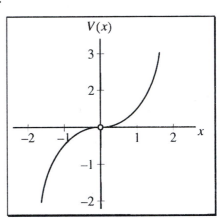

Potential energy of $\ddot{x} + x^2 = 0$

12. $\ddot{x} = \dfrac{1}{x^2}$

(a) $\text{KE} = \dfrac{1}{2}\dot{x}^2$, $V = -\displaystyle\int \dfrac{1}{x^2} dx = \dfrac{1}{x}$

$E(x,\,\dot{x}) = \text{KE} + V = \dfrac{1}{2}\dot{x}^2 + \dfrac{1}{x}$

(b)   To find the equilibrium points, we seek the solutions of the two equations

$$\frac{\partial E}{\partial x} = \frac{-1}{x^2} = 0, \ \frac{\partial E}{\partial \dot{x}} = \dot{x} = 0.$$

Because the first equation does not have a solution, there is no equilibrium point.

(c)   The graph of the potential energy $V(x)$ is shown. Note that $V(x)$ does not have any local maxima or minima, which corresponds to the absence of equilibrium points noted in part (b).

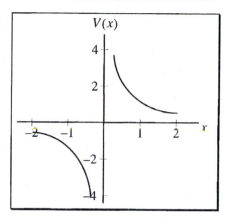

Potential energy of $\ddot{x} = 1/x^2$

■   **Conservative or Nonconservative?**

15. $\ddot{x} + kx = 0$

Conservative because it has the form

$$m\ddot{x} + F(x) = 0.$$

The total energy of this conservative system is

$$E(x, \dot{x}) = \frac{1}{2}m\dot{x} + \int F(x)\,dx = \frac{1}{2}\dot{x}^2 + \frac{1}{2}kx^2.$$

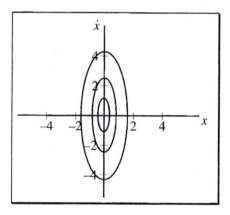

We draw contour curves for this surface over the $x\dot{x}$-plane to view the trajectories of the differential equation in the $x\dot{x}$ plane. The trajectories of $\ddot{x} + kx = 0$ are ellipses each with height $\sqrt{k}$ times its width.

18. $\ddot{\theta} + \sin\theta = 1$

Conservative because it can be written in the form

$$m\ddot{\theta} + F(\theta) = 0,$$

where $F(\theta) = \sin\theta - 1$. The total energy is

$$E(\theta, \dot{\theta}) = \frac{1}{2}\dot{\theta}^2 - \cos\theta - \theta.$$

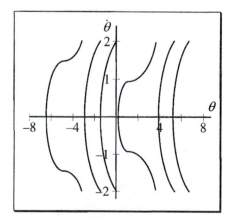

We can draw contour curves for this surface over the $\theta\dot{\theta}$-plane to view the trajectories of the differential equation in the $\theta\dot{\theta}$ plane.

■   **Computer Lab: Undamped Spring**

21.   IDE Lab

■   **Conversion of Equations**

24. $\ddot{\theta} + \dfrac{g}{L}\sin\theta = 0$

Letting $x_1 = \theta$, $x_2 = \dot{\theta}$, we have

$$\dot{x}_1 = x_2$$

$$\dot{x}_2 = -\frac{g}{L}\sin x_1.$$

This system is not linear, so there is no matrix form.

27. $t^2\ddot{x} + t\dot{x} + \left(t^2 - n^2\right)x = 0$

Letting $x_1 = x$, $x_2 = \dot{x}$, we have

$$\dot{x}_1 = x_2$$

$$\dot{x}_2 = -\frac{\left(t^2 - n^2\right)}{t^2}x_1 - \frac{1}{t}x_2.$$

In matrix form, this becomes

$$\begin{bmatrix} \dot{x}_1 \\ \dot{x}_2 \end{bmatrix} = \begin{bmatrix} 0 & 1 \\ -\dfrac{t^2 - n^2}{t^2} & -\dfrac{1}{t} \end{bmatrix} \begin{bmatrix} x_1 \\ x_2 \end{bmatrix}.$$

30. $\dfrac{d^4y}{dt^4} + 3\dfrac{d^3y}{dt^3} + 2\dfrac{d^2y}{dt^2} + \dfrac{dy}{dt} + 4y = 1$

If we introduce

$$x_1 = y$$

$$x_2 = \frac{dy}{dt}$$

$$x_3 = \frac{d^2y}{dt^2}$$

$$x_4 = \frac{d^3y}{dt^3}$$

we have the differential equations

$$\dot{x}_1 = x_2$$
$$\dot{x}_2 = x_3$$
$$\dot{x}_3 = x_4$$
$$\dot{x}_4 = -4x_1 - x_2 - 2x_3 - 3x_4 + 1$$

or in matrix form

$$
\begin{bmatrix} \dot{x}_1 \\ \dot{x}_2 \\ \dot{x}_3 \\ \dot{x}_4 \end{bmatrix} = \begin{bmatrix} 0 & 1 & 0 & 0 \\ 0 & 0 & 1 & 0 \\ 0 & 0 & 0 & 1 \\ -4 & -1 & -2 & -3 \end{bmatrix} \begin{bmatrix} x_1 \\ x_2 \\ x_3 \\ x_4 \end{bmatrix} + \begin{bmatrix} 0 \\ 0 \\ 0 \\ 1 \end{bmatrix}.
$$

■ Conversion of IVPs

33. $y'' + 3y' + 2z = e^{-t}$, $y(0) = 0$, $y'(0) = 1$

$z'' + y + 2z = 1$, $z(0) = 1$, $z'(0) = 0$

Letting $x_1 = y$, $x_2 = y'$, $x_3 = z$, $x_4 = z'$ yields

$$
\begin{array}{ll}
x_1' = x_2 & x_1(0) = 0 \\
x_2' = -3x_2 - 2x_3 + e^{-t} & x_2(0) = 1 \\
x_3' = x_4 & x_3(0) = 1 \\
x_4' = -x_1 - 2x_3 + 1 & x_4(0) = 0
\end{array}
$$

In matrix form this becomes

$$
\begin{bmatrix} x_1' \\ x_2' \\ x_3' \\ x_4' \end{bmatrix} = \begin{bmatrix} 0 & 1 & 0 & 0 \\ 0 & -3 & -2 & 0 \\ 0 & 0 & 0 & 1 \\ -1 & 0 & -2 & 0 \end{bmatrix} \begin{bmatrix} x_1 \\ x_2 \\ x_3 \\ x_4 \end{bmatrix} + \begin{bmatrix} 0 \\ e^{-t} \\ 0 \\ 1 \end{bmatrix}, \quad \begin{bmatrix} x_1(0) \\ x_2(0) \\ x_3(0) \\ x_4(0) \end{bmatrix} = \begin{bmatrix} 0 \\ 1 \\ 1 \\ 0 \end{bmatrix}.
$$

■ Conversion of Systems

36. $y''' = f(t, y, y', y'', z, z')$
$z'' = f(t, y, y', y'', z, z')$

Letting $x_1 = y$, $x_2 = y'$, $x_3 = y''$, $x_4 = z$, $x_5 = z'$ yields

$$
\begin{array}{l}
x_1' = x_2 \\
x_2' = x_3 \\
x_3' = f(t, x_1, x_2, x_3, x_4, x_5) \\
x_4' = x_5 \\
x_5' = g(t, x_1, x_2, x_3, x_4, x_5).
\end{array}
$$

■   **Solving Linear Systems**

39. $x_1' = 3x_1 - 2x_2$
$x_2' = 2x_1 - 2x_2$

From first DE: $x_2 = -\dfrac{1}{2}x_1' + \dfrac{3}{2}x_1$. Substituting in second DE yields a second order DE to solve for x_1.

$$\left(-\frac{1}{2}x_1' + \frac{3}{2}x_1\right)' = 2x_1 - 2\left(-\frac{1}{2}x_1' + \frac{3}{2}x_1\right)$$

$$-\frac{1}{2}x_1'' + \frac{3}{2}x_1' = 2x_1 + x_1' - 3x_1$$

$$x_1'' - x_1' - 2x_1 = 0$$

$$x_1 = c_1 e^{2t} + c_2 e^{-t}$$

To find x_2, substitute the solution for x_1 back into the first DE.

$$x_2 = -\frac{1}{2}x_1' + \frac{3}{2}x_1 = -\frac{1}{2}\left(2c_1 e^{2t} - c_2 e^{-t}\right) + \frac{3}{2}\left(c_1 e^{2t} + c_2 e^{-t}\right) = \frac{1}{2}c_1 e^{2t} + 2c_2 e^{-t}.$$

■   **Solving IVPs for Systems**

42. $x_1' = 6x_1 - 3x_2$
$x_2' = 2x_1 + x_2$

From first DE: $x_2 = 2x_1 - \dfrac{1}{3}x_1'$. Substituting in second DE yields a second order DE to solve for x_1.

$$\left(2x_1 - \frac{1}{3}x_1'\right)' = 2x_1 + \left(2x_1 - \frac{1}{3}x_1'\right)$$

$$x_1'' - 7x_1' + 12x_1 = 0$$

$$x_1 = c_1 e^{3t} + c_2 e^{4t}.$$

From first calculation, $x_2 = 2x_1 - \dfrac{1}{3}x_1'$, so

$$x_2 = 2\left(c_1 e^{3t} + c_2 e^{4t}\right) - \frac{1}{3}\left(3c_1 e^{3t} + 4c_2 e^{4t}\right) = c_1 e^{3t} + \frac{2}{3}c_2 e^{4t}.$$

Applying initial conditions:

$$x_1(0) = 2 \quad \Rightarrow \quad c_1 + c_2 = 2$$

$$x_2(0) = 3 \quad \Rightarrow \quad c_1 + \frac{2}{3}c_2 = 3$$

so

$$c_2 = -3 \text{ and } c_1 = 5.$$

The solution to the IVP is $x_1 = 5e^{3t} - 3e^{4t}$, $x_2 = 5e^{3t} - 2e^{4t}$.

■   **Coupled Mass-Spring System**

45.   Given the linear system

$$m\ddot{x}_1 = -k_1 x_1 + k_2 (x_2 - x_1) = -(k_1 + k_2) x_1 + k_2 x_2$$
$$m\ddot{x}_2 = -k_2 (x_2 - x_1) = k_2 x_1 - k_2 x_2,$$

we let

$$z_1 = x_1 \quad z_3 = x_2$$
$$z_2 = \dot{x}_1 \quad z_4 = \dot{x}_2.$$

We then have the first-order system

$$\dot{z}_1 = z_2$$
$$\dot{z}_2 = -\frac{k_1 + k_2}{m} z_1 + \frac{k_2}{m} z_3$$
$$\dot{z}_3 = z_4$$
$$\dot{z}_4 = \left(\frac{k_2}{m}\right) z_1 - \left(\frac{k_2}{m}\right) z_3.$$

In matrix form this becomes

$$\begin{bmatrix} \dot{z}_1 \\ \dot{z}_2 \\ \dot{z}_3 \\ \dot{z}_4 \end{bmatrix} = \begin{bmatrix} 0 & 1 & 0 & 0 \\ -\dfrac{(k_1 + k_2)}{m} & 0 & \dfrac{k_2}{m} & 0 \\ 0 & 0 & 0 & 1 \\ \dfrac{k_2}{m} & 0 & -\dfrac{k_2}{m} & 0 \end{bmatrix} \begin{bmatrix} z_1 \\ z_2 \\ z_3 \\ z_4 \end{bmatrix}.$$

■   **Suggested Journal Entry**

48.   Student Project

CHAPTER 5

Linear Transformations

5.1 Linear Transformations

Note: Many different arguments may be used to prove nonlinearity; our solutions to Problems 1–23 provide a sampling.

■ **Checking Linearity**

3. $T(x, y) = (xy, 2y)$

If we let $\vec{u} = (u_1, u_2)$, we have

$$cT(\vec{u}) = cT(u_1, u_2) = c(u_1u_2, 2u_2) = (cu_1u_2, 2cu_2)$$

and

$$T(c\vec{u}) = T(cu_1, cu_2) = (c^2u_1u_2, 2cu_2).$$

Hence $cT(\vec{u}) \neq T(c\vec{u})$, so T is not a linear transformation.

6. $T(x, y) = (x, 1, y, 1)$

Because T does not map the zero vector $[0, 0] \in \mathbb{R}^2$ into the zero vector $[0, 0, 0, 0] \in \mathbb{R}^4$, T is not a linear transformation.

9. $T(f) = tf'(t)$

If f and g are continuous functions on $[0, 1]$, then

$$T(f + g) = t[f(t) + g(t)]' = tf'(t) + tg'(t) = T(f) + T(g)$$

and

$$T(cf) = t(cf(t))' = ctf'(t) = cT(f).$$

Hence, T is a linear transformation.

12. $T\left(at^3 + bt^2 + ct + d\right) = a + b$

If we introduce the two vectors

$$\vec{p} = a_1 t^3 + b_1 t^2 + c_1 t + d_1$$
$$\vec{q} = a_2 t^3 + b_2 t^2 + c_2 t + d_2$$

then

$$T\left(\vec{p} + \vec{q}\right) = T\left(\left(a_1 + a_2\right)t^3 + \left(b_1 + b_2\right)t^2 + \left(c_1 + c_2\right)t + \left(d_1 + d_2\right)\right) = \left(a_1 + a_2\right) + \left(b_1 + b_2\right)$$
$$= \left(a_1 + b_1\right) + \left(a_2 + b_2\right) = T\left(\vec{p}\right) + T\left(\vec{q}\right)$$
$$T\left(c\vec{p}\right) = T\left(c\left(a_1 t^3 + b_1 t^2 + c_1 t + d_1\right)\right) = T\left(ca_1 t^3 + cb_1 t^2 + cc_1 t + cd_1\right) = ca_1 + cb_1$$
$$= c\left(a_1 + b_1\right) = cT\left(\vec{p}\right).$$

Hence, the derivative transformation defined on $\mathbb{P}_3$ is a linear transformation.

15. $T\begin{bmatrix} a & b \\ c & d \end{bmatrix} = \mathrm{Tr}\begin{bmatrix} a & b \\ c & d \end{bmatrix}$

Let $\mathbf{A} = \begin{bmatrix} a_{11} & a_{12} \\ a_{21} & a_{22} \end{bmatrix}, \ \mathbf{B} = \begin{bmatrix} b_{11} & b_{12} \\ b_{21} & b_{22} \end{bmatrix}$

so that $\mathbf{A} + \mathbf{B} = \begin{bmatrix} a_{11} + b_{11} & a_{12} + b_{12} \\ a_{21} + b_{21} & a_{22} + b_{22} \end{bmatrix}.$

Then $T\left(\mathbf{A} + \mathbf{B}\right) = \left(a_{11} + b_{11}\right) + \left(a_{22} + b_{22}\right) = \left(a_{11} + a_{22}\right) + \left(b_{11} + b_{22}\right) = T\left(\mathbf{A}\right) + T\left(\mathbf{B}\right)$

and $T\left(k\mathbf{A}\right) = T\begin{bmatrix} ka & kb \\ kc & kd \end{bmatrix} = ka + kd = k\left(a + d\right) = kT\left(\mathbf{A}\right).$

Hence, T is a linear transformation on $\mathbb{M}_{22}$.

■   **Linear Systems of DEs**

18. $T(x, y) = (x' - y, 2x + y')$

$$T\left((x_1, y_1) + (x_2, y_2)\right) = T(x_1 + x_2, y_1 + y_2) = \left((x_1 + x_2)' - (y_1 + y_2), 2(x_1 + x_2) + (y_1 + y_2)'\right)$$
$$= (x_1' + x_2' - y_1 - y_2, 2x_1 + 2x_2 + y_1' + y_2')$$
$$= (x_1' - y_1, 2x_1 + y_1') + (x_2' - y_2, 2x_2 + y_2') = T(x_1, y_1) + T(x_2, y_2)$$

$$T\left(c(x, y)\right) = T(cx, cy) = \left((cx)' - cy, 2(cx) + (cy)'\right)$$
$$= (cx' - cy, 2cx + cy') = \left(c(x' - y), c(2x + y')\right) = c(x' - y, 2x + y')$$
$$= cT(x, y)$$

■ **Laying Linearity on the Line**

21. $T(x) = ax + b$

$$T(kx) = a(kx) + b = akx + b \neq kT(x) = k(ax + b) = akx + kb$$

so $T(kx) \neq kT(x)$. Hence, T is not a linear transformation.

24. $T(x) = \sin x$

Because $\qquad T(kx) = \sin(kx)$ and $kT(x) = k \sin x,$

we have that $\qquad T(kx) \neq kT(x)$

so T is not a linear transformation. We could also simply note that

$$T\left(\frac{\pi}{2} + \frac{\pi}{2}\right) = T(\pi) = \sin(0) = 0$$

but $\qquad T\left(\frac{\pi}{2}\right) + T\left(\frac{\pi}{2}\right) = \sin\left(\frac{\pi}{2}\right) + \sin\left(\frac{\pi}{2}\right) = 1 + 1 = 2.$

■ **Geometry of a Linear Transformation**

27. Direct computation: the vector $[0, y]$ lies on the y-axis and $[2y, y]$ lies on the line $y = \dfrac{x}{2}$, so the

transformation maps vectors on the y-axis onto vectors on the line $y = \dfrac{x}{2}$.

■ **Geometric Interpretations in $\mathbb{R}^2$**

30. $T(x, y) = (x, 0)$

This map projects points to the x-axis. A matrix representation is

$$\begin{bmatrix} 1 & 0 \\ 0 & 0 \end{bmatrix}.$$

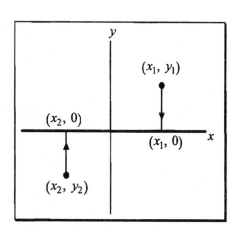

■ **Find the Standard Matrix**

33. $T(x, y) = x + 2y$

T maps the point $(x, y) \in \mathbb{R}^2$ into the real number $x + 2y \in \mathbb{R}$. In matrix form,

$$T(x, y) = \begin{bmatrix} 1 & 2 \end{bmatrix} \begin{bmatrix} x \\ y \end{bmatrix} = x + 2y.$$

36. $T(x, y) = (x + 2y, x - 2y, y)$

T maps   the   point $(x, y) \in \mathbb{R}^2$ in   two   dimensions   into   the   new   point $T(x, y) = (x + 2y, x - 2y, y) \in \mathbb{R}^3$. In matrix form, the linear transformation T can be written

$$T(x, y) = \begin{bmatrix} 1 & 2 \\ 1 & -2 \\ 0 & 1 \end{bmatrix} \begin{bmatrix} x \\ y \end{bmatrix} = \begin{bmatrix} x + 2y \\ x - 2y \\ y \end{bmatrix}.$$

39. $T(v_1, v_2, v_3) = (v_1 + 2v_2, v_3, -v_1 + 4v_2 + 3v_3)$

T maps $(v_1, v_2, v_3) \in \mathbb{R}^3$ into $(v_1 + 2v_2, v_3, -v_1 + 4v_2 + 3v_3) \in \mathbb{R}^3$. In matrix form,

$$T(v_1, v_2, v_3) = \begin{bmatrix} 1 & 2 & 0 \\ 0 & 0 & 1 \\ -1 & 4 & 3 \end{bmatrix} \begin{bmatrix} v_1 \\ v_2 \\ v_3 \end{bmatrix} = \begin{bmatrix} v_1 + 2v_2 \\ v_3 \\ -v_1 + 4v_2 + 3v_3 \end{bmatrix}.$$

■   **Mapping and Images**

42. $T(x, y) = (x + y, x)$

T maps a vector $[x, y] \in \mathbb{R}^2$ into the vector $[x + y, x] \in \mathbb{R}^2$. For $\vec{u} = [1, 0]$,

$$T(\vec{u}) = T([1, 0]) = [1 + 0, 1] = [1, 1].$$

Setting

$$\dot{T}([x, y]) = [x + y, x] = \vec{w} = [3, 1]$$

yields $x + y = 3$, $x = 1$, which has the solution $x = 1$, $y = 2$, or $[x, y] = [1, 2]$

45. $T(u_1, u_2) = (u_1, u_1 + u_2, u_1 - u_2)$

T maps a vector $[u_1, u_2] \in \mathbb{R}^2$ into the vector $[u_1, u_1 + u_2, u_1 - u_2] \in \mathbb{R}^3$. For $\vec{u} = [1, 1]$,

$$T(\vec{u}) = T([1, 1]) = [1, 1 + 1, 1 - 1] = [1, 2, 0].$$

Setting

$$T([u_1, u_2]) = [u_1, u_1 + u_2, u_1 - u_2] = \vec{w} = [1, 1, 0]$$

yields $u_1 = 1$, $u_1 + u_2 = 1$, $u_1 - u_2 = 0$, which has no solutions. In other words, no vectors $[u_1, u_2] \in \mathbb{R}^2$ map into $[1, 1, 0]$ under the linear transformation T.

48. $T\left(u_1,\ u_2,\ u_3\right)=\left(u_1,\ u_2,\ u_1+u_3\right)$

T maps a vector $\left[u_1,\ u_2,\ u_3\right]\in\mathbb{R}^3$ into $\left[u_1,\ u_2,\ u_1+u_3\right]\in\mathbb{R}^3$. For $\vec{\mathbf{u}}=\left[1,\ 2,\ 3\right]$,

$$T\left(\vec{\mathbf{u}}\right)=T\left(\left[1,\ 2,\ 1\right]\right)=\left[1,\ 2,\ 1+1\right]=\left[1,\ 2,\ 2\right].$$

Setting

$$T\left(\left[u_1,\ u_2,\ u_3\right]\right)=\left[u_1,\ u_2,\ u_1+u_3\right]=\vec{\mathbf{w}}=\left[0,\ 0,\ 1\right]$$

yields $u_1=0$, $u_2=0$, $u_1+u_3=1$, which yields $u_1=0$, $u_2=0$, $u_3=1$, so $[0,0,1]$ maps into itself.

■   **Transforming Areas**

51.   For the points

$$\left(0,\ 0\right),\ \left(1,\ 0\right),\ \left(1,\ 2\right),\ \left(0,\ 2\right),$$

the image is the parallelogram with vertices

$$\left(0,\ 0\right),\ \left(1,\ 2\right),\ \left(-1,\ 4\right),\ \left(-2,\ 2\right).$$

The original rectangle (shown in gray) has area 2; the new area is 6.

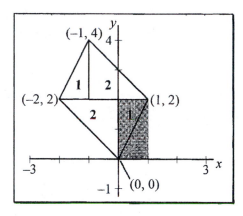

■   **Linear Transformations in the Plane**

54.   (a)   (B) shear; in the x direction.

(b)   (E) nonlinear; linear transformations map lines into straight lines.

(c)   (C) rotation; a 90-degree rotation in the counterclockwise direction.

(d)   (E) nonlinear; $\left(0,\ 0\right)$ must map into $\left(0,\ 0\right)$ in a linear transformation.

(e)   (A) scaling (dilation or contraction); contraction in both the x and y directions.

(f)   (B) shear; in the y-direction.

(g)   (D) reflection; through the x-axis.

■   **Finding the Matrices**

57. $\mathbf{L}=\begin{bmatrix}1 & 0\\ 0 & -1\end{bmatrix}$ describes (G), (reflection through the x-axis)

■   **Shear Transformation**

60.   (a)   The matrix

$$\begin{bmatrix} 1 & 0 \\ 1 & 1 \end{bmatrix}$$

produces a shear in the y-direction of one unit. Figure B is a shear in the y direction.

(b)   Figure A is a shear of -1 unit and would be carried out by the matrix

$$\begin{bmatrix} 1 & 0 \\ -1 & 1 \end{bmatrix}.$$

(c)   Figure C is a shear of 1 unit in the x direction; matrix is

$$\begin{bmatrix} 1 & 1 \\ 0 & 1 \end{bmatrix}.$$

■   **Pinwheel**

63.   (a)   A negative shear of 1 in the y-direction is

$$\begin{bmatrix} 1 & 0 \\ -1 & 1 \end{bmatrix}.$$

An easy way to see this is by observing how each point gets mapped. We have

$$\begin{bmatrix} 1 & 0 \\ -1 & 1 \end{bmatrix}\begin{bmatrix} x \\ y \end{bmatrix} = \begin{bmatrix} x \\ -x+y \end{bmatrix} = \begin{bmatrix} x \\ y \end{bmatrix} - \begin{bmatrix} 0 \\ x \end{bmatrix}.$$

Each point moves down by value of its x-coordinate. In other words, the further you are away in the x-direction from the x-axis the more the points move down (or up in case x is negative). Note that for the pinwheel the line that sticks out to the right is sheared down, whereas the line that sticks out to the left (in the NEGATIVE x region) is sheared up. Twelve rotations of $30°$ give the identity matrix.

(b) $\left(\mathbf{Rot}\left(30°\right)\right)^n = I$, only when n is a multiple of 12.

■ Reflections

66.   (a)   A reflection about the x-axis followed by a reflection through the y-axis would be

$$\mathbf{B}_y\mathbf{B}_x = \begin{bmatrix} 1 & 0 \\ 0 & -1 \end{bmatrix}\begin{bmatrix} -1 & 0 \\ 0 & 1 \end{bmatrix}$$

$$= \begin{bmatrix} -1 & 0 \\ 0 & -1 \end{bmatrix},$$

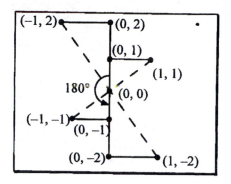

which is a reflection through the origin. The transformed r-shape is shown.

Reflection of r shape through the origin

(b)   An 180° rotation in the counterclockwise direction has matrix

$$\begin{bmatrix} \cos\pi & -\sin\pi \\ \sin\pi & \cos\pi \end{bmatrix} = \begin{bmatrix} -1 & 0 \\ 0 & -1 \end{bmatrix},$$

which is equivalent to the steps in part (a).

■ Anatomy of Another Transformation

69.   (a)   Solving for x, y in the system $\begin{bmatrix} 1 & 1 & -1 \\ 2 & 2 & -3 \end{bmatrix}\begin{bmatrix} x \\ y \\ z \end{bmatrix} = \begin{bmatrix} 0 \\ 0 \end{bmatrix}$, we find

$$\begin{aligned} x + y - z &= 0 \\ 2x + 2y - 3z &= 0, \end{aligned}$$

which we can solve getting the one-dimensional subspace of $\mathbb{R}^3$ of solutions

$$\begin{bmatrix} x \\ y \\ z \end{bmatrix} = \alpha\begin{bmatrix} 1 \\ -1 \\ 0 \end{bmatrix}, \; \alpha \text{ any real number.}$$

(b)   Setting $\begin{bmatrix} 1 & 1 & -1 \\ 2 & 2 & -3 \end{bmatrix}\begin{bmatrix} x \\ y \\ z \end{bmatrix} = \begin{bmatrix} 1 \\ 1 \end{bmatrix}$, we have

$$\begin{aligned} x + y - z &= 1 \\ 2x + 2y - 3z &= 1, \end{aligned}$$

which has solutions

$$\begin{bmatrix} x \\ y \\ z \end{bmatrix} = \begin{bmatrix} 2-\alpha \\ \alpha \\ 1 \end{bmatrix} = \begin{bmatrix} 2 \\ 0 \\ 1 \end{bmatrix} + \alpha\begin{bmatrix} -1 \\ 1 \\ 0 \end{bmatrix}$$

for α any real number. In other words, the vectors that map into $\begin{bmatrix} 1, & 1 \end{bmatrix}$ consist of a line in $\mathbb{R}^3$ passing through $(2, 0, 1)$ in the direction of $\begin{bmatrix} -1, & 1, & 0 \end{bmatrix}$.

(c)   We can write the matrix product as

$$\begin{bmatrix} 1 & 1 & -1 \\ 2 & 2 & -3 \end{bmatrix} \begin{bmatrix} x \\ y \\ z \end{bmatrix} = x \begin{bmatrix} 1 \\ 2 \end{bmatrix} + y \begin{bmatrix} 1 \\ 2 \end{bmatrix} + z \begin{bmatrix} -1 \\ -3 \end{bmatrix} = (x+y) \begin{bmatrix} 1 \\ 2 \end{bmatrix} + z \begin{bmatrix} -1 \\ -3 \end{bmatrix}.$$

The image of T consists of the span of the vectors $\begin{bmatrix} 1, & 2 \end{bmatrix}$ and $\begin{bmatrix} -1, & -3 \end{bmatrix}$, which is $\mathbb{R}^2$.

■   **Functionals**

72. $T(f) = -2 \int_0^1 f(t) dt$

$$T(f + g) = -2 \int_0^1 (f(t) + g(t)) dt = -2 \int_0^1 f(t) dt - 2 \int_0^1 g(t) dt = T(f) + T(g)$$

$$T(cf) = -2 \int_0^1 cf(t) dt = c \left(-2 \int_0^1 f(t) dt \right) = cT(f).$$

Hence, T is a linear functional.

■   **Further Linearity Checks**

75. $L_1(y) = \int_0^\infty e^{-st} f(t) dt$

$$L(f + g) = \int_0^\infty e^{-st} \left[f(t) + g(t) \right] dt = \int_0^\infty e^{-st} f(t) dt + \int_0^\infty e^{-st} g(t) dt = L(f) + L(g)$$

$$L(cf) = \int_0^\infty e^{-st} cf(t) dt = c \int_0^\infty e^{-st} f(t) dt = cL(f)$$

■   **Projections**

78.   The transformation $T : \mathbb{V} \rightarrow \mathbb{W}$ defined by $T(x, y, z) = (x, 0, 3x)$ can be represented by matrix multiplication as

$$\begin{bmatrix} 1 & 0 & 0 \\ 0 & 0 & 0 \\ 3 & 0 & 0 \end{bmatrix} \begin{bmatrix} x \\ y \\ z \end{bmatrix} = \begin{bmatrix} x \\ 0 \\ 3x \end{bmatrix} = x \begin{bmatrix} 1 \\ 0 \\ 3 \end{bmatrix}.$$

Hence, $\mathbb{W}$ is the line spanned by $\begin{bmatrix} 1, & 0, & 3 \end{bmatrix}$. Also T reduces to the identity on $\mathbb{W}$ because

$$\begin{bmatrix} 1 & 0 & 0 \\ 0 & 0 & 0 \\ 3 & 0 & 0 \end{bmatrix} \begin{bmatrix} x \\ 0 \\ 3x \end{bmatrix} = \begin{bmatrix} x \\ 0 \\ 3x \end{bmatrix}$$

and, hence T is a projection on $\mathbb{W}$.

■   **Rotational Transformations**

81.   Using the trigonometric identities

$$\cos(\theta + \alpha) = \cos\theta\cos\alpha - \sin\theta\sin\alpha$$
$$\sin(\theta + \alpha) = \sin\theta\cos\alpha + \cos\theta\sin\alpha$$

we have

$$\begin{bmatrix} \cos\theta & -\sin\theta \\ \sin\theta & \cos\theta \end{bmatrix}\begin{bmatrix} r\cos\alpha \\ r\sin\alpha \end{bmatrix} = \begin{bmatrix} r\cos\theta\cos\alpha - r\sin\theta\sin\alpha \\ r\sin\theta\cos\alpha + r\cos\theta\sin\alpha \end{bmatrix} = \begin{bmatrix} r\cos(\theta + \alpha) \\ r\sin(\theta + \alpha) \end{bmatrix}.$$

Hence, the original point $(r\cos\alpha,\ r\sin\alpha)$ is mapped into $(r\cos(\theta + \alpha),\ r\sin(\theta + \alpha))$.

■   **Computer Lab: Matrix Machine**

84. $\begin{bmatrix} 1 & 1 \\ 1 & 1 \end{bmatrix}$

(a)   Only $[0, 0]$ is not moved. All (nonzero) vectors are moved.

(b)   Only vectors on the line $y = x$ do not change direction.

(c)   Only $[0, 0]$ remains constant in magnitude. The magnitudes of all nonzero vectors change.

(d)   The nullspace consists of all vectors on the line $y = -x$.

(e)   No vectors map onto the vector $[1, 0]$.   (f)   The image is the line $y = x$.

87. $\begin{bmatrix} 2 & 0 \\ 0 & 3 \end{bmatrix}$

(a)   Only $[0, 0]$ is not moved.

(b)   The only vectors whose direction is not changed are those in the direction of $[1, 0]$ or $[0, 1]$.

(c)   Only $[0, 0]$ has unchanged magnitude.

(d)   The nullspace contains only $[0, 0]$.

(e)   The only vector that maps into $[1, 0]$ is $[0.5, 0]$.

(f)   The image is all of $\mathbb{R}^2$.

5.2   Properties of Linear Transformations

■   **Finding Kernels**

3. $T(x, y, z) = (z, y, 0)$

The transformation $T(x, y, z) = (x, y, 0)$ in matrix form is

$$T(x, y) = \begin{bmatrix} 1 & 0 & 0 \\ 0 & 1 & 0 \\ 0 & 0 & 0 \end{bmatrix} \begin{bmatrix} x \\ y \\ z \end{bmatrix}.$$

By solving

$$\begin{bmatrix} 1 & 0 & 0 \\ 0 & 1 & 0 \\ 0 & 0 & 0 \end{bmatrix} \begin{bmatrix} x \\ y \\ z \end{bmatrix} = \begin{bmatrix} 0 \\ 0 \\ 0 \end{bmatrix}$$

we get the kernel of T as $\{(0, 0, \alpha) \mid \alpha \text{ is any real number}\}$. In other words, the kernel consists of the z-axis in $\mathbb{R}^3$.

6. $D^2(f) = f''$

Setting $D^2(f) = \dfrac{d^2 f}{dt^2} = 0$, we get $f(t) = c_1 t + c_2$. Hence, the kernel of the second derivative transformation consists of the family of linear functions.

9. $T(\mathbf{A}) = \mathbf{A}^T$

Setting

$$T\left(\begin{bmatrix} a & b & c \\ d & e & f \end{bmatrix}\right) = \begin{bmatrix} a & d \\ b & e \\ c & f \end{bmatrix} = \begin{bmatrix} 0 & 0 \\ 0 & 0 \\ 0 & 0 \end{bmatrix}$$

we have that $\mathbf{A}^T = 0$ and, hence, $\mathbf{A}$ is also the zero vector. Hence, the kernel of T consists of the zero vector in $\mathbf{M}_{23}$, i.e., $\begin{bmatrix} 0 & 0 & 0 \\ 0 & 0 & 0 \end{bmatrix}$.

■   **Calculus Kernels**

12. $T\left(at^2 + bt + c\right) = 2at + b$

Setting

$$T\left(at^2 + bt + c\right) = 2at + b = 0,$$

which gives $a = b = 0$, c arbitrary. Hence, the kernel of T consists of all constant function.

15. $T\left(at^3 + bt^2 + ct + d\right) = 6at + 2b$

Setting

$$T\left(at^3 + bt^2 + ct + d\right) = 6at + b = 0$$

gives $a = b = 0$, c and d arbitrary. Hence, the kernel of T consists of all polynomials in $\mathbb{P}_3$ of the form $p(t) = ct + d$.

■   **Superposition Principle**

18.   Direct substitution

Substituting $y = t^2 - 2$ into the DE $y'' - y' - 2y = 6 - 2t - 2t^2$ yields

$$(t^2 - 2)'' - (t^2 - 2)' - 2(t^2 - 2) = 2 - 2t - 2t^2 + 4 = 6 - 2t - 2t^2$$

■   **Dissecting Transformations**

21. $\begin{bmatrix} 0 & 0 \\ 0 & 0 \end{bmatrix}$

Solving

$$T(x,\ y) = \begin{bmatrix} 0 & 0 \\ 0 & 0 \end{bmatrix}\begin{bmatrix} x \\ y \end{bmatrix} = \begin{bmatrix} 0 \\ 0 \end{bmatrix},$$

we see the solution space, or kernel of T, consists of all points in $\mathbb{R}^2$. The dim Ker (T) is 2. The image of T contains vectors of the form

$$\begin{bmatrix} 0 & 0 \\ 0 & 0 \end{bmatrix}\begin{bmatrix} x \\ y \end{bmatrix} = \begin{bmatrix} 0 \\ 0 \end{bmatrix},$$

which means the range contains only the zero vector $[0,\ 0]$ so the dim Im (T) is 0. We also know this fact because

$$\dim \text{Ker } (T) + \dim \text{Im}(T) = \dim \mathbb{R}^2 = 2.$$

The transformation is neither surjective (onto) nor injective (one-to-one).

24. $\begin{bmatrix} 1 & 2 \\ 4 & 1 \end{bmatrix}$

We have the linear transformation $T : \mathbb{R}^2 \to \mathbb{R}^2$ defined by $T(\vec{v}) = A\vec{v}$. The image of T is the set of all vectors

$$\begin{bmatrix} 1 & 2 \\ 4 & 1 \end{bmatrix}\begin{bmatrix} x \\ y \end{bmatrix} = x\begin{bmatrix} 1 \\ 4 \end{bmatrix} + y\begin{bmatrix} 2 \\ 1 \end{bmatrix}$$

for all x, y real numbers. But the vectors $[1, 4]$ and $[2, 1]$ are linearly independent, so this linear combination yields all vectors in $\mathbb{R}^2$. The image of this matrix transformation is $\mathbb{R}^2$. You can also show that the only vector that this matrix maps into the zero vector is $[0, 0]$, so the kernel consists of only $[0, 0]$ and the dim Ker (T) is 0. Note that $\dim \mathrm{Ker}(T) + \dim \mathrm{Im}(T) = \dim \mathbb{R}^2 = 2$. The transformation is injective and surjective.

27. $\begin{bmatrix} 1 & 1 & 1 \\ 1 & 2 & 1 \end{bmatrix}$

Solving $\begin{bmatrix} 1 & 1 & 1 \\ 1 & 2 & 1 \end{bmatrix}\begin{bmatrix} x \\ y \\ z \end{bmatrix} = \begin{bmatrix} 0 \\ 0 \end{bmatrix}$

we see the kernel of T is the line

$$\{[-\alpha,\ 0,\ \alpha] \mid \alpha \text{ any real number}\} \in \mathbb{R}^3.$$

Hence, the dim Ker $(T) = 1$. The image of T consists of vectors of the form

$$\begin{bmatrix} 1 & 1 & 1 \\ 1 & 2 & 1 \end{bmatrix}\begin{bmatrix} x \\ y \\ z \end{bmatrix} = x\begin{bmatrix} 1 \\ 1 \end{bmatrix} + y\begin{bmatrix} 1 \\ 2 \end{bmatrix} + z\begin{bmatrix} 1 \\ 1 \end{bmatrix} = (x+y)\begin{bmatrix} 1 \\ 1 \end{bmatrix} + y\begin{bmatrix} 1 \\ 2 \end{bmatrix}$$

which consists of all vectors in $\mathbb{R}^2$. Hence, the dim Im (T) is 2. We also know that, because $\dim \mathrm{Ker}(T) + \dim \mathrm{Im}(T) = \dim \mathbb{R}^2 = 3$, we could find the dim Im (T) from this equation as well.

The transformation is neither injective nor surjective.

30.
$$\begin{bmatrix} 1 & 3 & 1 \\ 2 & 2 & 1 \end{bmatrix}$$

We have the linear transformation $T : \mathbb{R}^3 \to \mathbb{R}^2$ defined by $T(\vec{v}) = A\vec{v}$. The image of T is the set of all vectors

$$\begin{bmatrix} 1 & 3 & 1 \\ 2 & 2 & 1 \end{bmatrix} \begin{bmatrix} x \\ y \\ z \end{bmatrix} = x\begin{bmatrix} 1 \\ 2 \end{bmatrix} + y\begin{bmatrix} 3 \\ 2 \end{bmatrix} + z\begin{bmatrix} 1 \\ 1 \end{bmatrix}$$

for all x, y, and z real numbers. There are two linearly independent vectors among the above three vectors, so this combination yields all vectors in $\mathbb{R}^2$. Hence, the image consists of $\mathbb{R}^2$. If we set $A\vec{v} = \vec{0}$ we will find a one-dimensional subspace

$$\left\{ [-\alpha, \ -\alpha, \ 4\alpha] \,\middle|\, \alpha \text{ any real number} \right\}$$

in $\mathbb{R}^3$. The dim Ker (T) is 1 and the dim Im (T) is 2. Thus, again we have the relationship $\dim\mathrm{Ker}(T) + \dim\mathrm{Im}(T) = \dim\mathbb{R}^2 = 3$. The transformation is not injective but it is surjective.

33.
$$\begin{bmatrix} 0 & 0 \\ 0 & 0 \\ 0 & 0 \end{bmatrix}$$

Solving
$$T(x, y) = \begin{bmatrix} 0 & 0 \\ 0 & 0 \\ 0 & 0 \end{bmatrix} \begin{bmatrix} x \\ y \end{bmatrix} = \begin{bmatrix} 0 \\ 0 \\ 0 \end{bmatrix}$$

we see the kernel of T consists of all of $\mathbb{R}^2$. The dim Ker (T) is 2. The image of T contains vectors of the form

$$\begin{bmatrix} 0 & 0 \\ 0 & 0 \\ 0 & 0 \end{bmatrix} \begin{bmatrix} x \\ y \end{bmatrix} = (x+y)\begin{bmatrix} 0 \\ 0 \\ 0 \end{bmatrix} = \begin{bmatrix} 0 \\ 0 \\ 0 \end{bmatrix},$$

which consists of the zero vector in $\mathbb{R}^3$, a zero dimensional subspace of $\mathbb{R}^3$. Hence, the rank of T is 0. The transformation is neither injective nor surjective.

36. $\begin{bmatrix} 1 & 1 & 1 \\ 1 & 2 & 1 \\ 2 & 3 & 2 \end{bmatrix}$

We have the linear transformation $T : \mathbb{R}^3 \to \mathbb{R}^3$ defined by $T(\vec{v}) = \mathbf{A}\vec{v}$. The image of T is the set of all vectors

$$\begin{bmatrix} 1 & 1 & 1 \\ 1 & 2 & 1 \\ 2 & 3 & 2 \end{bmatrix} \begin{bmatrix} x \\ y \\ z \end{bmatrix} = x \begin{bmatrix} 1 \\ 1 \\ 2 \end{bmatrix} + y \begin{bmatrix} 1 \\ 2 \\ 3 \end{bmatrix} + z \begin{bmatrix} 1 \\ 1 \\ 2 \end{bmatrix}$$

for all x, y, and z real numbers. But the determinate of this matrix is 0, which tells us the number of linearly independent columns is less than three. But we can see by inspection there are at least two linearly independent columns, so columns of $\mathbf{A}$ spans a two-dimensional subspace (i.e., a plane) of $\mathbb{R}^3$. To find the kernel of the transformation, we solve $\mathbf{A}\vec{v} = \vec{0}$, getting

$$\left\{ [-\alpha,\ 0,\ \alpha] \mid \alpha \text{ any real number} \right\},$$

which consists of a one-dimensional subspace (a line) in $\mathbb{R}^3$. Hence, the dim Ker (T) of the transformation is 1. The transformation is neither injective nor surjective.

39. $\begin{bmatrix} 1 & 2 & 0 \\ 0 & 1 & 1 \\ 0 & 0 & 1 \end{bmatrix}$

First we show that the transformation T defined by $T(\vec{v}) = \mathbf{A}\vec{v}$ is injective. The system

$$\begin{bmatrix} 1 & 2 & 0 \\ 0 & 1 & 1 \\ 0 & 0 & 1 \end{bmatrix} \begin{bmatrix} x \\ y \\ z \end{bmatrix} = \begin{bmatrix} 0 \\ 0 \\ 0 \end{bmatrix}$$

has only the zero solution because the determinant of the matrix is nonzero. Therefore dim Ker $(T) = 0$, so that dim Im$(T) = 3$. Therefore T is both injective and surjective.

■   **Transformations and Linear Dependence**

42.   Let T be the zero map. Then for any set $\{\vec{v}_1, \vec{v}_2, \vec{v}_3\}$ of vector in $\mathbb{R}^n$, the set $\{T(\vec{v}_1), T(\vec{v}_2), T(\vec{v}_3)\} = \{\vec{0}\}$ is linearly dependent.

45.   (a)   For any $p_1 \, p_2$ in $\mathbb{P}_2$,

$$T\big(p_1(t) + p_2(t)\big) = T\big((p_1 + p_2)(t)\big) = \begin{bmatrix} (p_1 + p_2)(0) \\ (p_1 + p_2)(1) \end{bmatrix} = \begin{bmatrix} p_1(0) + p_2(0) \\ p_1(1) + p_2(1) \end{bmatrix}$$

$$= \begin{bmatrix} p_1(0) \\ p_1(1) \end{bmatrix} + \begin{bmatrix} p_2(0) \\ p_2(1) \end{bmatrix} = T\big(p_1(t)\big) + T\big(p_2(t)\big).$$

Also $T\big(cp(t)\big) = \begin{bmatrix} (cp)(0) \\ (cp)(1) \end{bmatrix} = \begin{bmatrix} cp(0) \\ cp(1) \end{bmatrix} = c\begin{bmatrix} p(0) \\ p(1) \end{bmatrix} = cT\big(p(t)\big).$

Thus T is linear.

(b) $p(t) = a_2 t^2 + a_1 t + a_0$ is in Ker(T) if $\begin{bmatrix} a_2(0)^2 + a_1(0) + a_0 \\ a_2(1)^2 + a_2(1) + a_0 \end{bmatrix} = \begin{bmatrix} 0 \\ 0 \end{bmatrix}$,

that is, if $\begin{bmatrix} a_0 \\ a_2 + a_1 + a_0 \end{bmatrix} = \begin{bmatrix} 0 \\ 0 \end{bmatrix}$. This holds if $a_0 = 0$ and $a_2 + a_1 = 0$, that is, if $a_0 = 0$ and

$a_2 = -a_1$. A basis for Ker(T) is $\{t^2 - t\}$.

(c)   For $\begin{bmatrix} \alpha \\ \beta \end{bmatrix}$ in $\mathbb{R}^2$, $T\big(a_2 t^2 + (\beta - \alpha - a_2)t + \alpha\big) = \begin{bmatrix} \alpha \\ a_2 + \beta - \alpha - a_2 + \alpha \end{bmatrix} = \begin{bmatrix} \alpha \\ \beta \end{bmatrix}$, so T is

surjective.

$\left\{ \begin{bmatrix} 1 \\ 0 \end{bmatrix}, \begin{bmatrix} 0 \\ 1 \end{bmatrix} \right\}$ is a basis for Im(T).

■   **Kernels and Images**

48. $T: \mathbb{M}_{22} \to \mathbb{M}_{22}$, $T\begin{bmatrix} a & b \\ c & d \end{bmatrix} = \begin{bmatrix} a & b \\ b & c \end{bmatrix}$

(i)   Ker $(T) = \left\{ \begin{bmatrix} a & b \\ c & d \end{bmatrix} \in \mathbb{M}_{22} \,\middle|\, a = b = c = 0 \right\}$

(ii)   Im $(T) = \left\{ a\begin{bmatrix} 1 & 0 \\ 0 & 0 \end{bmatrix} + b\begin{bmatrix} 0 & 1 \\ 1 & 0 \end{bmatrix} + c\begin{bmatrix} 0 & 0 \\ 0 & 1 \end{bmatrix} \,\middle|\, a, b, c \in \mathbb{R} \right\}$

51. $T: \mathbb{R}^2 \to \mathbb{R}^3$, $T(x, y) = (x + y, 0, x - y)$

   (i) $(x, y) \in$ Ker $(T) \Leftrightarrow \begin{cases} x + y = 0 \\ x - y = 0 \end{cases} \Leftrightarrow x = y = 0,$

   so Ker $(T) = \{\vec{0}\}$.

   (ii)   Im $(T) = \left\{ x \begin{bmatrix} 1 \\ 0 \\ 1 \end{bmatrix} + y \begin{bmatrix} 1 \\ 0 \\ -1 \end{bmatrix} \middle| x, y, \in \mathbb{R} \right\}$

■   **Examples of Matrices**

54.   Set $\begin{bmatrix} a & b & c \\ d & e & f \\ g & h & i \end{bmatrix} \begin{bmatrix} 1 \\ 0 \\ 1 \end{bmatrix} = \begin{bmatrix} 0 \\ 0 \\ 0 \end{bmatrix}$ so that $\begin{matrix} a + c = 0 \\ d + f = 0 \\ g + i = 0 \end{matrix}$

   Set $\begin{bmatrix} a & b & c \\ d & e & f \\ g & h & i \end{bmatrix} \begin{bmatrix} 0 \\ 1 \\ 2 \end{bmatrix} = \begin{bmatrix} 0 \\ 0 \\ 0 \end{bmatrix}$ so that $\begin{matrix} b + 2c = 0 \\ e + 2f = 0 \\ h + 2i = 0 \end{matrix}$

   Therefore for $\mathbf{A} = \begin{bmatrix} 1 & 2 & -1 \\ 1 & 1 & -1 \\ 1 & 2 & -1 \end{bmatrix}$ and $T(\vec{x}) = \mathbf{A}\vec{x}$

   Ker T is spanned by $\left\{ \begin{bmatrix} 1 \\ 0 \\ 1 \end{bmatrix}, \begin{bmatrix} 0 \\ 1 \\ 2 \end{bmatrix} \right\}$

   (As a check, dim Ker(T) + dim Im(T) = 2 + rank A = 3 = dim $\mathbb{R}^3$)

■   **True/False Questions**

57.   True. Elementary row operations will not change the solutions of $\mathbf{A}\vec{x} = \vec{0}$.

60.   True. Im $(T) =$ span $\left\{ \begin{bmatrix} 1 \\ 1 \end{bmatrix} \right\}$.

■   **Still Investigating**

63.   The dim Im (T): $\mathbb{R}^3 \to \mathbb{R}^4$ is the number of linearly independent columns of the matrix $\mathbf{A}$, which is 3. Hence, the transformation T is not surjective.

   Also, because dim Ker(T) + dim Im(T) = dim $\mathbb{R}^3$ = 3, we see that dim Ker $(T) = 0$, which means that T is injective.

■ Review of Nonhomogeneous Algebraic Systems

66. $x + y = 1$. A particular solution of the nonhomogeneous equation $x + y = 1$ is $x = \dfrac{1}{2}$, $y = \dfrac{1}{2}$.

The general solution of the homogeneous equation $x + y = 0$ is

$$\{(-c,\ c)\,|\,c \text{ is an arbitrary constant}\}.$$

Hence, the general solution of the nonhomogeneous equation is

$$\begin{bmatrix} x \\ y \end{bmatrix} = \frac{1}{2}\begin{bmatrix} 1 \\ 1 \end{bmatrix} + c\begin{bmatrix} -1 \\ 1 \end{bmatrix}.$$

69.

$$\begin{array}{rcrcr} x & - & 2y & = & 5 \\ 2x & + & 4y & = & -5 \end{array}$$

Solving the original nonhomogeneous system, we find a particular solution of $x = \dfrac{10}{8}$, $y = -\dfrac{15}{8}$.

Then, solving the corresponding homogeneous system, we find the only solution to be $x = 0$, $y = 0$. Hence, the general solution of the original nonhomogeneous system is

$$\begin{bmatrix} x \\ y \end{bmatrix} = \frac{1}{8}\begin{bmatrix} 10 \\ -15 \end{bmatrix} + c\begin{bmatrix} 0 \\ 0 \end{bmatrix} = \frac{1}{8}\begin{bmatrix} 10 \\ -15 \end{bmatrix} \text{ for any real number } c.$$

■ Review of Nonhomogeneous First-Order DEs

72. $y' - y = 3$. By inspection, we find a particular solution of $y_p(t) = -3$. We then solve the corresponding linear homogeneous equation $y' - y = 0$, and get $y_h(t) = ce^t$. Hence, the general solution of the nonhomogeneous equation is

$$y(t) = y_h(t) + y_p(t) = ce^t - 3.$$

75. $y' + \dfrac{1}{t^2}y = \dfrac{2}{t^2}$. By inspection, we find a particular solution $y_p(t) = 2$. We then solve the corresponding homogeneous equation

$$y' + \frac{2}{t^2}y = 0$$

by separating variables and get

$$y_h(t) = ce^{2/t}.$$

Hence, the general solution of the nonhomogeneous equation is

$$y(t) = y_h + y_p = ce^{2/t} + 2.$$

■   **Review of Nonhomogeneous Second-Order DEs**

78. $y'' + y' - 2y = 2t - 3$

Using the method of undetermined coefficients, we find a particular solution

$$y_p(t) = -t + 1.$$

We then solve the corresponding homogeneous equation

$$y'' + y' - 2y = 0$$

and get

$$y_h(t) = c_1 e^t + c_2 e^{-2t}.$$

Hence, the general solution of the nonhomogeneous equation is

$$y(t) = y_h(t) + y_p(t) = c_1 e^t + c_2 e^{-2t} - t + 1.$$

81. $y'' + y = 2t$

By inspection, we find a particular solution

$$y_p(t) = 2t.$$

We then solve the corresponding homogeneous equation

$$y'' + y = 0$$

and get

$$y_h(t) = c_1 \cos t + c_2 \sin t.$$

Hence, the general solution of the nonhomogeneous equation is

$$y(t) = y_h + y_p = c_1 \cos t + c_2 \sin t + 2t.$$

5.3   Eigenvalues and Eigenvectors

■   **Computing Eigenstuff**

3. $\begin{bmatrix} 1 & 2 \\ 1 & 2 \end{bmatrix}$. The characteristic equation is

$$p(\lambda) = |\mathbf{A} - \lambda\mathbf{I}| = \begin{vmatrix} 1-\lambda & 2 \\ 1 & 2-\lambda \end{vmatrix} = \lambda^2 - 3\lambda = 0,$$

which has roots of $\lambda_1 = 0$, $\lambda_2 = 3$. To find the eigenvectors, we substitute λ_i into the system $\mathbf{A}\vec{v} = \lambda_i\vec{v}$ and solve for $\vec{v}$. For $\lambda_1 = 0$, substitution yields

$$\begin{bmatrix} 1 & 2 \\ 1 & 2 \end{bmatrix}\begin{bmatrix} x \\ y \end{bmatrix} = 0\begin{bmatrix} x \\ y \end{bmatrix} = \begin{bmatrix} 0 \\ 0 \end{bmatrix}.$$

Solving this 2×2 system for x and y yields the equation $x + 2y = 0$, which implies all vectors $\vec{v}_1 = \alpha[2, \ -1]$ in $\mathbb{R}^2$ are eigenvectors for any real number α.

We find the eigenvectors corresponding to $\lambda_2 = 3$ by solving $\mathbf{A}\vec{v} = 3\vec{v}$, yielding the equations

$$x + 2y = 3x, \ x + 2y = 3y,$$

which gives $x = y$.

Hence $\vec{v}_2 = \alpha[1, \ 1]$ where α is any real number.

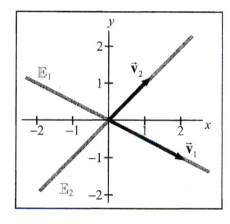

6. $\begin{bmatrix} 3 & 2 \\ -2 & -3 \end{bmatrix}$. The characteristic equation is

$$p(\lambda) = |\mathbf{A} - \lambda\mathbf{I}| = \begin{vmatrix} 3-\lambda & 2 \\ -2 & -3-\lambda \end{vmatrix} = \lambda^2 - 5 = 0,$$

which has roots of $\lambda_1 = \sqrt{5}$, $\lambda_2 = -\sqrt{5}$. To find the eigenvectors, we substitute λ_i into the system $\mathbf{A}\vec{v} = \lambda_i\vec{v}$ and solve for $\vec{v}$.

To find eigenvectors for $\lambda_1 = \sqrt{5}$, substitution yields

$$\begin{bmatrix} 3 & 2 \\ -2 & -3 \end{bmatrix}\begin{bmatrix} x \\ y \end{bmatrix} = \sqrt{5}\begin{bmatrix} x \\ y \end{bmatrix}.$$

continued on the next page

Solving this 2×2 system for x and y yields one independent equation $\left(3-\sqrt{5}\right)x+2y=0$,

which gives an eigenvector

$$\vec{v}_1 = \left[1, \ \frac{1}{2}\sqrt{5}-\frac{3}{2}\right].$$

Likewise for the second eigenvalue, $\lambda_2 = -\sqrt{5}$,

we find eigenvector

$$\vec{v}_2 = \left[1, \ -\frac{1}{2}\sqrt{5}-\frac{3}{2}\right].$$

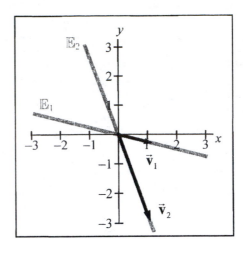

9. $\quad A = \begin{bmatrix} 1 & 4 \\ -4 & 11 \end{bmatrix}$

$$\det(A-\lambda I) = \det\begin{bmatrix} 1-\lambda & 4 \\ -4 & 11-\lambda \end{bmatrix} = (1-\lambda)(11-\lambda)+16$$

$$= \lambda^2 - 12\lambda + 27$$
$$= (\lambda-3)(\lambda-9) = 0 \qquad \Rightarrow \lambda_1 = 3, \lambda_2 = 9$$

To find eigenvectors for $\lambda_1 = 3$:

$$\begin{bmatrix} -2 & 4 & | & 0 \\ -4 & 8 & | & 0 \end{bmatrix} \rightarrow \begin{bmatrix} 1 & -2 & | & 0 \\ -4 & 8 & | & 0 \end{bmatrix} \begin{array}{c} \text{RREF} \\ \rightarrow \end{array} \begin{bmatrix} 1 & -2 & | & 0 \\ 0 & 0 & | & 0 \end{bmatrix} \Rightarrow \begin{array}{l} v_1 - 2v_2 = 0 \\ v_2 \text{ free} \end{array}$$

$$\mathbb{E}_1 = \text{span} \left\{ \begin{bmatrix} 2 \\ 1 \end{bmatrix} \right\}$$

To find eigenvectors for $\lambda_2 = 9$:

$$\begin{bmatrix} -8 & 4 & | & 0 \\ -4 & 2 & | & 0 \end{bmatrix} \rightarrow \begin{bmatrix} 1 & -1/2 & | & 0 \\ -4 & 2 & | & 0 \end{bmatrix} \begin{array}{c} \text{RREF} \\ \rightarrow \end{array} \begin{bmatrix} 1 & -1/2 & | & 0 \\ 0 & 0 & | & 0 \end{bmatrix} \Rightarrow \begin{array}{l} v_1 - \dfrac{1}{2}v_2 = 0 \\ v_2 \text{ free} \end{array}$$

$$\mathbb{E}_2 = \text{span} \left\{ \begin{bmatrix} 1 \\ 2 \end{bmatrix} \right\}$$

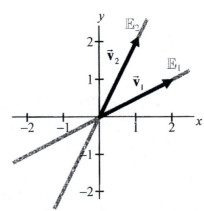

12. $\begin{bmatrix} 1 & 1 \\ 0 & 1 \end{bmatrix}$. The characteristic equation is

$$p(\lambda) = |\mathbf{A} - \lambda\mathbf{I}| = \begin{vmatrix} 1-\lambda & 1 \\ 0 & 1-\lambda \end{vmatrix} = (\lambda - 1)^2,$$

which has roots of $\lambda_1 = 1, 1$.

To find eigenvectors for $\lambda_1 = 1$:

$$\begin{bmatrix} 1 & 1 \\ 0 & 1 \end{bmatrix}\begin{bmatrix} x \\ y \end{bmatrix} = 1\begin{bmatrix} x \\ y \end{bmatrix}.$$

Solving this 2×2 system for x and y, yields

$$x + y = x$$
$$y = y$$

which implies x arbitrary, $y = 0$. Hence, there is only one independent eigenvector:

$$\vec{\mathbf{v}}_1 = \begin{bmatrix} 1 \\ 0 \end{bmatrix}$$

And $\mathbb{E}_1 = \mathrm{span}\left\{\begin{bmatrix} 1 \\ 0 \end{bmatrix}\right\}$ is its eigenspace.

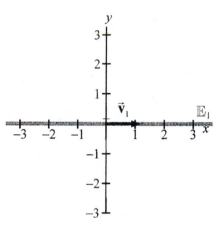

15. $\begin{bmatrix} 2 & -1 \\ 1 & 4 \end{bmatrix}$ $\lambda^2 + 6\lambda + 9 = 0$, so $\lambda = 3, 3$.

To find eigenvectors for $\lambda = 3$:

$$\begin{bmatrix} 2-3 & -1 \\ 1 & 4-3 \end{bmatrix}\begin{bmatrix} v_1 \\ v_2 \end{bmatrix} = \begin{bmatrix} 0 \\ 0 \end{bmatrix} \Rightarrow -v_1 - v_2 = 0 \Rightarrow \begin{matrix} v_2 = -v_1 \\ v_1 \text{ free} \end{matrix} \Rightarrow \vec{\mathbf{v}} = \begin{bmatrix} 1 \\ -1 \end{bmatrix}$$

$\mathbb{E} = \mathrm{span}\left\{\begin{bmatrix} 1 \\ -1 \end{bmatrix}\right\}$

Dim $\mathbb{E} = 1$.

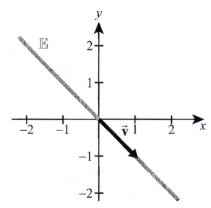

■   **When Shortcut Fails**

18.   In all the three matrices given the shortcut $\vec{v} = \begin{bmatrix} -b \\ a - \lambda \end{bmatrix}$ fails for $\lambda = 3$, because it gives

$\begin{bmatrix} 0 \\ 3-3 \end{bmatrix} = \begin{bmatrix} 0 \\ 0 \end{bmatrix}$ which cannot be an eigenvector.

If matrix element $b = 0$, the eigenvector system (*) in Problem 17 gives

$$(a - \lambda)v_1 = 0$$
$$cv_1 + (d - \lambda)v_2 = 0.$$

There are several possibilities for solving this system, depending on which factor of the first equation is zero.

•   If $v_1 = 0$, the second equation says one possibility is that $v_2 = 0$ and $\vec{v} = \begin{bmatrix} 0 \\ 0 \end{bmatrix}$, but a zero

vector cannot be an eigenvector.

The only other option is that $d = \lambda$ in which case v_2 can be anything, so $\vec{v} = \begin{bmatrix} 0 \\ 1 \end{bmatrix}$.

•   If $a = \lambda$, then v_1 can be anything and the second equation gives $\vec{v} = \begin{bmatrix} d - \lambda \\ -c \end{bmatrix}$.

The second equation of the system then determines the outcome.

(a) $\begin{bmatrix} 3 & 0 \\ 5 & 3 \end{bmatrix}$ has eigenvalue 3.

The first eigenvector equation, $3v_1 = 3v_1$ says v_1 can be anything.

But the second equation, $5v_1 + (3 - 3)v_2 = 0$, requires that $v_1 = 0$; then v_2 can be anything, so $\vec{v} = \begin{bmatrix} 0 \\ 1 \end{bmatrix}$.

(b) $\begin{bmatrix} 3 & 0 \\ 5 & 2 \end{bmatrix}$ has eigenvalues 3 and 2.

For $\lambda_1 = 3$, the first equation says $3v_1 = 3v_1$ so v_1 can be anything.

But the second equation, $5v_1 + 2v_2 = 3v_2$, requires $v_2 = 5v_1$ so $\vec{v}_1 = \begin{bmatrix} 1 \\ 5 \end{bmatrix}$ is an eigenvector.

For $\lambda_2 = 2$, the first equation $3v_1 = 2v_1$ says $v_1 = 0$.

Then the second equation $5v_1 + 2v_2 = 2v_2$ says v_2 can be anything, so $\vec{v}_2 = \begin{bmatrix} 0 \\ 1 \end{bmatrix}$.

(c) $\begin{bmatrix} 3 & 0 \\ 0 & 3 \end{bmatrix}$ has a double eigenvalue 3.

$$3v_1 = 3v_1$$
$$3v_2 = 3v_2$$

so both v_1 and v_2 can be anything! Any vector is an eigenvector .The eigenspace is two dimensional, all of $\mathbb{R}^2$ spanned by $\begin{bmatrix} 0 \\ 1 \end{bmatrix}$ and $\begin{bmatrix} 1 \\ 0 \end{bmatrix}$ (or any other two linearly independent vectors).

21.   The characteristic equation of the given matrix is

$$p(\lambda) = |\mathbf{A} - \lambda\mathbf{I}| = \begin{vmatrix} 1-\lambda & 2 & 2 \\ 2 & -\lambda & 3 \\ 2 & 3 & -\lambda \end{vmatrix} = \lambda^3 - \lambda^2 - 17\lambda - 15 = 0,$$

which has eigenvalues $\lambda_1 = 5$, $\lambda_2 = -3$, and $\lambda_3 = -1$. To find the eigenvector corresponding to $\lambda_1 = 5$, we substitute 5 into the system $\mathbf{A\vec{v}} = \lambda\vec{v}$ and solve for **v**. The system

$$\begin{bmatrix} 1 & 2 & 2 \\ 2 & 0 & 3 \\ 2 & 3 & 0 \end{bmatrix} \begin{bmatrix} x \\ y \\ z \end{bmatrix} = 5 \begin{bmatrix} x \\ y \\ z \end{bmatrix}$$

gives $x = c$, $y = c$, $z = c$, where c is any real number. Hence, we have the eigenvector $\vec{v}_1 = [1,\ 1,\ 1]$. By a similar argument, the eigenvectors corresponding to $\lambda_2 = -3$ and $\lambda_3 = -1$ are

$$\lambda_2 = -3 \Rightarrow \vec{v}_2 = [0,\ -1,\ 1]$$
$$\lambda_3 = -1 \Rightarrow \vec{v}_3 = [-2,\ 1,\ 1]$$

Each eigenvalue corresponds to a one-dimensional eigenspace in $\mathbb{R}^3$.

24.   The characteristic equation of the given matrix is

$$p(\lambda) = |\mathbf{A} - \lambda\mathbf{I}| = \begin{vmatrix} 1-\lambda & 0 & 0 \\ -1 & 3-\lambda & 0 \\ 3 & 2 & -2-\lambda \end{vmatrix} = (\lambda-1)(\lambda-3)(\lambda+2),$$

which has roots $\lambda_1 = 1$, $\lambda_2 = 3$, and $\lambda_3 = -2$. To find the eigenvector corresponding to $\lambda_1 = 1$, we substitute into the system $\mathbf{A\vec{v}} = \lambda_1\vec{v}$ and solve for $\vec{v}$. Doing this yields an eigenvector $\vec{v}_1 = [6,\ 3,\ 8]$. By a similar analysis, we substitute $\lambda_2 = 3$ and $\lambda_3 = -2$ into $\mathbf{A\vec{v}} = \lambda\vec{v}$ for λ yielding eigenvectors $\vec{v}_2 = [0,\ 5,\ 2]$, $\vec{v}_3 = [0,\ 0,\ 1]$. Each eigenvalue corresponds to a one-dimensional eigenspace.

27. $A = \begin{bmatrix} 1 & 0 & 0 \\ -4 & 3 & 0 \\ -4 & 2 & 1 \end{bmatrix}$

$\det(A - \lambda I) = \det \begin{bmatrix} 1-\lambda & 0 & 0 \\ -4 & 3-\lambda & 0 \\ -4 & 2 & 1-\lambda \end{bmatrix} = (1-\lambda)^2(3-\lambda) \Rightarrow \lambda_{1,2} = 1, \ 1, \ \lambda_3 = 3.$

To find eigenvectors for double eigenvector $\lambda_{1,2} = 1$:

$\begin{bmatrix} 0 & 0 & 0 \\ -4 & 2 & 0 \\ -4 & 2 & 0 \end{bmatrix} \xrightarrow[\rightarrow]{RREF} \begin{bmatrix} 1 & -1/2 & 0 \\ 0 & 0 & 0 \\ 0 & 0 & 0 \end{bmatrix} \Rightarrow \begin{matrix} v_1 - \dfrac{1}{2}v_2 = 0 \\ v_2, v_3 \ \text{free} \end{matrix}$

$\mathbb{E}_{1,2} = \text{span} \left\{ \begin{bmatrix} 1 \\ 2 \\ 0 \end{bmatrix}, \begin{bmatrix} 0 \\ 0 \\ 1 \end{bmatrix} \right\}, \ \dim \mathbb{E}_{1,2} = 2.$

To find eigenvectors for $\lambda_3 = 3$:

$\begin{bmatrix} -2 & 0 & 0 \\ -4 & 0 & 0 \\ -4 & 2 & -2 \end{bmatrix} \rightarrow \begin{bmatrix} 1 & 0 & 0 \\ 0 & 0 & 0 \\ 0 & 2 & -2 \end{bmatrix} \xrightarrow[\rightarrow]{RREF} \begin{bmatrix} 1 & 0 & 0 \\ 0 & 1 & -1 \\ 0 & 0 & 0 \end{bmatrix} \Rightarrow \begin{matrix} v_1 = 0 \\ v_2 - v_3 = 0 \\ v_3 \ \text{free} \end{matrix}$

$\mathbb{E}_3 = \text{span} \left\{ \begin{bmatrix} 0 \\ 1 \\ 1 \end{bmatrix} \right\}, \ \dim \mathbb{E}_3 = 1.$

30. $A = \begin{bmatrix} 0 & 0 & 2 \\ -1 & 1 & 2 \\ -1 & 0 & 3 \end{bmatrix}$

$\det(A - \lambda I) = \det \begin{bmatrix} -\lambda & 0 & 2 \\ -1 & 1-\lambda & 2 \\ -1 & 0 & 3-\lambda \end{bmatrix} = -\lambda \begin{vmatrix} 1-\lambda & 2 \\ 0 & 3-\lambda \end{vmatrix} + 2 \begin{vmatrix} -1 & 1-\lambda \\ -1 & 0 \end{vmatrix}$

$= \lambda \big((1-\lambda)(3-\lambda) \big) + 2(1-\lambda)$

$= (1-\lambda)\big[-\lambda(3-\lambda) + 2 \big]$

$= (1-\lambda)(\lambda^2 - 3\lambda + 2) = (1-\lambda)(\lambda - 2)(\lambda - 1) \Rightarrow \lambda_{1,2} = 1, \ 1, \ \lambda_3 = 2.$

To find eigenvectors for double eigenvalue $\lambda_{1,2} = 1$:

$\begin{bmatrix} -1 & 0 & 2 \\ -1 & 0 & 2 \\ -1 & 0 & 2 \end{bmatrix} \xrightarrow[\rightarrow]{RREF} \begin{bmatrix} 1 & 0 & -2 \\ 0 & 0 & 0 \\ 0 & 0 & 0 \end{bmatrix} \begin{matrix} v_1 - 2v_3 = 0 \\ v_2, v_3 \ \text{free} \end{matrix}$

$$\mathbb{E}_{1,2} = \text{span} \left\{ \begin{bmatrix} 0 \\ 1 \\ 0 \end{bmatrix}, \begin{bmatrix} 2 \\ 0 \\ 1 \end{bmatrix} \right\}, \ \dim \mathbb{E}_{1,2} = 2.$$

To find eigenvectors for $\lambda_3 = 2$:

$$\begin{bmatrix} -2 & 0 & 2 \\ -1 & -1 & 2 \\ -1 & 0 & 1 \end{bmatrix} \rightarrow \begin{bmatrix} 1 & 0 & -1 \\ 0 & -1 & 1 \\ 0 & 0 & 0 \end{bmatrix} \ \overset{\text{RREF}}{\rightarrow} \ \begin{bmatrix} 1 & 0 & -1 \\ 0 & 1 & -1 \\ 0 & 0 & 0 \end{bmatrix} \Rightarrow \ \begin{matrix} v_1 - v_3 = 0 \\ v_2 - v_3 = 0 \\ v_3 \ \text{free} \end{matrix}$$

$$\mathbb{E}_3 = \text{span} \left\{ \begin{bmatrix} 1 \\ 1 \\ 1 \end{bmatrix} \right\}, \ \dim \mathbb{E}_3 = 1.$$

33. $\quad \mathbf{A} = \begin{bmatrix} 2 & 0 & 1 & 2 \\ 0 & 2 & 0 & 0 \\ 0 & 0 & 6 & 0 \\ 0 & 0 & 1 & 4 \end{bmatrix}$

$$\det(\mathbf{A} - \lambda \mathbf{I}) = \det \begin{bmatrix} 2-\lambda & 0 & 1 & 2 \\ 0 & 2-\lambda & 0 & 0 \\ 0 & 0 & 6-\lambda & 0 \\ 0 & 0 & 1 & 4-\lambda \end{bmatrix}$$

$$= (2-\lambda) \begin{vmatrix} 2-\lambda & 0 & 0 \\ 0 & 6-\lambda & 0 \\ 0 & 1 & 4-\lambda \end{vmatrix}$$

$$= (2-\lambda^2)(4-\lambda)(6-\lambda) \Rightarrow \lambda_{1,2} = 2, \ 2, \ \lambda_3 = 4, \ \lambda_4 = 6.$$

To find eigenvectors for double eigenvalue $\lambda_{1,2} = 2$:

$$\begin{bmatrix} 0 & 0 & 1 & 2 \\ 0 & 0 & 0 & 0 \\ 0 & 0 & 4 & 0 \\ 0 & 0 & 1 & 2 \end{bmatrix} \rightarrow \begin{bmatrix} 0 & 0 & 1 & 2 \\ 0 & 0 & 4 & 0 \\ 0 & 0 & 0 & 0 \\ 0 & 0 & 0 & 0 \end{bmatrix} \overset{\text{RREF}}{\rightarrow} \begin{bmatrix} 0 & 0 & 1 & 2 \\ 0 & 0 & 0 & 1 \\ 0 & 0 & 0 & 0 \\ 0 & 0 & 0 & 0 \end{bmatrix} \Rightarrow \begin{matrix} v_3 + 2v_4 = 0 \\ v_4 = 0 \\ v_1, v_2 \ \text{free} \end{matrix}$$

$$\mathbb{E}_{1,2} = \text{span} \left\{ \begin{bmatrix} 1 \\ 0 \\ 0 \\ 0 \end{bmatrix}, \begin{bmatrix} 0 \\ 1 \\ 0 \\ 0 \end{bmatrix} \right\}, \ \dim \mathbb{E}_{1,2} = 2.$$

continued on the next page

To find eigenvectors for $\lambda_3 = 4$:

$$\begin{bmatrix} -2 & 0 & 1 & 0 \\ 0 & -2 & 0 & 0 \\ 0 & 0 & 2 & 0 \\ 0 & 0 & 1 & 0 \end{bmatrix} \xrightarrow[\rightarrow]{\text{RREF}} \begin{bmatrix} 1 & 0 & -1/2 & -1 \\ 0 & 1 & 0 & 0 \\ 0 & 0 & 1 & 0 \\ 0 & 0 & 0 & 0 \end{bmatrix} \Rightarrow \begin{array}{l} v_2 - \dfrac{1}{2}v_3 - v_4 = 0 \\ v_2 = 0 \\ v_3 = 0 \\ v_4 \ \text{free} \end{array}$$

$$\mathbb{E}_3 = \text{span} \left\{ \begin{bmatrix} 1 \\ 0 \\ 0 \\ 1 \end{bmatrix} \right\}, \ \dim \mathbb{E}_3 = 1.$$

To find eigenvectors for $\lambda_4 = 6$:

$$\begin{bmatrix} -4 & 0 & 1 & 2 \\ 0 & -4 & 0 & 0 \\ 0 & 0 & 0 & 0 \\ 0 & 0 & 1 & -2 \end{bmatrix} \xrightarrow[\rightarrow]{\text{RREF}} \begin{bmatrix} 1 & 0 & -1/4 & -1/2 \\ 0 & 1 & 0 & 0 \\ 0 & 0 & 1 & -2 \\ 0 & 0 & 0 & 0 \end{bmatrix} \Rightarrow \begin{array}{l} v_1 - \dfrac{1}{4}v_3 - \dfrac{1}{2}v_4 = 0 \\ v_2 = 0 \\ v_3 - 2v_4 = 0 \\ v_4 \ \text{free} \end{array}$$

$$\mathbb{E}_4 = \text{span} \left\{ \begin{bmatrix} 1 \\ 0 \\ 2 \\ 1 \end{bmatrix} \right\}, \ \dim \mathbb{E}_4 = 1.$$

■   **Distinct Eigenvalues Extended**

36.   We wish to show that for any three distinct eigenvalues of a 3×3 matrix **A**, their eigenvectors are linearly independent. Assume λ_1, λ_2, λ_3 are distinct eigenvalues and let

$$c_1 \vec{\mathbf{v}}_1 + c_2 \vec{\mathbf{v}}_2 + c_3 \vec{\mathbf{v}}_3 = 0.$$

Multiplying by **A** yields

$$\mathbf{A}\left(c_1 \vec{\mathbf{v}}_1 + c_2 \vec{\mathbf{v}}_2 + c_3 \vec{\mathbf{v}}_3 \right) = c_1 \lambda_1 \vec{\mathbf{v}}_1 + c_2 \lambda_2 \vec{\mathbf{v}}_2 + c_3 \lambda_3 \vec{\mathbf{v}}_3 = 0.$$

Subtracting $\lambda_3 \left(c_1 \vec{\mathbf{v}}_1 + c_2 \vec{\mathbf{v}}_2 + c_3 \vec{\mathbf{v}}_3 \right)$ (whose value is zero) from this equation yields

$$c_1 \left(\lambda_1 - \lambda_3 \right) \vec{\mathbf{v}}_1 + c_2 \left(\lambda_2 - \lambda_3 \right) \vec{\mathbf{v}}_2 = 0.$$

We have seen that the eigenvectors of any two distinct eigenvalues are linearly independent so $c_1 = c_2 = 0$ in the preceding equation. But if this is true, the equation

$$c_1\vec{v}_1 + c_2\vec{v}_2 + c_3\vec{v}_3 = 0$$

shows that $c_3 = 0$ also and, hence, we have proven the result for three distinct eigenvalues. We can continue this process indefinitely for any number of vectors.

■ **Invertible Matrices**

39. One example of the result in Problem 38 is

$$\mathbf{A} = \begin{bmatrix} 2 & 0 \\ 0 & 3 \end{bmatrix}, \text{ with eigenvalues 2, 3, and for which}$$

$$\mathbf{A}^{-1} = \begin{bmatrix} \dfrac{1}{2} & 0 \\ 0 & \dfrac{1}{3} \end{bmatrix}, \text{ with eigenvalues } \dfrac{1}{2} \text{ and } \dfrac{1}{3}.$$

■ **Eigenvalues and Inversion**

42. We have seen that if a matrix $\mathbf{A}$ has an inverse, none of its eigenvalues are zero. Hence, for an eigenvalue λ we can write

$$\mathbf{A}\vec{v} = \lambda\vec{v}$$

so

$$\mathbf{A}^{-1}\mathbf{A}\vec{v} = \mathbf{A}^{-1}\lambda\vec{v}$$

or

$$\vec{v} = \lambda\mathbf{A}^{-1}\vec{v}.$$

Thus

$$\mathbf{A}^{-1}\vec{v} = \dfrac{1}{\lambda}\vec{v},$$

which shows that the eigenvalues of $\mathbf{A}$ and $\mathbf{A}^{-1}$ are reciprocals and that they have the same eigenvectors. (Creating an example is left to the reader.)

■   **Triangular Matrices**

45.
$$\begin{bmatrix} 1 & 0 & 3 \\ 0 & 4 & 1 \\ 0 & 0 & 2 \end{bmatrix}$$

The characteristic polynomial is

$$p(\lambda) = |\mathbf{A} - \lambda\mathbf{I}| = \begin{vmatrix} 1-\lambda & 0 & 3 \\ 0 & 4-\lambda & 1 \\ 0 & 0 & 2-\lambda \end{vmatrix} = (\lambda-1)(\lambda-4)(\lambda-2) = 0.$$

Hence, the eigenvalues are 1, 2, 4, which are the elements on the diagonal of the matrix.

■   **Eigenvalues of a Transpose**

48.   By Problem 40 in Section 3.1,

$$|\mathbf{A} - \lambda\mathbf{I}|^\mathrm{T} = \mathbf{A}^\mathrm{T} - (\lambda\mathbf{I})^\mathrm{T} = \mathbf{A}^\mathrm{T} - \lambda\mathbf{I}$$

so that by Problem 47 in this section

$$|\mathbf{A} - \lambda\mathbf{I}| = |\mathbf{A} - \lambda\mathbf{I}|^\mathrm{T} = |\mathbf{A}^\mathrm{T} - \lambda\mathbf{I}|.$$

Therefore $\mathbf{A}$ and $\mathbf{A}^\mathrm{T}$ have the same characteristic polynomial and hence, the same eigenvalues.

■   **Another Eigenspace**

51.   The matrix representation for the linear transformation $T\left(ax^2 + bx + c\right) = bx + c$ is

$$\mathbf{T} = \begin{bmatrix} 0 & 0 & 0 \\ 0 & 1 & 0 \\ 0 & 0 & 1 \end{bmatrix} \text{ because } \begin{bmatrix} 0 & 0 & 0 \\ 0 & 1 & 0 \\ 0 & 0 & 1 \end{bmatrix}\begin{bmatrix} a \\ b \\ c \end{bmatrix} = \begin{bmatrix} 0 \\ b \\ c \end{bmatrix} \begin{matrix} \leftarrow x^2 \\ \leftarrow x \\ \leftarrow 1 \end{matrix}.$$

Because $\mathbf{T}$ is a diagonal matrix, the eigenvalues are 0, 1, 1.

The eigenvector corresponding to $\lambda_1 = 0$ satisfies

$$\begin{bmatrix} 0 & 0 & 0 \\ 0 & 1 & 0 \\ 0 & 0 & 1 \end{bmatrix}\begin{bmatrix} x_1 \\ x_2 \\ x_3 \end{bmatrix} = \begin{bmatrix} 0 \\ 0 \\ 0 \end{bmatrix},$$

so $\vec{\mathbf{v}}_1$ can be $\begin{bmatrix} \alpha, & 0, & 0 \end{bmatrix}$ for any real number α.

The eigenvector corresponding to the multiple eigenvalue $\lambda_2 = \lambda_3 = 1$ satisfies the equation

$$\begin{bmatrix} 0 & 0 & 0 \\ 0 & 1 & 0 \\ 0 & 0 & 1 \end{bmatrix}\begin{bmatrix} x_1 \\ x_2 \\ x_3 \end{bmatrix} = \begin{bmatrix} x_1 \\ x_2 \\ x_3 \end{bmatrix}.$$

The only condition these equations specify on x_1, x_2, x_3 is $x_1 = 0$. Hence, we have the two-dimensional eigenspace

$$E_{\lambda_2, \lambda_3} = \left\{ \alpha \begin{bmatrix} 0 \\ 0 \\ 1 \end{bmatrix} + \beta \begin{bmatrix} 0 \\ 1 \\ 0 \end{bmatrix} \middle| \alpha, \beta \text{ any real numbers} \right\}.$$

■ **Looking for Matrices**

54.
$$\begin{bmatrix} a & b \\ c & d \end{bmatrix} \begin{bmatrix} 1 \\ 1 \end{bmatrix} = \lambda \begin{bmatrix} 1 \\ 1 \end{bmatrix} \Rightarrow = \begin{bmatrix} a+b \\ c+d \end{bmatrix} = \begin{bmatrix} \lambda \\ \lambda \end{bmatrix} \Rightarrow \begin{cases} a+b = \lambda \\ c+d = \lambda \end{cases}$$

Thus, matrices of the form $\begin{bmatrix} \lambda - b & b \\ \lambda - d & d \end{bmatrix}$ have $\begin{bmatrix} 1 \\ 1 \end{bmatrix}$ as an eigenvector with eigenvalue λ.

57. For $\lambda = -1$,

$$\begin{bmatrix} a & b \\ c & d \end{bmatrix} \begin{bmatrix} 0 \\ 2 \end{bmatrix} \qquad \text{and} \qquad \begin{bmatrix} a & 0 \\ c & -1 \end{bmatrix} \begin{bmatrix} 1 \\ -1 \end{bmatrix} = \begin{bmatrix} a \\ c+1 \end{bmatrix} = \begin{bmatrix} -1 \\ 1 \end{bmatrix} \Rightarrow \begin{matrix} a = -1 \\ c = 0 \end{matrix}$$

$$= \begin{bmatrix} 2b \\ 2d \end{bmatrix} = \begin{bmatrix} 0 \\ -2 \end{bmatrix} \Rightarrow \begin{matrix} b = 0 \\ d = -1 \end{matrix} \qquad \begin{bmatrix} a-b \\ c-d \end{bmatrix} \begin{bmatrix} -1-b \\ -d \end{bmatrix} = \begin{bmatrix} -1 \\ 1 \end{bmatrix} \Rightarrow \begin{matrix} b = 0 \\ d = -1 \end{matrix}$$

Hence $\begin{bmatrix} -1 & 0 \\ 0 & -1 \end{bmatrix}$ is the only matrix with double eigenvalue -1 and the given eigenvectors.

■ **Linear Transformations in the Plane**

60.
$$A = \begin{bmatrix} \cos \pi/4 & \sin \pi/4 \\ -\sin \pi/4 & \cos \pi/4 \end{bmatrix} = \begin{bmatrix} \sqrt{2}/2 & \sqrt{2}/2 \\ -\sqrt{2}/2 & \sqrt{2}/2 \end{bmatrix} = \frac{\sqrt{2}}{2} \begin{bmatrix} 1 & 1 \\ -1 & 1 \end{bmatrix}.$$

$$|A - \lambda I| = \begin{vmatrix} \sqrt{2}/2 - \lambda & \sqrt{2}/2 \\ -\sqrt{2}/2 & \sqrt{2}/2 - \lambda \end{vmatrix} = \left(\sqrt{2}/2 - \lambda \right)\left(\sqrt{2}/2 - \lambda \right) + \frac{1}{2} = \lambda^2 - \sqrt{2}\lambda + 1 = 0$$

Hence, $\lambda_1 = \sqrt{2}/2 + \sqrt{2}/2 i$, $\lambda_2 = \sqrt{2}/2 - \sqrt{2}/2 i$, or $\lambda = \dfrac{\sqrt{2}}{2}(1 \pm i)$.

To find eigenvectors for $\dfrac{\sqrt{2}}{2}(1+i)$:

$$\frac{\sqrt{2}}{2} \begin{bmatrix} 1 & 1 \\ -1 & 1 \end{bmatrix} \begin{bmatrix} v_1 \\ v_2 \end{bmatrix} = \frac{\sqrt{2}}{2}(1+i) \begin{bmatrix} v_1 \\ v_2 \end{bmatrix} \Rightarrow \begin{bmatrix} v_1 - v_2 \\ -v_1 + v_2 \end{bmatrix} = \begin{bmatrix} v_1 + iv_1 \\ v_2 + iv_2 \end{bmatrix}$$

$$\Rightarrow \begin{matrix} v_2 = iv_1 \\ -v_1 = iv_2 \end{matrix} \Rightarrow \vec{v}_1 = \begin{bmatrix} 1 \\ -i \end{bmatrix}$$

Eigenvectors for $\lambda_2 = \dfrac{\sqrt{2}}{2}(1-i)$ are complex conjugates of $\vec{v}_1$, so $\vec{v}_2 = \begin{bmatrix} 1 \\ i \end{bmatrix}$.

■   **Cayley-Hamilton**

63. $\begin{bmatrix} 1 & 1 \\ 4 & 1 \end{bmatrix}$

The characteristic equation is

$$p(\lambda)=|\mathbf{A}-\lambda\mathbf{I}| = \begin{vmatrix} 1-\lambda & 1 \\ 4 & 1-\lambda \end{vmatrix} = \lambda^2 - 2\lambda - 3 = 0$$

Substituting **A** into this polynomial, we can easily verify

$$p(\mathbf{A})=\mathbf{A}^2 - 2\mathbf{A} - 3\mathbf{I} = \begin{bmatrix} 1 & 1 \\ 4 & 1 \end{bmatrix}^2 - 2\begin{bmatrix} 1 & 1 \\ 4 & 1 \end{bmatrix} - 3\begin{bmatrix} 1 & 0 \\ 0 & 1 \end{bmatrix} = \begin{bmatrix} 0 & 0 \\ 0 & 0 \end{bmatrix}.$$

66. $\begin{bmatrix} 1 & 1 & 2 \\ 0 & 2 & 3 \\ 1 & 0 & 4 \end{bmatrix}$

The characteristic equation is

$$p(\lambda)=|\mathbf{A}-\lambda\mathbf{I}| = \begin{vmatrix} 1-\lambda & 1 & 2 \\ 0 & 2-\lambda & 3 \\ 1 & 0 & 4-\lambda \end{vmatrix} = \lambda^3 - 7\lambda^2 + 12\lambda - 7 = 0.$$

Substituting **A** for λ into this polynomial, we can easily verify

$$p(\mathbf{A})=\mathbf{A}^3 - 7\mathbf{A}^2 + 12\mathbf{A} - 7\mathbf{I} = \begin{bmatrix} 0 & 0 & 0 \\ 0 & 0 & 0 \\ 0 & 0 & 0 \end{bmatrix}.$$

■   **Inverses by Cayley-Hamilton**

69.   The matrix $\mathbf{A} = \begin{bmatrix} a & b \\ c & d \end{bmatrix}$ has the characteristic equation

$$\lambda^2 - (a+d)\lambda + (ad-bc) = 0, \text{ so}$$

$$\mathbf{A}^2 - (a+d)\mathbf{A} + (ad-bc)\mathbf{I} = \mathbf{0}.$$

Premultiplying by $\mathbf{A}^{-1}$ yields the equation $\mathbf{A} - (a+d)\mathbf{I} + (ad-bc)\mathbf{A}^{-1} = \mathbf{0}$.

Solving for $\mathbf{A}^{-1}$

$$\mathbf{A}^{-1} = \frac{a+d}{ad-bc}\mathbf{I} - \frac{1}{ad-bc}\mathbf{A} = \frac{1}{\det\mathbf{A}}\big((tr\mathbf{A})\mathbf{I} - \mathbf{A}\big).$$

(a) $\begin{bmatrix} 3 & 2 \\ -2 & -3 \end{bmatrix}$

Using the preceding formula yields

$$\mathbf{A}^{-1} = \frac{0}{-5}\mathbf{I} - \frac{1}{-5}\begin{bmatrix} 3 & 2 \\ -2 & -3 \end{bmatrix} = \frac{1}{5}\begin{bmatrix} 3 & 2 \\ -2 & -3 \end{bmatrix} = \begin{bmatrix} \dfrac{3}{5} & \dfrac{2}{5} \\ -\dfrac{2}{5} & -\dfrac{3}{5} \end{bmatrix}.$$

(b) $\begin{bmatrix} 3 & 5 \\ -1 & -1 \end{bmatrix}$

$$\mathbf{A}^{-1} = \frac{2}{2}\mathbf{I} - \frac{1}{2}\begin{bmatrix} 3 & 5 \\ -1 & -1 \end{bmatrix} = \frac{1}{2}\begin{bmatrix} -1 & -5 \\ 1 & 3 \end{bmatrix} = \begin{bmatrix} -\dfrac{1}{2} & -\dfrac{5}{2} \\ \dfrac{1}{2} & \dfrac{3}{2} \end{bmatrix}.$$

■   **Eigenvalues and Conversion**

72. $y'' - y' - 2y = 0$ has characteristic equation $r^2 - r - 2 = 0$ with roots $r = -1$ and 2.

On the other hand, $y_1 = y$, $y_2 = y'$ yields the first-order system $y_1' = y_2$, $y_2' = 2y_1 + y_2$, which in matrix form is

$$\begin{bmatrix} y_1' \\ y_2' \end{bmatrix} = \begin{bmatrix} 0 & 1 \\ 2 & 1 \end{bmatrix}\begin{bmatrix} y_1 \\ y_2 \end{bmatrix}.$$

The coefficient matrix of this system has the characteristic polynomial

$$p(\lambda) = \begin{vmatrix} -\lambda & 1 \\ 2 & 1-\lambda \end{vmatrix} = -\lambda(1-\lambda) - 2 = (\lambda+1)(\lambda-2),$$

so the roots of the characteristic equation are the same as the eigenvalues of the companion matrix.

75. $y''' - 2y'' - 5y' + 6y = 0$ has characteristic equation $r^3 - 2r^2 - 5r + 6 = 0$.

On the other hand $y_1 = y$, $y_2 = y'$, $y_3 = y''$ yields $y_1' = y_2$, $y_2' = y_3$, $y_3' = -6y_1 + 5y_2 + 2y_3$ which in matrix form is

$$\begin{bmatrix} y_1' \\ y_2' \\ y_3' \end{bmatrix} = \begin{bmatrix} 0 & 1 & 0 \\ 0 & 0 & 1 \\ -6 & 5 & 2 \end{bmatrix}\begin{bmatrix} y_1 \\ y_2 \\ y_3 \end{bmatrix}.$$

continued on the next page

The coefficient matrix of this system has the characteristic polynomial

$$p(\lambda) = \begin{vmatrix} -\lambda & 1 & 0 \\ 0 & -\lambda & 1 \\ -6 & 5 & 2-\lambda \end{vmatrix} = (\lambda - 1)(\lambda - 3)(\lambda + 2),$$

which is the same as the characteristic polynomial of the third order DE so the eigenvalues of the companion matrix are the same as the roots of the characteristic polynomial.

■   Eigenfunction Boundary-Value Problems

78. $y'' + \lambda y = 0$

If $\lambda = 0$, then $y(t) = at + b$, so that $y'(t) = a$.

$$y(-\pi) = y(\pi) \implies -a\pi + b = a\pi + b \implies a = 0.$$

There are non zero solutions if $y(t) =$ a nonzero constant.

If $\lambda > 0$, then $y(t) = c_1 \cos\sqrt{\lambda}t + c_2 \sin\sqrt{\lambda}t$, $y'(t) = -c_1\sqrt{\lambda}\sin\sqrt{\lambda}t + c_2\sqrt{\lambda}\cos\sqrt{\lambda}t$.

$$y(-\pi) = y(\pi) \Rightarrow c_1\cos\left(-\pi\sqrt{\lambda}\right) + c_2\sin\left(-\pi\sqrt{\lambda}\right) = c_1\cos\left(\pi\sqrt{\lambda}\right) + c_2\sin\left(\pi\sqrt{\lambda}\right)$$

$$\Rightarrow c_2\sin\left(-\pi\sqrt{\lambda}\right) = c_2\sin\left(\pi\sqrt{\lambda}\right)$$

$$\Rightarrow -\sin\left(\pi\sqrt{\lambda}\right) = \sin\left(\pi\sqrt{\lambda}\right)$$

$$\Rightarrow \sin\left(\pi\sqrt{\lambda}\right) = 0$$

$$\Rightarrow \lambda = n^2, \text{ for } n \text{ any nonzero integer.}$$

$$y'(-\pi) = y'(\pi) \Rightarrow c_1\sqrt{\lambda}\sin\left(-\pi\sqrt{\lambda}\right) - c_2\sqrt{\lambda}\cos\left(-\pi\sqrt{\lambda}\right) = -c_1\sqrt{\lambda}\sin\left(\pi\sqrt{\lambda}\right) + c_2\sqrt{\lambda}\cos\left(\pi\sqrt{\lambda}\right)$$

$$\Rightarrow -c_1\sqrt{\lambda}\sin\left(-\pi\sqrt{\lambda}\right) = -c_1\sqrt{\lambda}\sin\left(\pi\sqrt{\lambda}\right)$$

$$\Rightarrow c_1\sqrt{\lambda}\sin\left(\pi\sqrt{\lambda}\right) = -c_1\sqrt{\lambda}\sin\left(\pi\sqrt{\lambda}\right)$$

$$\Rightarrow \sin\left(\pi\sqrt{\lambda}\right) = 0$$

$$\Rightarrow \lambda = n^2, \text{ for } n \text{ any nonzero integer.}$$

5.4   Coordinates and Diagonalization

■   **Changing Coordinates I**

3. $\vec{u}_S = \mathbf{M}_B \vec{u}_B$, $\mathbf{M}_B \begin{bmatrix} 3 \\ -1 \end{bmatrix} = \begin{bmatrix} 3 & -4 \\ -2 & 3 \end{bmatrix} \begin{bmatrix} 3 \\ -1 \end{bmatrix} = \begin{bmatrix} 13 \\ -9 \end{bmatrix}$, $\mathbf{M}_B \begin{bmatrix} 2 \\ 2 \end{bmatrix} = \begin{bmatrix} 3 & -4 \\ -2 & 3 \end{bmatrix} \begin{bmatrix} 2 \\ 2 \end{bmatrix} = \begin{bmatrix} -2 \\ 2 \end{bmatrix}$,

$\mathbf{M}_B \begin{bmatrix} 1 \\ 0 \end{bmatrix} = \begin{bmatrix} 3 & -4 \\ -2 & 3 \end{bmatrix} \begin{bmatrix} 1 \\ 0 \end{bmatrix} = \begin{bmatrix} 3 \\ -2 \end{bmatrix}$

■   **Changing Coordinates II**

6. $\vec{u}_S = \mathbf{M}_S \vec{u}_B$, $\mathbf{M}_B \begin{bmatrix} 2 \\ 2 \end{bmatrix} = \begin{bmatrix} 1 & -1 \\ -1 & 2 \end{bmatrix} \begin{bmatrix} 2 \\ 2 \end{bmatrix} = \begin{bmatrix} 0 \\ 2 \end{bmatrix}$, $\mathbf{M}_B \begin{bmatrix} 1 \\ -1 \end{bmatrix} = \begin{bmatrix} 1 & -1 \\ -1 & 2 \end{bmatrix} \begin{bmatrix} 1 \\ -1 \end{bmatrix} = \begin{bmatrix} 2 \\ -3 \end{bmatrix}$,

$\mathbf{M}_B \begin{bmatrix} 1 \\ 0 \end{bmatrix} = \begin{bmatrix} 1 & -1 \\ -1 & 2 \end{bmatrix} \begin{bmatrix} 1 \\ 0 \end{bmatrix} = \begin{bmatrix} 1 \\ -1 \end{bmatrix}$

■   **Changing Coordinates III**

9. $\vec{u}_S = \mathbf{M}_B \vec{u}_B$, $\mathbf{M}_B \begin{bmatrix} 1 \\ 0 \\ -1 \end{bmatrix} = \begin{bmatrix} 1 & 1 & 1 \\ 0 & 1 & 1 \\ 0 & 0 & 1 \end{bmatrix} \begin{bmatrix} 1 \\ 0 \\ -1 \end{bmatrix} = \begin{bmatrix} 0 \\ -1 \\ -1 \end{bmatrix}$, $\mathbf{M}_B \begin{bmatrix} 1 \\ 1 \\ 3 \end{bmatrix} = \begin{bmatrix} 1 & 1 & 1 \\ 0 & 1 & 1 \\ 0 & 0 & 1 \end{bmatrix} \begin{bmatrix} 1 \\ 1 \\ 3 \end{bmatrix} = \begin{bmatrix} 5 \\ 4 \\ 3 \end{bmatrix}$,

$\mathbf{M}_B \begin{bmatrix} -2 \\ 1 \\ 1 \end{bmatrix} = \begin{bmatrix} 1 & 1 & 1 \\ 0 & 1 & 1 \\ 0 & 0 & 1 \end{bmatrix} \begin{bmatrix} -2 \\ 1 \\ 1 \end{bmatrix} = \begin{bmatrix} 0 \\ 2 \\ 1 \end{bmatrix}$

■   **Changing Coordinates IV**

12. $\vec{u}_S = \mathbf{M}_B \vec{u}_B$, $\mathbf{M}_B \begin{bmatrix} -1 \\ -1 \\ -4 \end{bmatrix} = \begin{bmatrix} 1 & 0 & 2 \\ 0 & 0 & 1 \\ 0 & 1 & -1 \end{bmatrix} \begin{bmatrix} -1 \\ -1 \\ -4 \end{bmatrix} = \begin{bmatrix} -9 \\ -4 \\ 3 \end{bmatrix}$, $\mathbf{M}_B \begin{bmatrix} 1 \\ -1 \\ 3 \end{bmatrix} = \begin{bmatrix} 1 & 0 & 2 \\ 0 & 0 & 1 \\ 0 & 1 & -1 \end{bmatrix} \begin{bmatrix} 1 \\ -1 \\ 3 \end{bmatrix} = \begin{bmatrix} 7 \\ 3 \\ -4 \end{bmatrix}$,

$\mathbf{M}_B \begin{bmatrix} 3 \\ 1 \\ 4 \end{bmatrix} = \begin{bmatrix} 1 & 0 & 2 \\ 0 & 0 & 1 \\ 0 & 1 & -1 \end{bmatrix} \begin{bmatrix} 3 \\ 1 \\ 4 \end{bmatrix} = \begin{bmatrix} 11 \\ 4 \\ -3 \end{bmatrix}$

■   **Polynomial Coordinates I**

15.   For the basis $N = \{2x^2 - x,\ x^2,\ x^2 + 1\}$ the vectors; $\vec{u}(x)$, $\vec{v}(x)$, $\vec{w}(x)$, whose coordinate vectors are, respectively, $[1,\ 0,\ 2]$, $[-2,\ 2,\ 3]$, $[-1,\ -1,\ 0]$, are

$$u(x) = 1(2x^2 - x) + 0(x^2) + 2(x^2 + 1) = 4x^2 - x + 2 \qquad \text{so } \vec{u}_S = \begin{bmatrix} 4 \\ -1 \\ 2 \end{bmatrix}.$$

$$v(x) = -2(2x^2 - x) + 2(x^2) + 3(x^2 + 1) = x^2 + 2x + 3 \qquad \text{so } \vec{v}_S = \begin{bmatrix} 1 \\ 2 \\ 3 \end{bmatrix}.$$

$$w(x) = -1(2x^2 - x) - 1(x^2) + 0(x^2 + 1) = -3x^2 + x \qquad \text{so } \vec{w}_S = \begin{bmatrix} -3 \\ 1 \\ 0 \end{bmatrix}.$$

■   **Polynomial Coordinates II**

18.   Here, we have $\vec{u}_p = M_Q \vec{u}_Q$, using M_Q from Problem 16. So the standardized representations of $\vec{u}(x)$, $\vec{v}(x)$, $\vec{w}(x)$ are

$$M_Q \begin{bmatrix} 1 \\ -1 \\ 0 \\ 2 \end{bmatrix} = \begin{bmatrix} 1 & 1 & 0 & 0 \\ 0 & 0 & 1 & 1 \\ 0 & 1 & 0 & 0 \\ 0 & 0 & 0 & 1 \end{bmatrix} \begin{bmatrix} 1 \\ -1 \\ 0 \\ 2 \end{bmatrix} = \begin{bmatrix} 0 \\ 2 \\ -1 \\ 2 \end{bmatrix} = [\vec{u}]_p$$

$$M_Q \begin{bmatrix} -2 \\ 0 \\ -2 \\ 0 \end{bmatrix} = \begin{bmatrix} 1 & 1 & 0 & 0 \\ 0 & 0 & 1 & 1 \\ 0 & 1 & 0 & 0 \\ 0 & 0 & 0 & 1 \end{bmatrix} \begin{bmatrix} -2 \\ 0 \\ -2 \\ 0 \end{bmatrix} = \begin{bmatrix} -2 \\ -2 \\ 0 \\ 0 \end{bmatrix} = [\vec{v}]_p$$

$$M_Q \begin{bmatrix} 3 \\ -1 \\ 4 \\ 2 \end{bmatrix} = \begin{bmatrix} 1 & 1 & 0 & 0 \\ 0 & 0 & 1 & 1 \\ 0 & 1 & 0 & 0 \\ 0 & 0 & 0 & 1 \end{bmatrix} \begin{bmatrix} 3 \\ -1 \\ 4 \\ 2 \end{bmatrix} = \begin{bmatrix} 2 \\ 6 \\ -1 \\ 2 \end{bmatrix} = [\vec{w}]_p .$$

We can check our results by observing that $\vec{u}(x)$, $\vec{v}(x)$, $\vec{w}(x)$ can be written in both coordinate systems as with the coordinates shown bold:

$$\vec{u}(x) = \mathbf{1}(x^3) + (\mathbf{-1})(x^3 + x) + \mathbf{0}(x^2) + \mathbf{2}(x^2 + 1) = \mathbf{0}(x^3) + (\mathbf{2})x^2 + (\mathbf{-1})(x) + \mathbf{2}(1)$$

$$\vec{v}(x) = \mathbf{-2}(x^3) + (\mathbf{0})(x^3 + x) + (\mathbf{-2})(x^2) + \mathbf{0}(x^2 + 1) = (\mathbf{-2})(x^3) + (\mathbf{-2})x^2 + (\mathbf{0})(x) + \mathbf{0}(1)$$

$$\vec{w}(x) = \mathbf{3}(x^3) + (\mathbf{-1})(x^3 + x) + \mathbf{4}(x^2) + \mathbf{2}(x^2 + 1) = \mathbf{2}(x^3) + (\mathbf{6})x^2 + (\mathbf{-1})(x) + \mathbf{2}(1)$$

■   **Matrix Representations for Polynomial Transformations**

21. $T\left(f\left(t\right)\right)=f'''(t)$ and $f(t)=at^4+bt^3+ct^2+dt+e.$

We first write $T\left(f\right)=f'''=24at+6b$. We then apply the matrix that sends the coordinates of f into the coordinates of f''':

$$\begin{bmatrix} 0 & 0 & 0 & 0 & 0 \\ 0 & 0 & 0 & 0 & 0 \\ 0 & 0 & 0 & 0 & 0 \\ 24 & 0 & 0 & 0 & 0 \\ 0 & 6 & 0 & 0 & 0 \end{bmatrix}\begin{bmatrix} a \\ b \\ c \\ d \\ e \end{bmatrix}=\begin{bmatrix} 0 \\ 0 \\ 0 \\ 24a \\ 6b \end{bmatrix}\begin{matrix} \leftarrow t^4 \\ \leftarrow t^3 \\ \leftarrow t^2 \\ \leftarrow t \\ \leftarrow 1 \end{matrix}.$$

(a)   We can use this matrix to find the third derivative of $g\left(t\right)=t^4-t^3+t^2-t+1$ by multiplying the coordinate vector of $g\left(t\right)$, which is $\begin{bmatrix} 1, & -1, & 1, & -1, & 1 \end{bmatrix}$ to find

$$\begin{bmatrix} 0 & 0 & 0 & 0 & 0 \\ 0 & 0 & 0 & 0 & 0 \\ 0 & 0 & 0 & 0 & 0 \\ 24 & 0 & 0 & 0 & 0 \\ 0 & 6 & 0 & 0 & 0 \end{bmatrix}\begin{bmatrix} 1 \\ -1 \\ 1 \\ -1 \\ 1 \end{bmatrix}=\begin{bmatrix} 0 \\ 0 \\ 0 \\ 24 \\ -6 \end{bmatrix}\begin{matrix} \leftarrow t^4 \\ \leftarrow t^3 \\ \leftarrow t^2 \\ \leftarrow t \\ \leftarrow 1 \end{matrix}.$$

This means that the third derivative of $g\left(t\right)$ is $g'''=24t-6$ and $T\left(g(t)\right)=[0,\ 0,\ 0,\ 24,\ -6]$. We find the third derivatives in parts (b), (c), and (d) in the same way.

(b) $\begin{bmatrix} 0, 0, 0, 24, 0 \end{bmatrix}$ (c) $\begin{bmatrix} 0, 0, 0, -96, 18 \end{bmatrix}$ (d) $\begin{bmatrix} 0, 0, 0, 24, 0 \end{bmatrix}$

24. $T\left(f\left(t\right)\right)=f''(t)+f\left(t\right)$

We write

$$T\left(f\right)=f''+f=at^4+bt^3+\left(c+12a\right)t^2+\left(d+6b\right)t+\left(e+2c\right).$$

We use the matrix that sends the coordinates of f into the coordinates of $f''+f$:

$$\begin{bmatrix} 1 & 0 & 0 & 0 & 0 \\ 0 & 1 & 0 & 0 & 0 \\ 12 & 0 & 1 & 0 & 0 \\ 0 & 6 & 0 & 1 & 0 \\ 0 & 0 & 2 & 0 & 1 \end{bmatrix}\begin{bmatrix} a \\ b \\ c \\ d \\ e \end{bmatrix}=\begin{bmatrix} a \\ b \\ c+12a \\ d+6b \\ e+2c \end{bmatrix}\begin{matrix} \leftarrow t^4 \\ \leftarrow t^3 \\ \leftarrow t^2 \\ \leftarrow t \\ \leftarrow 1 \end{matrix}.$$

continued on the next page

(a)   We evaluate $g''(t) + g(t)$ for the function $g(t) = t^4 - t^3 + t^2 - t + 1$ by multiplying the matrix by the coordinate vector of $g(t)$, which is $\begin{bmatrix} 1, & -1, & 1, & -1, & 1 \end{bmatrix}$ to find

$$\begin{bmatrix} 1 & 0 & 0 & 0 & 0 \\ 0 & 1 & 0 & 0 & 0 \\ 12 & 0 & 1 & 0 & 0 \\ 0 & 6 & 0 & 1 & 0 \\ 0 & 0 & 2 & 0 & 1 \end{bmatrix} \begin{bmatrix} 1 \\ -1 \\ 1 \\ -1 \\ 1 \end{bmatrix} = \begin{bmatrix} 1 \\ -1 \\ 13 \\ -7 \\ 3 \end{bmatrix} \begin{array}{l} \leftarrow t^4 \\ \leftarrow t^3 \\ \leftarrow t^2 \\ \leftarrow t \\ \leftarrow 1 \end{array}.$$

This means that $g''(t) + g(t) = t^4 - t^3 + 13t^2 - 7t + 3$, and $T(g(t)) = \begin{bmatrix} 1, -1, 13, -7, 3 \end{bmatrix}$.

(b) $\begin{bmatrix} 1, 0, 14, 0, 8 \end{bmatrix}$ (c) $\begin{bmatrix} -4, 3, -48, 18, 0 \end{bmatrix}$ (d) $\begin{bmatrix} 1, 0, 4, 0, 0 \end{bmatrix}$

■   **Diagonalization**

27. $A = \begin{bmatrix} 1 & 2 \\ 2 & 1 \end{bmatrix}$

The matrix has two independent eigenvectors, which are the columns of the matrix

$$P = \begin{bmatrix} -1 & 1 \\ 1 & 1 \end{bmatrix}$$

$$P^{-1} = \frac{1}{2} \begin{bmatrix} -1 & 1 \\ 1 & 1 \end{bmatrix}.$$

Hence, the eigenvalues will be the diagonal elements of the matrix

$$P^{-1}AP = \frac{1}{2} \begin{bmatrix} -1 & 1 \\ 1 & 1 \end{bmatrix} \begin{bmatrix} 1 & 2 \\ 2 & 1 \end{bmatrix} \begin{bmatrix} -1 & 1 \\ 1 & 1 \end{bmatrix} = \begin{bmatrix} -1 & 0 \\ 0 & 3 \end{bmatrix}.$$

Note that **P** is not unique.

30. $A = \begin{bmatrix} 0 & -1 \\ 1 & 0 \end{bmatrix}$

The matrix **P** that diagonalizes the given matrix **A** is the matrix of eigenvectors of **A**.

$$P = \begin{bmatrix} 1 & 1 \\ -i & i \end{bmatrix}, \quad P^{-1} = \frac{1}{2} \begin{bmatrix} 1 & i \\ 1 & -i \end{bmatrix}, \quad P^{-1}AP = \frac{1}{2} \begin{bmatrix} 1 & i \\ 1 & -i \end{bmatrix} \begin{bmatrix} 0 & -1 \\ 1 & 0 \end{bmatrix} \begin{bmatrix} 1 & 1 \\ -i & i \end{bmatrix} = \begin{bmatrix} i & 0 \\ 0 & -i \end{bmatrix}.$$

33.   The matrix $\mathbf{A} = \begin{bmatrix} 4 & -2 \\ 1/2 & 2 \end{bmatrix}$ has repeated eigenvalue $\lambda = 3, 3$

For $\lambda = 3$: $\begin{bmatrix} 1 & -2 & | & 0 \\ 1/2 & -1 & | & 0 \end{bmatrix}$ $\underrightarrow{\text{RREF}}$ $\begin{bmatrix} 1 & -2 & | & 0 \\ 0 & 0 & | & 0 \end{bmatrix}$ $\Rightarrow$ $\begin{matrix} v_1 - 2v_2 = 0 \\ v_2 \text{ free} \end{matrix}$

Therefore there is only one linearly independent eigenvector $\begin{bmatrix} 2 \\ 1 \end{bmatrix}$.

Since dim $\mathbb{E} = 1 < 2$, $\mathbf{A}$ is not diagonalizable.

36. $\mathbf{A} = \begin{bmatrix} 0 & 1 & -1 \\ 0 & -1 & 1 \\ 0 & 0 & 0 \end{bmatrix}$ has eigenvalues $-1, 0, 0$ and three linearly independent eigenvectors to use as

columns of $\mathbf{P}$, a matrix that diagonalizes the given matrix $\mathbf{A}$.

$$\mathbf{P} = \begin{bmatrix} -1 & 0 & 1 \\ 1 & 1 & 0 \\ 0 & 1 & 0 \end{bmatrix}, \ \mathbf{P}^{-1} = \begin{bmatrix} 0 & 1 & -1 \\ 0 & 0 & 1 \\ 1 & 1 & -1 \end{bmatrix}, \ \mathbf{P}^{-1}\mathbf{A}\mathbf{P} = \begin{bmatrix} -1 & 0 & 0 \\ 0 & 0 & 0 \\ 0 & 0 & 0 \end{bmatrix}.$$

We can confirm that the eigenvectors (columns of $\mathbf{P}$) are linearly independent because the determinant of $\mathbf{P}$ is nonzero.

39. $\mathbf{A} = \begin{bmatrix} 3 & -1 & 1 \\ 7 & -5 & 1 \\ 6 & -6 & 2 \end{bmatrix}$

The eigenvalue $\lambda_1 = -4$ has eigenvector $\begin{bmatrix} 0, 1, 1 \end{bmatrix}$. However, the double eigenvalue $\lambda_2 = \lambda_3 = 2$, has only one linearly independent eigenvector $\begin{bmatrix} 1, 1, 0 \end{bmatrix}$. This matrix cannot be diagonalized because it has only two linearly independent eigenvectors.

42. $\mathbf{A} = \begin{bmatrix} 4 & 2 & 3 \\ 2 & 1 & 2 \\ -1 & 2 & 0 \end{bmatrix}$

We find eigenvalues $1, 5, -1$, and use their linearly independent eigenvectors to form the matrix $\mathbf{P}$ that diagonalizes $\mathbf{A}$.

$$\mathbf{P} = \begin{bmatrix} -1 & 2 & 1 \\ 0 & 1 & 2 \\ 1 & 0 & -3 \end{bmatrix}, \ \mathbf{P}^{-1} = \begin{bmatrix} -\dfrac{1}{2} & 1 & \dfrac{1}{2} \\ \dfrac{1}{3} & \dfrac{1}{3} & \dfrac{1}{3} \\ -\dfrac{1}{6} & \dfrac{1}{3} & -\dfrac{1}{6} \end{bmatrix}, \ \mathbf{P}^{-1}\mathbf{A}\mathbf{P} = \begin{bmatrix} 1 & 0 & 0 \\ 0 & 5 & 0 \\ 0 & 0 & -1 \end{bmatrix}.$$

45. $A = \begin{bmatrix} 0 & 0 & 2 \\ -1 & 1 & 2 \\ -1 & 0 & 3 \end{bmatrix}$ The eigenvalues are $\lambda_1 = 1, 1$ and $\lambda_2 = 2$.

To find eigenvectors for $\lambda_1 = 1$:

$$\begin{bmatrix} -1 & 0 & 2 & | & 0 \\ -1 & 0 & 2 & | & 0 \\ -1 & 0 & 2 & | & 0 \end{bmatrix} \underline{\text{RREF}} \begin{bmatrix} 1 & 0 & -2 & | & 0 \\ 0 & 0 & 0 & | & 0 \\ 0 & 0 & 0 & | & 0 \end{bmatrix} \Rightarrow \begin{array}{c} v_1 - 2v_3 = 0 \\ v_2, v_3 \text{ free} \end{array}$$

Two (linearly independent) eigenvectors for $\lambda_1 = 1$ are $\begin{bmatrix} 0 \\ 1 \\ 0 \end{bmatrix}$ and $\begin{bmatrix} 2 \\ 0 \\ 1 \end{bmatrix}$.

At this point, we know that A is diagonalizable.

To find an eigenvector for $\lambda_2 = 2$:

$$\begin{bmatrix} -2 & 0 & 2 & | & 0 \\ -1 & -1 & 2 & | & 0 \\ -1 & 0 & 1 & | & 0 \end{bmatrix} \underline{\text{RREF}} \begin{bmatrix} 1 & 0 & -1 & | & 0 \\ 0 & 1 & -1 & | & 0 \\ 0 & 0 & 0 & | & 0 \end{bmatrix} \Rightarrow \begin{array}{c} v_1 - v_3 = 0 \\ v_2 - v_3 = 0 \\ v_3 \text{ free} \end{array}$$

An eigenvector for $\lambda_2 = 2$ is $\begin{bmatrix} 1 \\ 1 \\ 1 \end{bmatrix}$.

The matrix $P = \begin{bmatrix} 0 & 2 & 1 \\ 1 & 0 & 1 \\ 0 & 1 & 1 \end{bmatrix}$ diagonalizes A, and $P^{-1}AP = \begin{bmatrix} 1 & 0 & 0 \\ 0 & 1 & 0 \\ 0 & 0 & 2 \end{bmatrix}$.

48. $A = \begin{bmatrix} 2 & 0 & 1 & 2 \\ 0 & 2 & 0 & 0 \\ 0 & 0 & 6 & 0 \\ 0 & 0 & 1 & 4 \end{bmatrix}$ The eigenvalues are $\lambda_1 = 2, 2, \lambda_2 = 4, \lambda_3 = 6$.

To find eigenvectors for $\lambda_1 = 2$:

$$\begin{bmatrix} 0 & 0 & 1 & 2 \\ 0 & 0 & 0 & 0 \\ 0 & 0 & 4 & 0 \\ 0 & 0 & 1 & 2 \end{bmatrix} \underline{\text{RREF}} \begin{bmatrix} 0 & 0 & 1 & 2 \\ 0 & 0 & 0 & 1 \\ 0 & 0 & 0 & 0 \\ 0 & 0 & 0 & 0 \end{bmatrix} \Rightarrow \begin{array}{c} v_3 + 2v_4 = 0 \\ v_4 = 0 \\ v_1, v_2 \text{ free} \end{array}$$

Two (linearly independent) eigenvectors for $\lambda_1 = 2$ are $\begin{bmatrix} 1 \\ 0 \\ 0 \\ 0 \end{bmatrix}$ and $\begin{bmatrix} 0 \\ 1 \\ 0 \\ 0 \end{bmatrix}$.

(Note: **A** is diagonalizable.)

To find an eigenvector for $\lambda_2 = 4$:

$$\begin{bmatrix} -2 & 0 & 1 & 2 \\ 0 & -2 & 0 & 0 \\ 0 & 0 & 2 & 0 \\ 0 & 0 & 1 & 0 \end{bmatrix} \xrightarrow{\text{RREF}} \begin{bmatrix} 1 & 0 & -1/2 & -1 \\ 0 & 1 & 0 & 0 \\ 0 & 0 & 1 & 0 \\ 0 & 0 & 0 & 0 \end{bmatrix} \Rightarrow \begin{array}{c} v_1 - \frac{1}{2}v_3 - v_4 = 0 \\ v_2 = 0 = v_3 \\ v_4 \text{ free} \end{array}$$

An eigenvector for $\lambda_2 = 4$ is $\begin{bmatrix} 1 \\ 0 \\ 0 \\ 1 \end{bmatrix}$.

To find an eigenvector for $\lambda_3 = 6$:

$$\begin{bmatrix} -4 & 0 & 1 & 2 \\ 0 & -4 & 0 & 0 \\ 0 & 0 & 0 & 0 \\ 0 & 0 & 1 & -2 \end{bmatrix} \xrightarrow{\text{RREF}} \begin{bmatrix} 1 & 0 & -1/4 & -1/2 \\ 0 & 1 & 0 & 0 \\ 0 & 0 & 1 & -2 \\ 0 & 0 & 0 & 0 \end{bmatrix} \Rightarrow \begin{array}{c} v_1 - \frac{1}{4}v_3 - \frac{1}{2}v_4 = 0 \\ v_2 = 0 \\ v_3 - 2v_4 = 0 \\ v_4 \text{ free} \end{array}$$

An eigenvector for $\lambda_3 = 6$ is $\begin{bmatrix} 1 \\ 0 \\ 2 \\ 1 \end{bmatrix}$.

The matrix $\mathbf{P} = \begin{bmatrix} 1 & 0 & 1 & 1 \\ 0 & 1 & 0 & 0 \\ 0 & 0 & 0 & 2 \\ 0 & 0 & 1 & 1 \end{bmatrix}$ diagonalizes **A**, and $\mathbf{P}^{-1}\mathbf{AP} = \begin{bmatrix} 2 & 0 & 0 & 0 \\ 0 & 2 & 0 & 0 \\ 0 & 0 & 4 & 0 \\ 0 & 0 & 0 & 6 \end{bmatrix}$.

■   **Constructing Counterexamples**

51. $A = \begin{bmatrix} 1 & 1 \\ 0 & 1 \end{bmatrix}$ is invertible because $|A| = 1 \neq 0$, but A is not diagonalizable.

That is, $\lambda = 1, 1$ but $\vec{v} = \begin{bmatrix} 1 \\ 0 \end{bmatrix}$ is the only linearly independent eigenvector.

■   **Computer Lab: Diagonalization**

54.   (a)   The given matrix has an eigenvalue of 1 with multiplicity 4 and only one linearly independent eigenvector $(1, 0, 0, 0)$. Hence, it cannot be diagonalized.

(b)   For $\begin{bmatrix} -2 & 1 & 1 & 0 & 0 \\ 1 & -2 & 1 & 0 & 0 \\ 1 & 1 & -2 & 0 & 0 \\ 0 & 0 & 0 & 1 & 1 \\ 0 & 0 & 0 & 4 & 1 \end{bmatrix}$, $\lambda = 0, -3, -3, 3, -1$, $P = \begin{bmatrix} 1 & -1 & -1 & 0 & 0 \\ 1 & 0 & 1 & 0 & 0 \\ 1 & 1 & 0 & 0 & 0 \\ 0 & 0 & 0 & 1 & -1 \\ 0 & 0 & 0 & 2 & 2 \end{bmatrix}$.

(c)   For $\begin{bmatrix} 3 & 0 & 0 & 0 \\ 0 & 1 & 1 & 0 \\ 0 & 1 & 1 & 0 \\ 0 & 0 & 0 & 5 \end{bmatrix}$, $\lambda = 3, 0, 2, 5$, $P = \begin{bmatrix} 1 & 0 & 0 & 0 \\ 0 & 1 & 1 & 0 \\ 0 & -1 & 1 & 0 \\ 0 & 0 & 0 & 1 \end{bmatrix}$.

■   **Computer Lab: Similarity Challenge**

57.   (a)   We need to show that A is similar to B where

$$A = \begin{bmatrix} 1 & 2 & -3 \\ 2 & 0 & 1 \\ 1 & -3 & 1 \end{bmatrix} \text{ and } B = \begin{bmatrix} 1 & -19 & 58 \\ 1 & 12 & -27 \\ 5 & 15 & -11 \end{bmatrix}$$

Both matrices have the same characteristic polynomial $p(\lambda) = -\lambda^3 + 2\lambda^2 - 3\lambda + 19$. There is one real eigenvalue λ_1 and a complex conjugate pair of eigenvalues λ_2 and λ_3. Because all three eigenvalues are distinct, there are three linearly independent eigenvectors so that A and B are similar to the same diagonal matrix D:

$$D = \begin{bmatrix} \lambda_1 & 0 & 0 \\ 0 & \lambda_2 & 0 \\ 0 & 0 & \lambda_3 \end{bmatrix}.$$

(b) A and B both have trace 2 and determinant of 19.

■ **When Diagonalization Fails**

60.   Let $\mathbf{A} = \begin{bmatrix} a & b \\ c & d \end{bmatrix}$ with double eigenvalue λ and only one linearly independent eigenvector

$\vec{\mathbf{v}} = \begin{bmatrix} v_1 \\ v_2 \end{bmatrix}$, where $v_2 \neq 0$.

Let $\mathbf{Q} = \begin{bmatrix} v_1 & 1 \\ v_2 & 0 \end{bmatrix}$

$$\mathbf{AQ} = \begin{bmatrix} a & b \\ c & d \end{bmatrix}\begin{bmatrix} v_1 & 1 \\ v_2 & 0 \end{bmatrix} = \begin{bmatrix} av_1 + bv_2 & a \\ cv_1 + dv_2 & c \end{bmatrix} = \begin{bmatrix} \lambda v_1 & a \\ \lambda v_2 & c \end{bmatrix}$$

because $av_1 + bv_2 = \lambda v_1$ and $cv_1 + dv_2 = \lambda v_2$, by the fact that $\mathbf{A}\vec{\mathbf{v}} = \lambda\vec{\mathbf{v}}$.

$$\mathbf{Q}^{-1} = -\frac{1}{v_2}\begin{bmatrix} 0 & -1 \\ -v_2 & v_1 \end{bmatrix}$$

$$\mathbf{Q}^{-1}\mathbf{AQ} = -\frac{1}{v_2}\begin{bmatrix} 0 & -1 \\ -v_2 & v_1 \end{bmatrix}\begin{bmatrix} \lambda v_1 & a \\ \lambda v_2 & c \end{bmatrix} = -\frac{1}{v_2}\begin{bmatrix} -\lambda v_2 & -c \\ 0 & -av_2 + cv_1 \end{bmatrix}$$

$$= \begin{bmatrix} \lambda & -c/v_2 \\ 0 & a - cv_1/v_2 \end{bmatrix} \sim \mathbf{A}$$

By Section 5.3, Problem 40(a), $\mathbf{Q}^{-1}\mathbf{AQ} \sim \mathbf{A}$, and they will have the same eigenvalues. By Section 5.3, Problems 43–46, a triangular matrix has its eigenvalues on the main diagonal, so the lower right element must also be λ. Hence $\lambda = a - \dfrac{cv_1}{v_2}$.

■ **Suggested Journal Entry I**

63.   Student Project

CHAPTER 6

Linear Systems of Differential Equations

6.1 Theory of Linear DE Systems

■ Breaking Out Systems

3.
$$x_1' = 4x_1 + 3x_2 + e^{-t}$$
$$x_2' = -x_1 - x_2$$

■ Checking It Out

6.
$$\vec{x}' = \begin{bmatrix} 4 & -1 \\ 2 & 1 \end{bmatrix} \vec{x}$$

By substitution, we verify that $\vec{u}(t) = \begin{bmatrix} e^{3t} \\ e^{3t} \end{bmatrix}$ and $\vec{v}(t) = \begin{bmatrix} e^{2t} \\ 2e^{2t} \end{bmatrix}$ satisfy the system.

The fundamental matrix $\mathbf{X}(t) = \begin{bmatrix} e^{3t} & e^{2t} \\ e^{3t} & 2e^{2t} \end{bmatrix}$.

The general solution $\vec{x}(t) = c_1 \begin{bmatrix} e^{3t} \\ e^{3t} \end{bmatrix} + c_2 \begin{bmatrix} e^{2t} \\ 2e^{2t} \end{bmatrix}$.

■   **Uniqueness in the Phase Plane**

9.   The direction field of $x' = y$, $y' = -x$ is shown.

We have drawn three distinct trajectories for the six initial conditions $(x(0), y(0)) =$

$$(1,0), (2,0), (3,0), (0,1), (0,2), (0,3).$$

Note that although the trajectories may (and do) coincide if one starts at a point lying on another, they never cross each other.

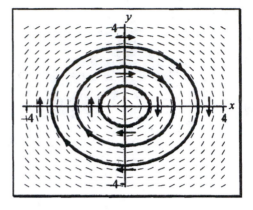

However, if we plot coordinate $x = x(t)$ or $y = y(t)$ for these same six initial conditions we get the six intersecting curves shown in the tx and ty planes.

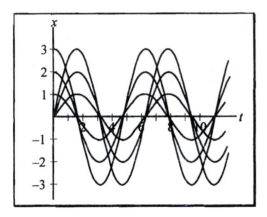

Intersecting solutions $x = x(t)$

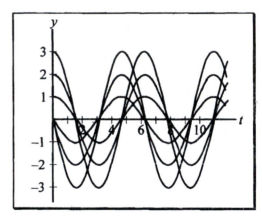

Intersecting solutions $y = y(t)$

■   **Euler's Method Numerics**

12.   (a)   The IVP studied in Example 5,

$$x'' + 0.1x = 0, \ x(0) = 1, \ x'(0) = 0,$$

can be solved numerically with a spreadsheet using the following coding:

	A	**B**	**C**	**D**	**E**
1	t	x	y	$\dfrac{dx}{dt}$	$\dfrac{dy}{dt}$
2	0	1	0	= C2	= –0.1 * B2
3	= A2 + 0.1	= B2 + 0.1 * D2	= C2 + 0.1 * E2	= C3	= –0.1 * B3

Row 3 can now be dragged down to produce the following values on $0 \le t \le 1$.

t	x	y	$\dfrac{dx}{dt}$	$\dfrac{dy}{dt}$
0.0	1.0000	0.0000	0.0000	–0.1000
0.1	1.0000	–0.0100	–0.0100	–0.1000
0.2	0.9990	–0.0200	–0.0200	–0.0999
0.3	0.9970	–0.0300	–0.0300	–0.0997
0.4	0.9940	–0.0400	–0.0400	–0.0994
0.5	0.9900	–0.0499	–0.0499	–0.0990
0.6	0.9850	–0.0598	–0.0598	–0.0985
0.7	0.9790	–0.0697	–0.0697	–0.0979
0.8	0.9721	–0.0794	–0.0794	–0.0972
0.9	0.9641	–0.0892	–0.0892	–0.0964
1.0	0.9552	–0.0988	–0.0988	–0.0955

continued on the next four pages

If the domain is continued to $t = 40$, then the graphs that correspond to Figure 6.1.1 look like the following.

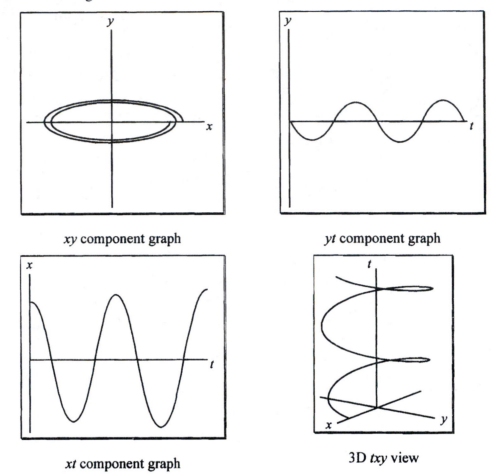

xy component graph   yt component graph

xt component graph   3D txy view

Note that the xy trajectory does not close as in Figure 6.1.1; with Euler's method, a smaller step size would do better.

(b)   The IVP studied in Example 6

$$x'' + 0.05x' + 0.1x = 0, \quad x(-5) = -0.1, \quad x'(-5) = 0.5,$$

can be solved numerically with a spreadsheet using the following coding:

	A	B	C	D	E
1	t	x	y	$\dfrac{dx}{dt}$	$\dfrac{dy}{dt}$
2	0	1	0	= C2	= –0.1 * B2 – 0.05 * C2
3	= A2 + 0.1	= B2 + 0.1 * D2	= C2 + 0.1 * E2	= C3	= –0.1 * B3 – 0.05 * C3

Row 3 can now be dragged down to produce the following values on $-5 \leq t \leq -4$.

t	x	y	$\dfrac{dx}{dt}$	$\dfrac{dy}{dt}$
−5.0	−0.1000	0.5000	0.5000	−0.0150
−4.9	−0.0500	0.4985	0.4985	−0.0199
−4.8	−0.0001	0.4965	0.4965	−0.0248
−4.7	0.0495	0.4940	0.4940	−0.0297
−4.6	0.0989	0.4911	0.4911	−0.0344
−4.5	0.1480	0.4876	0.4876	−0.0392
−4.4	0.1968	0.4837	0.4837	−0.0439
−4.3	0.2451	0.4793	0.4793	−0.0485
−4.2	0.2931	0.4745	0.4745	−0.0530
−4.1	0.3405	0.4692	0.4692	−0.0575
−4.0	0.3874	0.4634	0.4634	−0.0619

If the domain is continued to $t = 25$, then the graphs that correspond to Figure 6.1.2 look like the following.

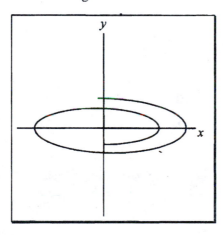

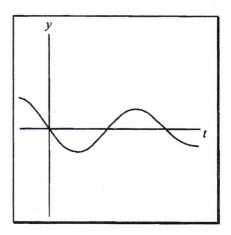

xy phase portrait   *yt* component graph

continued on the next page

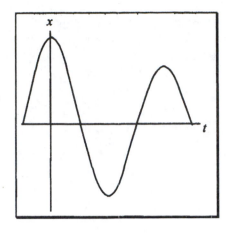

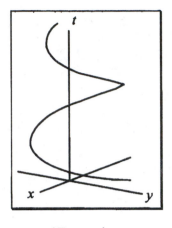

xt component graph   **3D txy view**

(c)   The system of Example 9,

$$
\dot{\mathbf{x}} = \begin{bmatrix} t & 0 & 0 & \sqrt{t^2+1} \\ 0 & 1 & 2 & -1 \\ 1 & -1 & 3 & 0 \\ 0 & 1 & 1 & t \end{bmatrix} \vec{x} + \begin{bmatrix} \cos t \\ \sin t \\ t^3 \\ e^{t^2} \end{bmatrix}, \quad \text{or} \quad \begin{aligned} \dot{\mathbf{x}}_1 &= tx_1 + \sqrt{t^2+1}\,x_4 + \cos t \\ \dot{\mathbf{x}}_2 &= x_2 + 2x_3 - x_4 + \sin t \\ \dot{\mathbf{x}}_3 &= x_1 - x_2 + 3x_3 + t^3 \\ \dot{\mathbf{x}}_4 &= x_2 + x_3 + tx_4 + e^{t^2} \end{aligned},
$$

can be solved numerically with a spreadsheet using the following coding. We choose stepsize $h = 0.01$, in hope of a good approximation, and enter the initial conditions in boldface. There are so many columns for a 4D-system that we have to break our display into two lines.

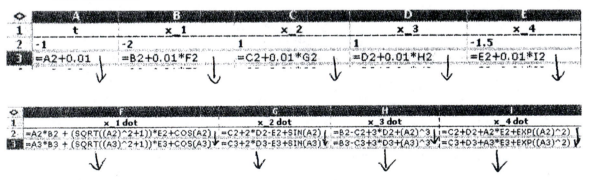

Dragging down Row 3 results in the following value for the first 100 steps, or $-1 \le t \le 0$.

	A	B	C	D	E	F	G	H	I
1	t	x_1	x_2	x_3	x_4	x_1 dot	x_2 dot	x_3 dot	x_4 dot
2	-1.00	-2.00	1.00	1.00	-1.50	0.42	3.66	-1.00	6.22
3	-0.99	-2.00	1.04	0.99	-1.44	0.50	3.62	-1.03	6.11
4	-0.98	-1.99	1.07	0.98	-1.38	0.58	3.58	-1.07	6.01
5	-0.97	-1.98	1.11	0.97	-1.32	0.66	3.54	-1.10	5.92
6	-0.96	-1.98	1.14	0.96	-1.26	0.73	3.50	-1.13	5.82
7	-0.95	-1.97	1.18	0.95	-1.20	0.80	3.46	-1.17	5.73
8	-0.94	-1.96	1.21	0.94	-1.14	0.87	3.42	-1.20	5.64
9	-0.93	-1.95	1.25	0.92	-1.09	0.93	3.38	-1.24	5.55
10	-0.92	-1.95	1.28	0.91	-1.03	1.00	3.34	-1.27	5.47
11	-0.91	-1.94	1.31	0.90	-0.98	1.06	3.30	-1.31	5.39
12	-0.90	-1.92	1.35	0.88	-0.92	1.11	3.26	-1.35	5.31
⋮	⋮	⋮	⋮	⋮	⋮	⋮	⋮	⋮	⋮
98	-0.04	-0.06	1.36	-3.07	1.48	2.48	-6.30	-10.62	-0.76
99	-0.03	-0.03	1.30	-3.17	1.47	2.47	-6.55	-10.86	-0.92
100	-0.02	-0.01	1.23	-3.28	1.46	2.46	-6.82	-11.09	-1.08
101	-0.01	0.02	1.16	-3.39	1.45	2.45	-7.09	-11.33	-1.24
102	0.00	0.04	1.09	-3.51	1.44	2.44	-7.36	-11.58	-1.41

Choosing to plot columns B, C, D, E as "lines" gives the following "chart" for component graphs, or time series.

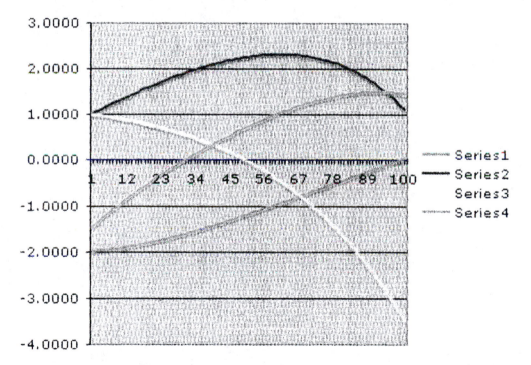

These curves agree with the graphs shown in Example 9, Figure 6.1.7, for the behaviors between $t_0 = -1$ and $t_f = 0$.

■   **Computer Check**

15.   Student Lab Project with IDE.

■   **The Wronskian**

When the Wronskian is not zero, the vectors are linearly independent and form a fundamental set.

(If the Wronskian of two solutions is nonzero on any interval it will always be nonzero on that interval.)

18. $W[\vec{x}_1, \vec{x}_2] = \begin{vmatrix} e^{3t} & 2e^{-t} \\ e^{3t} & -3e^{-t} \end{vmatrix} = -5e^{2t} \neq 0$, so the vectors form a fundamental set.

21. $W[\vec{x}_1, \vec{x}_2] = \begin{vmatrix} e^t \cos t & e^t \sin t \\ -e^t \sin t & e^t \cos t \end{vmatrix} = e^{2t}(\cos^2 t + \sin^2 t) = e^{2t} \neq 0$; the vectors form a fundamental set.

■   **Suggested Journal Entry II**

24.   Student Project

6.2   Linear Systems with Real Eigenvalues

■ Sketching Second-Order DEs

3. $x'' + x = 1$

(a)   Letting $y = x'$, we write the equation as the first-order system

$$x' = y$$
$$y' = -x + 1.$$

(b)   The equilibrium point is $(x, y) = (1, 0)$.

(c) h − nullcline $x = 1$
       v − nullcline $y = 0$

   (See figure.)

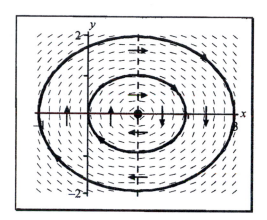

(d)   From the direction field, the equilibrium point $(x, y) = (1, 0)$ is stable.

(e)   A mass-spring system with this equation shows no damping and steady forcing; hence, periodic motion about an equilibrium to the right of the origin.

■ Matching Games

6.   C

■ Solutions in General

9. $\vec{x}' = \begin{bmatrix} -4 & 2 \\ 2 & -1 \end{bmatrix} \vec{x}$

The characteristic equation of the system is $p(\lambda) = \begin{vmatrix} -4 - \lambda & 2 \\ 2 & -1 - \lambda \end{vmatrix} = \lambda^2 + 5\lambda = 0$,

which has solutions $\lambda_1 = 0$, $\lambda_2 = -5$. Finding the eigenvectors corresponding to each eigenvalue yields

$$\lambda_1 = 0 \Rightarrow \vec{v}_1 = \begin{bmatrix} 1 \\ 2 \end{bmatrix}, \quad \lambda_2 = -5 \Rightarrow \vec{v}_2 = \begin{bmatrix} -2 \\ 1 \end{bmatrix}.$$

Hence, the general solution is $\vec{x}(t) = c_1 \begin{bmatrix} 1 \\ 2 \end{bmatrix} + c_2 e^{-5t} \begin{bmatrix} -2 \\ 1 \end{bmatrix}$.

12. $\vec{x}' = \begin{bmatrix} 10 & -5 \\ 8 & -12 \end{bmatrix} \vec{x}$

The characteristic equation of the system is

$$p(\lambda) = \begin{vmatrix} 10 - \lambda & -5 \\ 8 & -12 - \lambda \end{vmatrix} = \lambda^2 + 2\lambda - 80 = 0,$$

which has solutions $\lambda_1 = -10$, $\lambda_2 = 8$.

Finding the eigenvectors corresponding to each eigenvalue yields

$$\lambda_1 = -10 \Rightarrow \vec{v}_1 = \begin{bmatrix} 1 \\ 4 \end{bmatrix}, \lambda_2 = 8 \Rightarrow \vec{v}_2 = \begin{bmatrix} 5 \\ 2 \end{bmatrix}.$$

Hence, the general solution is $\vec{x}(t) = c_1 e^{-10t} \begin{bmatrix} 1 \\ 4 \end{bmatrix} + c_2 e^{8t} \begin{bmatrix} 5 \\ 2 \end{bmatrix}$.

15. $\vec{x}' = \begin{bmatrix} 1 & 0 \\ -2 & 2 \end{bmatrix} \vec{x}$

The characteristic equation of the system is $p(\lambda) = \begin{vmatrix} 1 - \lambda & 0 \\ -2 & 2 - \lambda \end{vmatrix} = (\lambda - 1)(\lambda - 2) = 0$,

which has solutions $\lambda_1 = 1$, $\lambda_2 = 2$. The eigenvectors corresponding to each eigenvalue are

$$\lambda_1 = 1 \Rightarrow \vec{v}_1 = \begin{bmatrix} 1 \\ 2 \end{bmatrix}, \lambda_2 = 2 \Rightarrow \vec{v}_2 = \begin{bmatrix} 0 \\ 1 \end{bmatrix}.$$

Hence, the general solution is $\vec{x}(t) = c_1 e^t \begin{bmatrix} 1 \\ 2 \end{bmatrix} + c_2 e^{2t} \begin{bmatrix} 0 \\ 1 \end{bmatrix}$.

18. $\vec{x}' = \begin{bmatrix} 4 & 3 \\ -4 & -4 \end{bmatrix} \vec{x}$

The characteristic equation of the system is $p(\lambda) = \begin{vmatrix} 4 - \lambda & 3 \\ -4 & -4 - \lambda \end{vmatrix} = \lambda^2 - 4 = 0$,

which has solutions $\lambda_1 = 2$, $\lambda_2 = -2$.

Finding the eigenvectors corresponding to each eigenvalue yields

$$\lambda_1 = 2 \Rightarrow \vec{v}_1 = \begin{bmatrix} -3 \\ 2 \end{bmatrix}, \lambda_2 = -2 \Rightarrow \vec{v}_2 = \begin{bmatrix} 1 \\ -2 \end{bmatrix}.$$

Hence, the general solution is $\vec{x}(t) = c_1 e^{2t} \begin{bmatrix} -3 \\ 2 \end{bmatrix} + c_2 e^{-2t} \begin{bmatrix} 1 \\ -2 \end{bmatrix}$.

21. $\vec{x}' = \begin{bmatrix} 4 & -3 \\ 8 & -6 \end{bmatrix} \vec{x}$

The characteristic equation of the system is $p(\lambda) = \begin{vmatrix} 4-\lambda & -3 \\ 8 & -6-\lambda \end{vmatrix} = \lambda^2 + 2\lambda = 0$,

which has solutions $\lambda_1 = 0$, $\lambda_2 = -2$.

The eigenvectors corresponding to each eigenvalue are

$$\lambda_1 = 0 \Rightarrow \vec{v}_1 = \begin{bmatrix} 3 \\ 4 \end{bmatrix}, \quad \lambda_2 = -2 \Rightarrow \vec{v}_2 = \begin{bmatrix} 1 \\ 2 \end{bmatrix}.$$

Hence, the general solution is $\vec{x}(t) = c_1 \begin{bmatrix} 3 \\ 4 \end{bmatrix} + c_2 e^{-2t} \begin{bmatrix} 1 \\ 2 \end{bmatrix}$.

■   **Repeated Eigenvalues**

24. $\vec{x}' = \begin{bmatrix} 3 & 2 \\ -8 & -5 \end{bmatrix} \vec{x}$

The characteristic equation of the system is $p(\lambda) = \begin{vmatrix} 3-\lambda & 2 \\ -8 & -5-\lambda \end{vmatrix} = \lambda^2 + 2\lambda + 1 = 0$,

which has solutions $\lambda = -1, \ -1$ with one linearly independent eigenvector, $\vec{v} = \begin{bmatrix} 1 \\ -2 \end{bmatrix}$.

The general solution is, therefore,

$$\vec{x}(t) = c_1 e^{-t} \begin{bmatrix} 1 \\ -2 \end{bmatrix} + c_2 \left\{ t e^{-t} \begin{bmatrix} 1 \\ -2 \end{bmatrix} + e^{-t} \begin{bmatrix} u_1 \\ u_2 \end{bmatrix} \right\}$$

where $\vec{u}$ is a *generalized eigenvector* satisfying $(\mathbf{A} - \lambda \mathbf{I}) \vec{u} = \vec{v}$, or

$$\begin{bmatrix} 4 & 2 \\ -8 & -4 \end{bmatrix} \begin{bmatrix} u_1 \\ u_2 \end{bmatrix} = \begin{bmatrix} 1 \\ -2 \end{bmatrix},$$

which has one linearly independent equation, $4u_1 + 2u_2 = 1$. Hence,

$$\vec{u} = \begin{bmatrix} u_1 \\ u_2 \end{bmatrix} = \begin{bmatrix} k \\ \dfrac{1}{2} - 2k \end{bmatrix} = \begin{bmatrix} 0 \\ \dfrac{1}{2} \end{bmatrix} + k \begin{bmatrix} 1 \\ -2 \end{bmatrix}$$

$$\vec{x}(t) = c_1 e^{-t} \begin{bmatrix} 1 \\ -2 \end{bmatrix} + c_2 \left\{ t e^{-t} \begin{bmatrix} 1 \\ -2 \end{bmatrix} + k e^{-t} \begin{bmatrix} 1 \\ -2 \end{bmatrix} + e^{-t} \begin{bmatrix} 0 \\ \dfrac{1}{2} \end{bmatrix} \right\}.$$

Because the term involving k is a multiple of the first term, we have

$$\vec{x}(t) = c \, e^{-t} \begin{bmatrix} 1 \\ -2 \end{bmatrix} + c_2 e^{-t} \left\{ t \begin{bmatrix} 1 \\ -2 \end{bmatrix} + \begin{bmatrix} 0 \\ \dfrac{1}{2} \end{bmatrix} \right\}.$$

■   **Solutions in Particular**

27. $\vec{x}' = \begin{bmatrix} 2 & 0 \\ 0 & 3 \end{bmatrix} \vec{x}, \ \vec{x}(0) = \begin{bmatrix} 5 \\ 4 \end{bmatrix}$

The characteristic equation of the system is $p(\lambda) = \begin{vmatrix} 2-\lambda & 0 \\ 0 & 3-\lambda \end{vmatrix} = (\lambda-2)(\lambda-3) = 0$,

which has the solutions $\lambda_1 = 2$ and $\lambda_2 = 3$.

Finding the eigenvectors corresponding to each eigenvalue yields

$$\lambda_1 = 2 \Rightarrow \vec{v}_1 = \begin{bmatrix} 1 \\ 0 \end{bmatrix}, \ \lambda_2 = 3 \Rightarrow \vec{v}_2 = \begin{bmatrix} 0 \\ 1 \end{bmatrix}.$$

The general solution is $\vec{x}(t) = c_1 e^{2t} \begin{bmatrix} 1 \\ 0 \end{bmatrix} + c_2 e^{3t} \begin{bmatrix} 0 \\ 1 \end{bmatrix}$.

Substituting the initial conditions $\vec{x}(0) = \begin{bmatrix} 5 \\ 4 \end{bmatrix}$ yields $c_1 = 5$ and $c_2 = 4$.

The solution of the IVP is $\vec{x}(t) = 5e^{2t} \begin{bmatrix} 1 \\ 0 \end{bmatrix} + 4e^{3t} \begin{bmatrix} 0 \\ 1 \end{bmatrix} = \begin{bmatrix} 5e^{2t} \\ 4e^{3t} \end{bmatrix}$.

30. $\vec{x}' = \begin{bmatrix} -3 & 2 \\ 1 & -2 \end{bmatrix} \vec{x}, \ \vec{x}(0) = \begin{bmatrix} -1 \\ 6 \end{bmatrix}$

The characteristic equation of the system is $p(\lambda) = \begin{vmatrix} -3-\lambda & 2 \\ 1 & -2-\lambda \end{vmatrix} = \lambda^2 + 5\lambda + 4 = 0$,

which has the solutions $\lambda_1 = -4$ and $\lambda_2 = -1$.

Finding the eigenvectors corresponding to each eigenvalue yields

$$\lambda_1 = -4 \Rightarrow \vec{v}_1 = \begin{bmatrix} -2 \\ 1 \end{bmatrix}, \ \lambda_2 = -1 \Rightarrow \vec{v}_2 = \begin{bmatrix} 1 \\ 1 \end{bmatrix}.$$

The general solution is $\vec{x}(t) = c_1 e^{-4t} \begin{bmatrix} -2 \\ 1 \end{bmatrix} + c_2 e^{-t} \begin{bmatrix} 1 \\ 1 \end{bmatrix}$.

Substituting the initial conditions $\vec{x}(0) = \begin{bmatrix} -1 \\ 6 \end{bmatrix}$ yields

$$-2c_1 + c_2 = -1$$
$$c_1 + c_2 = 6$$

which gives $c_1 = \dfrac{7}{3}, \ c_2 = \dfrac{11}{3}$. The solution of the IVP is $\vec{x}(t) = \left(\dfrac{7}{3}\right) e^{-4t} \begin{bmatrix} -2 \\ 1 \end{bmatrix} + \left(\dfrac{11}{3}\right) e^{-t} \begin{bmatrix} 1 \\ 1 \end{bmatrix}$.

33. $\vec{x}' = \begin{bmatrix} 1 & -1 \\ 2 & 4 \end{bmatrix} \vec{x}, \quad \vec{x}(0) = \begin{bmatrix} 1 \\ 0 \end{bmatrix}$

The characteristic equation of the system is $p(\lambda) = \begin{vmatrix} 1-\lambda & -1 \\ 2 & 4-\lambda \end{vmatrix} = \lambda^2 - 5\lambda + 6 = 0$,

which has the solutions $\lambda_1 = 2$ and $\lambda_2 = 3$.

Finding the eigenvectors corresponding to each eigenvalue yields

$$\lambda_1 = 2 \Rightarrow \vec{v}_1 = \begin{bmatrix} -1 \\ 1 \end{bmatrix}, \ \lambda_2 = 3 \Rightarrow \vec{v}_2 = \begin{bmatrix} 1 \\ -2 \end{bmatrix}.$$

The general solution is $\vec{x}(t) = c_1 e^{2t} \begin{bmatrix} -1 \\ 1 \end{bmatrix} + c_2 e^{3t} \begin{bmatrix} 1 \\ -2 \end{bmatrix}$.

Substituting the initial conditions $\vec{x}(0) = \begin{bmatrix} 1 \\ 0 \end{bmatrix}$ yields

$$-c_1 + c_2 = 1$$
$$c_1 - 2c_2 = 0$$

which gives $c_1 = -2$ and $c_2 = -1$. The solution of the IVP is $\vec{x}(t) = -2e^{2t} \begin{bmatrix} -1 \\ 1 \end{bmatrix} - e^{3t} \begin{bmatrix} 1 \\ -2 \end{bmatrix}$.

■   Repeated Eigenvalue Theory

36. $A = \begin{bmatrix} a & b \\ c & d \end{bmatrix}$

(a)   A has characteristic equation
$$\lambda^2 - (a+d)\lambda + (ad - bc) = 0.$$

$$\lambda = \frac{(a+d) \pm \sqrt{(a+d)^2 - 4(ad-bc)}}{2}$$

There is a double eigenvalue if and only if the discriminant

$$(a+d)^2 - 4(ad-bc) = 0 \ (a-d)^2 + 4bc = 0$$

in which case, $\lambda = \frac{1}{2}(a+d)$.

(b)   If $a = d$, then $bc = 0$ which implies that either b or $c = 0$, possibly both.

The double eigenvalue $\lambda = a = d$. To find the eigenvectors,

$$(\mathbf{A} - \lambda\mathbf{I})\bar{\mathbf{v}} = \begin{bmatrix} 0 & b \\ c & 0 \end{bmatrix}\begin{bmatrix} v_1 \\ v_2 \end{bmatrix} = \begin{bmatrix} 0 \\ 0 \end{bmatrix} \Rightarrow bv_2 = 0 \text{ and } cv_1 = 0.$$

Case 1: $b = 0, c \neq 0$ Then v_2 is free and $v_1 = 0$ and $\left\{ \begin{bmatrix} 0 \\ 1 \end{bmatrix} \right\}$ is a basis.

Case 2: $b \neq 0, c = 0$ Then v_1 is free and $v_2 = 0$ and $\left\{ \begin{bmatrix} 1 \\ 0 \end{bmatrix} \right\}$ is a basis.

Case 3: $b = c = 0$ Then both v_1 and v_2 are free and $\left\{ \begin{bmatrix} 1 \\ 0 \end{bmatrix}, \begin{bmatrix} 0 \\ 1 \end{bmatrix} \right\}$ is a basis.

Therefore the requirement that *both* the off-diagonal elements be zero, i.e., when $\lambda = a = d$, implies that $\mathbf{A}$ must be a multiple of the identity matrix.

(c)   Assume $a \neq d$: With zero discriminant, the double eigenvalue is now $\frac{1}{2}(a + d)$.

To find the eigenvectors, set

$$(\mathbf{A} - \lambda\mathbf{I})\bar{\mathbf{v}} = \begin{bmatrix} a - \frac{1}{2}(a+d) & b \\ c & d - \frac{1}{2}(a+d) \end{bmatrix}\begin{bmatrix} v_1 \\ v_2 \end{bmatrix} = \begin{bmatrix} 0 \\ 0 \end{bmatrix}.$$

The resulting system,

$$\frac{1}{2}(a - d)v_1 + bv_2 = 0$$

$$cv_1 + \frac{1}{2}(d - a)v_2 = 0$$

gives $v_1 = \frac{2b}{d - a}v_2$ so $\left\{ \begin{bmatrix} 2b \\ d - a \end{bmatrix} \right\}$ is a basis for the eigenspace.

An alternate basis vector is $\begin{bmatrix} d - a \\ 2c \end{bmatrix}$.

(d)   Assume $a \neq d$.

Because there is only one eigenvector for $\lambda = \frac{1}{2}(a+d)$, we need a generalized eigenvector $\bar{\mathbf{u}}$.

To find $\bar{\mathbf{u}}$, set $(\mathbf{A} - \lambda\mathbf{I})\bar{\mathbf{u}} = \bar{\mathbf{v}}$

$$\left.\begin{array}{c} \frac{1}{2}(a-d)u_1 + bu_2 = 2b \\[2mm] cu_1 + \frac{1}{2}(d-a)u_2 = d-a \end{array}\right\} \quad \Rightarrow \quad u_1 = 0,\ u_2 = 2.$$

Therefore $\bar{\mathbf{u}} = \begin{bmatrix} 0 \\ 2 \end{bmatrix}$ is a generalized eigenvector.

The general solution of the system is $\bar{\mathbf{x}}(t) = c_1 e^{\lambda t}\begin{bmatrix} 2b \\ d-a \end{bmatrix} + c_2 e^{\lambda t}\left(t\begin{bmatrix} 2b \\ d-a \end{bmatrix} + \begin{bmatrix} 0 \\ 2 \end{bmatrix}\right)$.

■ One Independent Eigenvector

39. $\mathbf{A} = \begin{bmatrix} 0 & 0 & 1 \\ 1 & 0 & -3 \\ 0 & 1 & 3 \end{bmatrix}$

(a)   The eigenvalue is $\lambda = 1$, with an algebraic multiplicity of 3. We find the eigenvector(s) by substituting $\lambda = 1$ into the equation $\mathbf{A}\bar{\mathbf{v}} = \lambda\bar{\mathbf{v}}$ and solving for the vector $\bar{\mathbf{v}}$. Doing this yields the single eigenvector $c[1,\ -2,\ 1]$.

(b)   From the eigenvalue and eigenvector, one solution has been found:

$$\bar{\mathbf{x}}_1(t) = ce^t\begin{bmatrix} 1 \\ -2 \\ 1 \end{bmatrix}.$$

(c)   Now we solve for a second solution of the form $\bar{\mathbf{x}}_2(t) = te^t\bar{\mathbf{v}} + e^t\bar{\mathbf{u}}$, where $\bar{\mathbf{v}} = [1,\ -2,\ 1]$ is the first eigenvector, and $\bar{\mathbf{u}} = [u_1,\ u_2,\ u_3]$ is an unknown vector. Substituting $\bar{\mathbf{x}}_2(t)$ into the system $\bar{\mathbf{x}}' = \mathbf{A}\bar{\mathbf{x}}$ and comparing coefficients of te^t and e^t yields equations for u_1, u_2, u_3, giving $u_1 = -1$, $u_2 = 1$, $u_3 = 0$. Hence, we obtain as a second solution

$$\bar{\mathbf{x}}_2(t) = te^t\begin{bmatrix} 1 \\ -2 \\ 1 \end{bmatrix} + e^t\begin{bmatrix} -1 \\ 1 \\ 0 \end{bmatrix}.$$

(d)   To find a third (linearly independent) solution, we try the specific form

$$\vec{x}_3(t) = \frac{1}{2}t^2 e^t \vec{v} + t e^t \vec{u} + e^t \vec{w}$$

where $\vec{v}$ and $\vec{u}$ are vectors previously found and $\vec{w}$ is the unknown vector. Substituting $\vec{x}_3(t)$ into the system results in the system of equations $(\mathbf{A} - \mathbf{I})\vec{w} = \vec{u}$. We then find $\vec{w} = [w_1, w_2, w_3]$. Solving this system yields $w_1 = 1$, $w_2 = 0$, $w_3 = 0$. Hence, we obtain as a third solution

$$\vec{x}_3(t) = \frac{1}{2}t^2 e^t \begin{bmatrix} 1 \\ -2 \\ 1 \end{bmatrix} + t e^t \begin{bmatrix} -1 \\ 1 \\ 0 \end{bmatrix} + e^t \begin{bmatrix} 1 \\ 0 \\ 0 \end{bmatrix}.$$

■   Spatial Particulars

42. $\vec{x}' = \begin{bmatrix} 1 & -1 & 0 \\ 0 & -1 & 3 \\ -1 & 1 & 0 \end{bmatrix} \vec{x}, \quad \vec{x}(0) = \begin{bmatrix} 0 \\ 0 \\ 1 \end{bmatrix}$

We find the eigenvalues and eigenvectors of the coefficient matrix by the usual procedure, obtaining

$$\lambda_1 = 0 \Rightarrow \vec{v}_1 = \begin{bmatrix} 3 \\ 3 \\ 1 \end{bmatrix}, \quad \lambda_2 = 2 \Rightarrow \vec{v}_2 = \begin{bmatrix} -1 \\ 1 \\ 1 \end{bmatrix}, \quad \lambda_3 = -2 \Rightarrow \vec{v}_3 = \begin{bmatrix} -1 \\ -3 \\ 1 \end{bmatrix}.$$

Hence, the general solution is $\vec{x}(t) = c_1 \begin{bmatrix} 3 \\ 3 \\ 1 \end{bmatrix} + c_2 e^{2t} \begin{bmatrix} -1 \\ 1 \\ 1 \end{bmatrix} + c_3 e^{-2t} \begin{bmatrix} -1 \\ -3 \\ 1 \end{bmatrix}.$

Substituting this vector into the initial condition $\vec{x}(0) = [0, 0, 1]$ yields the three equations

$$\begin{aligned} 3c_1 & - c_2 & - & c_3 & = & 0 \\ 3c_1 & + c_2 & - & 3c_3 & = & 0 \\ c_1 & + c_2 & + & c_3 & = & 0 \end{aligned}$$

with the solution $c_1 = \frac{1}{4}$, $c_2 = \frac{3}{8}$, $c_3 = \frac{3}{8}$.

Hence, the IVP has the solution $\vec{x}(t) = \frac{1}{4}\begin{bmatrix} 3 \\ 3 \\ 1 \end{bmatrix} + \frac{3}{8}e^{2t}\begin{bmatrix} -1 \\ 1 \\ 1 \end{bmatrix} + \frac{3}{8}e^{-2t}\begin{bmatrix} -1 \\ -3 \\ 1 \end{bmatrix}.$

■   **Adjoint Systems**

45. $\vec{x}' = \begin{bmatrix} 0 & 1 \\ 1 & 0 \end{bmatrix} \vec{x}$

(a)   The negative transpose of the given matrix is simply the matrix with –1s in the place of 1s, hence the adjoint system is

$$\vec{w}' = -\mathbf{A}^T \vec{w} = \begin{bmatrix} 0 & -1 \\ -1 & 0 \end{bmatrix} \vec{w}.$$

(b)   The first equality is simply the product rule for matrix derivatives. Using the adjoint system, yields

$$\vec{w}'^T = \left(-\mathbf{A}^T \vec{w}\right)^T = -\vec{w}^T \mathbf{A},$$

and hence,

$$\vec{w}'^T \vec{x} + \vec{w}^T \vec{x}' = -\vec{w}^T \vec{x}' + \vec{w}^T \vec{x}' = 0.$$

(c)   The characteristic equation of the matrix is simply $\lambda^2 - 1 = 0$, and hence, the eigenvalues are +1, –1. The eigenvector corresponding to +1 can easily be found and is [1, 1]. Likewise, the eigenvector for –1 is [1, -1]. Hence,

$$\vec{x}(t) = c_1 e^t \begin{bmatrix} 1 \\ 1 \end{bmatrix} + c_2 e^{-t} \begin{bmatrix} 1 \\ -1 \end{bmatrix}.$$

Substituting in the initial condition $\vec{x}(0)$ = [1, 0], yields $c_1 = c_2 = \dfrac{1}{2}$.

The IVP solution is $\vec{x}(t) = \left(\dfrac{1}{2}\right) e^t \begin{bmatrix} 1 \\ 1 \end{bmatrix} + \left(\dfrac{1}{2}\right) e^{-t} \begin{bmatrix} 1 \\ -1 \end{bmatrix} = \left(\dfrac{1}{2}\right) \begin{bmatrix} e^t + e^{-t} \\ e^t - e^{-t} \end{bmatrix}.$

(d)   In the adjoint system the eigenvalues are also 1 and –1, but the eigenvectors are reversed, [1,-1] and [1,1], respectively. Hence

$$\vec{w}(t) = k_1 e^t \begin{bmatrix} 1 \\ -1 \end{bmatrix} + k_2 e^{-t} \begin{bmatrix} 1 \\ 1 \end{bmatrix}.$$

If the initial conditions are $\vec{w}(0)$ = [0, 1], then

$k_1 = \dfrac{1}{2}$, $k_2 = -\dfrac{1}{2}$. So the solution of this IVP is

$$\vec{w}(t) = \dfrac{1}{2} \begin{bmatrix} -e^t + e^{-t} \\ e^t + e^{-t} \end{bmatrix}.$$

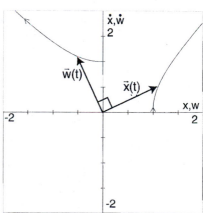

(e)   Part (b) shows that the vectors $\vec{w}(t)$ and $\vec{x}(t)$ have a constant dot product.

Note that the initial conditions $\vec{x}(0) = [1, 0]$ and $\vec{w}(0) = [0, 1]$ are orthogonal vectors, so this constant is zero. Hence the two resulting trajectories will be orthogonal for all $t > 0$. As trajectories evolve, the vector $\vec{w}(t)$ for the adjoint system is always orthogonal to the vector $\vec{x}(t)$ for the original system, as shown for a typical t value.

■   Computer Labs: Predicting Phase Portraits

For each of the linear systems (47–50) a few trajectories in the phase plane have been drawn. The analytic solutions are then computed.

48. $x' = 0$, $y' = -y$

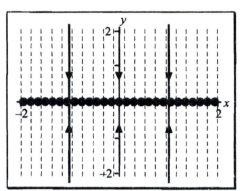

Solve each of these equations individually, obtaining

$$x = c_1 \text{ and } y = c_2 e^{-t}.$$

Eliminating t yields the trajectories

$$x = c,$$

which is the family of vertical lines. For any starting point (x_0, y_0) the solution moves asymptotically towards $(x_0, 0)$. The x-axis is composed entirely of stable equilibrium points.

■   Radioactive Decay Chain

51.   (a)   The amount of iodine is decreasing via radioactive decay; hence, $\dfrac{dI}{dt} = -k_1 I$, where k_1 is

the decay constant of iodine. Work in Chapter 2 showed that the decay constant is $\ln 2$ divided by the half-life of the material; hence,

$$k_1 = \frac{\ln 2}{6.7} \approx 0.1034548.$$

The amount of xenon $x(t)$ is increasing with the decay of iodine, but decreasing with its own radioactive decay, hence, the equation

$$\frac{dx}{dt} = k_1 I - k_2 x, \quad \text{where} \quad k_2 = \frac{\ln 2}{9.2} \approx 0.0753421.$$

(b)   In matrix form, the equations become

$$\begin{bmatrix} I' \\ x' \end{bmatrix} = \begin{bmatrix} -k_1 & 0 \\ k_1 & -k_2 \end{bmatrix} \begin{bmatrix} I \\ x \end{bmatrix}.$$

The eigenvalues of this triangular matrix can easily be seen and their eigenvectors calculated as

$$\lambda_1 = -k_1, \quad \vec{v}_1 = \begin{bmatrix} k_2 - k_1 \\ k_1 \end{bmatrix}; \quad \lambda_2 = -k_2, \quad \vec{v}_2 = \begin{bmatrix} 0 \\ 1 \end{bmatrix}.$$

Hence, the solution is $\vec{x}(t) = c_1 e^{-k_1 t} \begin{bmatrix} k_2 - k_1 \\ k_1 \end{bmatrix} + c_2 e^{-k_2 t} \begin{bmatrix} 0 \\ 1 \end{bmatrix}.$

■   **Mixing and Homogeneity**

54.   Instead of pouring pure water into Tank A, pour in a brine solution of $\dfrac{1}{2}$ lb/gal. Then the equations would be

$$x_1' = 2 - \frac{6x_1}{100} + \frac{2x_2}{100}, \quad x_2' = \frac{6x_1}{100} - \frac{6x_2}{100}, \quad \vec{x}' = \begin{bmatrix} -\dfrac{6}{100} & \dfrac{2}{100} \\ \dfrac{6}{100} & -\dfrac{6}{100} \end{bmatrix} \vec{x} + \begin{bmatrix} 2 \\ 0 \end{bmatrix}.$$

■   **Electrical Circuits**

57. $R_1 = 4$ ohms, $R_3 = 6$ ohms, $L_1 = 1$ henry, $L_2 = 2$ henries.

Using the fact that $I_3 = I_1 - I_2$, we obtain from Kirchoff's 2nd Law

(Loop 1) $I_1' + 4I_1 + 6(I_1 - I_2) = 0$

(Loop 2) $2I_2' - 6(I_1 - I_2) = 0$

so we have

$$\begin{matrix} I_1' = -10I_1 + 6I_2 \\ I_2' = 3I_1 - 3I_2 \end{matrix} \quad \text{or} \quad \begin{bmatrix} I_1 \\ I_2 \end{bmatrix}' = \begin{bmatrix} -10 & 6 \\ 3 & -3 \end{bmatrix} \begin{bmatrix} I_1 \\ I_2 \end{bmatrix}.$$

The eigenvalues and eigenvectors are

$$\lambda_1 = -1, \quad \vec{v}_1 = \begin{bmatrix} 2 \\ 3 \end{bmatrix}; \quad \lambda_2 = -12, \quad \vec{v}_2 = \begin{bmatrix} 3 \\ -1 \end{bmatrix}.$$

The general solution of our system is $\begin{bmatrix} I_1 \\ I_2 \end{bmatrix} = c_1 e^{-t} \begin{bmatrix} 2 \\ 3 \end{bmatrix} + c_2 e^{-12t} \begin{bmatrix} 3 \\ -1 \end{bmatrix},$ so

$$I_1(t) = 2c_1 e^{-t} + 3c_2 e^{-12t}$$

$$I_2(t) = 3c_1 e^{-t} - c_2 e^{-12t}$$

$$I_3(t) = -c_1 e^{-t} + 4c_2 e^{-12t}$$

6.3   Linear Systems with Nonreal Eigenvalues

> *For all problems in 6.3, $\lambda = \alpha \pm \beta i$ and $\vec{\mathbf{v}} = \vec{\mathbf{p}} \pm \vec{\mathbf{q}} i$.*

■   **Solutions in General**

3.
$$\vec{\mathbf{x}}' = \begin{bmatrix} 1 & 2 \\ -2 & 1 \end{bmatrix} \vec{\mathbf{x}}$$

The characteristic equation for the matrix is $\lambda^2 - 2\lambda + 5 = 0$, which yields complex eigenvalues $\lambda = 1 \pm 2i$. Substituting these values into $\mathbf{A}\vec{\mathbf{v}} = \lambda\vec{\mathbf{v}}$, respective eigenvectors $\vec{\mathbf{v}} = \begin{bmatrix} 1, & \pm i \end{bmatrix}$.

Therefore,

$$\alpha = 1, \ \beta = 2, \ \vec{\mathbf{p}} = \begin{bmatrix} 1, & 0 \end{bmatrix}, \ \vec{\mathbf{q}} = \begin{bmatrix} 0, & 1 \end{bmatrix}.$$

Two linearly independent solutions result:

$$\vec{\mathbf{x}}_1(t) = e^{\alpha t} \cos \beta t \, \vec{\mathbf{p}} - e^{\alpha t} \sin \beta t \, \vec{\mathbf{q}} = e^t \cos 2t \begin{bmatrix} 1 \\ 0 \end{bmatrix} - e^t \sin 2t \begin{bmatrix} 0 \\ 1 \end{bmatrix},$$

$$\vec{\mathbf{x}}_2(t) = e^{\alpha t} \sin \beta t \, \vec{\mathbf{p}} + e^{\alpha t} \cos \beta t \, \vec{\mathbf{q}} = e^t \sin 2t \begin{bmatrix} 1 \\ 0 \end{bmatrix} + e^t \cos 2t \begin{bmatrix} 0 \\ 1 \end{bmatrix}.$$

The general solution is $\vec{\mathbf{x}}(t) = c_1 \vec{\mathbf{x}}_1(t) + c_2 \vec{\mathbf{x}}_2(t) = c_1 e^t \begin{bmatrix} \cos 2t \\ -\sin 2t \end{bmatrix} + c_2 e^t \begin{bmatrix} \sin 2t \\ \cos 2t \end{bmatrix}.$

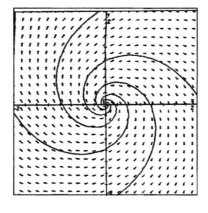

6. $\vec{\mathbf{x}}' = \begin{bmatrix} 2 & -4 \\ 2 & -2 \end{bmatrix} \vec{\mathbf{x}}$

The eigenvalues are $\lambda = \pm 2i$ with corresponding eigenvectors $\vec{\mathbf{v}} = [1 \pm i,\ 1]$. Therefore,

$$\alpha = 0,\ \beta = 2,\ \vec{\mathbf{p}} = [1,\ 1],\ \vec{\mathbf{q}} = [1,\ 0].$$

Two linearly independent solutions result:

$$\vec{\mathbf{x}}_1(t) = e^{\alpha t} \cos \beta t \vec{\mathbf{p}} - e^{\alpha t} \sin \beta t \vec{\mathbf{q}} = \cos 2t \begin{bmatrix} 1 \\ 1 \end{bmatrix} - \sin 2t \begin{bmatrix} 1 \\ 0 \end{bmatrix}$$

$$\vec{\mathbf{x}}_2(t) = e^{\alpha t} \sin \beta t \vec{\mathbf{p}} + e^{\alpha t} \cos \beta t \vec{\mathbf{q}} = \sin 2t \begin{bmatrix} 1 \\ 1 \end{bmatrix} + \cos 2t \begin{bmatrix} 1 \\ 0 \end{bmatrix}.$$

The general solution is

$$\vec{\mathbf{x}}(t) = c_1 \vec{\mathbf{x}}_1(t) + c_2 \vec{\mathbf{x}}_2(t) = c_1 \begin{bmatrix} \cos 2t - \sin 2t \\ \cos 2t \end{bmatrix} + c_2 \begin{bmatrix} \cos 2t + \sin 2t \\ \sin 2t \end{bmatrix}.$$

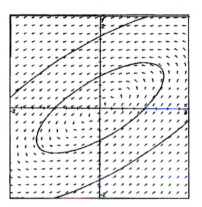

9. $\vec{\mathbf{x}}' = \begin{bmatrix} 1 & -1 \\ 5 & -3 \end{bmatrix} \vec{\mathbf{x}}$

The eigenvalues are $\lambda = -1 \pm i$, with complex eigenvectors $\vec{\mathbf{v}} = [2 \pm i,\ 5]$. Therefore,

$$\alpha = -1,\ \beta = 1,\ \vec{\mathbf{p}} = [2,\ 5],\ \vec{\mathbf{q}} = [1,\ 0].$$

Two linearly independent solutions result:

$$\vec{\mathbf{x}}_1(t) = e^{\alpha t} \cos \beta t \vec{\mathbf{p}} - e^{\alpha t} \sin \beta t \vec{\mathbf{q}} = e^{-t} \cos t \begin{bmatrix} 2 \\ 5 \end{bmatrix} - e^{-t} \sin t \begin{bmatrix} 1 \\ 0 \end{bmatrix},$$

$$\vec{\mathbf{x}}_2(t) = e^{\alpha t} \sin \beta t \vec{\mathbf{p}} + e^{\alpha t} \cos \beta t \vec{\mathbf{q}} = e^{-t} \sin t \begin{bmatrix} 2 \\ 5 \end{bmatrix} + e^{-t} \cos t \begin{bmatrix} 1 \\ 0 \end{bmatrix}.$$

The general solution is

$$\vec{\mathbf{x}}(t) = c_1 \vec{\mathbf{x}}_1(t) + c_2 \vec{\mathbf{x}}_2(t) = c_1 e^{-t} \begin{bmatrix} 2\cos t - \sin t \\ 5\cos t \end{bmatrix} + c_2 e^{-t} \begin{bmatrix} \cos t + 2\sin t \\ 5\sin t \end{bmatrix}.$$

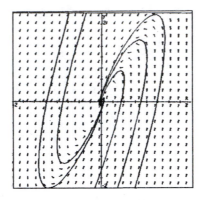

12. $\vec{\mathbf{x}}' = \begin{bmatrix} 2 & 4 \\ -2 & -2 \end{bmatrix} \vec{\mathbf{x}}$ The eigenvalues are $\lambda = \pm 2i$, with complex eigenvectors $\vec{\mathbf{v}} = [2, \ -1 \pm i]$.

Therefore,

$$\alpha = 0, \ \beta = 2, \ \vec{\mathbf{p}} = [2, \ -1], \ \vec{\mathbf{q}} = [0, \ 1].$$

Two linearly independent solutions result:

$$\vec{\mathbf{x}}_1(t) = e^{\alpha t} \cos \beta t \, \vec{\mathbf{p}} - e^{\alpha t} \sin \beta t \, \vec{\mathbf{q}} = \cos 2t \begin{bmatrix} 2 \\ -1 \end{bmatrix} - \sin 2t \begin{bmatrix} 0 \\ 1 \end{bmatrix},$$

$$\vec{\mathbf{x}}_2(t) = e^{\alpha t} \sin \beta t \, \vec{\mathbf{p}} + e^{\alpha t} \cos \beta t \, \vec{\mathbf{q}} = \sin 2t \begin{bmatrix} 2 \\ -1 \end{bmatrix} + \cos 2t \begin{bmatrix} 0 \\ 1 \end{bmatrix}.$$

The general solution is $\vec{\mathbf{x}}(t) = c_1 \vec{\mathbf{x}}_1(t) + c_2 \vec{\mathbf{x}}_2(t) = c_1 \begin{bmatrix} 2\cos 2t \\ -\cos 2t - \sin 2t \end{bmatrix} + c_2 \begin{bmatrix} 2\sin 2t \\ -\sin 2t + \cos 2t \end{bmatrix}.$

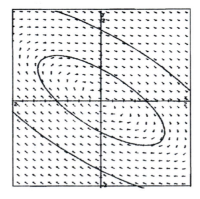

■   **Solutions in Particular**

15. $\vec{\mathbf{x}}' = \begin{bmatrix} -3 & 2 \\ -1 & -1 \end{bmatrix} \vec{\mathbf{x}}$, $\vec{\mathbf{x}}(0) = \begin{bmatrix} 1 \\ 1 \end{bmatrix}$

The coefficient matrix has eigenvalues $\lambda = -2 \pm i$, and corresponding eigenvectors $\vec{\mathbf{v}} = [1 \mp i, \ 1]$.

Hence, two linearly independent solutions are

$$\vec{\mathbf{x}}_1(t) = e^{\alpha t} \cos \beta t \, \vec{\mathbf{p}} - e^{\alpha t} \sin \beta t \, \vec{\mathbf{q}} = e^{-2t} \cos t \begin{bmatrix} 1 \\ 1 \end{bmatrix} - e^{-2t} \sin t \begin{bmatrix} -1 \\ 0 \end{bmatrix},$$

$$\vec{\mathbf{x}}_2(t) = e^{\alpha t} \sin \beta t \, \vec{\mathbf{p}} + e^{\alpha t} \cos \beta t \, \vec{\mathbf{q}} = e^{-2t} \sin t \begin{bmatrix} 1 \\ 1 \end{bmatrix} + e^{-2t} \cos 2t \begin{bmatrix} -1 \\ 0 \end{bmatrix}.$$

Substituting the initial conditions into $\vec{\mathbf{x}}(t) = c_1 \vec{\mathbf{x}}_1(t) + c_2 \vec{\mathbf{x}}_2(t)$:

$$\vec{\mathbf{x}}(0) = c_1 \vec{\mathbf{x}}_1(0) + c_2 \vec{\mathbf{x}}_2(0) = c_1 \begin{bmatrix} 1 \\ 1 \end{bmatrix} + c_2 \begin{bmatrix} -1 \\ 0 \end{bmatrix} = \begin{bmatrix} 1 \\ 1 \end{bmatrix}$$

yields $c_1 = 1$ and $c_2 = 0$.

The solution is, therefore, $\vec{\mathbf{x}}(t) = \vec{\mathbf{x}}_1(t) = e^{-2t} \begin{bmatrix} \cos t + \sin t \\ \cos t \end{bmatrix}$.

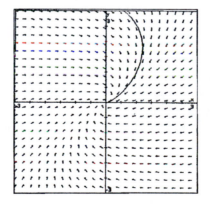

■   **Rotation Direction**

18.   For $\vec{\mathbf{x}} = \mathbf{A}\vec{\mathbf{x}}$ with $\mathbf{A} = \begin{bmatrix} a & b \\ c & d \end{bmatrix}$ and nonreal eigenvalues, the off-diagonal elements b and c must

be nonzero and of opposite sign. (See Problem 17(a).) We also know that nonreal eigenvalues give solutions with a rotation factor (see text equation (13)), so it will be sufficient to have a qualitative look at the vector field, determined by

$$x' = ax + by$$
$$y' = cx + dy,$$

for some sample points.

For example, if b is negative and c is positive,

- Along the positive y-axis (where $x = 0$), x' points *left*, not right (regardless of whether y' points up or down).

- Along the positive x-axis (where $y = 0$), y' points *up*, not down (regardless of whether x' points right or left).

Some sample possible phase-plane vectors are drawn in the first figure, and they show that the rotation is *counterclockwise*.

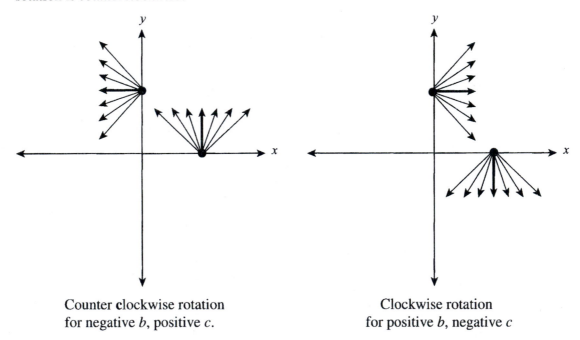

Counter clockwise rotation   Clockwise rotation
for negative b, positive c.   for positive b, negative c

By similar reasoning, if b is positive and c is negative, rotation is *clockwise*, as shown by second figure.

■   **"Boxing" the Ellipse**

From Problem 20, $\vec{x}' = \begin{bmatrix} a & b \\ c & d \end{bmatrix} \vec{x}, \quad \lambda = \alpha \pm \beta i, \quad \vec{v} = \begin{bmatrix} -b \\ a-\lambda \end{bmatrix}.$

21. $\vec{x}' = \begin{bmatrix} 4 & -5 \\ 5 & -4 \end{bmatrix} \vec{x}$ $\lambda = \pm 3i, \quad \vec{v} = \begin{bmatrix} 5 \\ 4 \mp 3i \end{bmatrix}.$

Therefore, $\alpha = 0$, $\beta = 3$, $\vec{p} = [5, 4]$, $\vec{q} = [0, -3]$.

See graph.

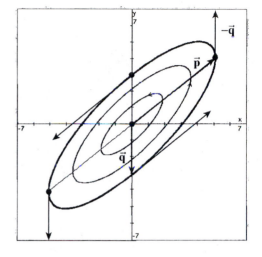

24. $\vec{x}' = \begin{bmatrix} -1 & 1 \\ -2 & 1 \end{bmatrix} \vec{x}$ $\lambda = \pm i, \quad \vec{v} = \begin{bmatrix} -1 \\ -1 \mp i \end{bmatrix}.$

Therefore, $\alpha = 0$, $\beta = 1$, $\vec{p} = [-1, -1]$, $\vec{q} = [0, -1]$.

See graph.

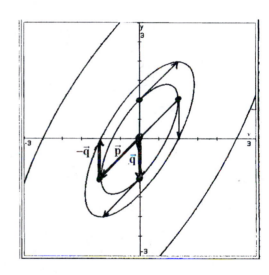

■ **Axes for Ellipses**

27. $\vec{x}' = \begin{bmatrix} -1 & -1 \\ 5 & 1 \end{bmatrix} \vec{x}$ $\qquad \lambda = \pm 2i, \ \vec{v} = \begin{bmatrix} 1 \\ -1 \mp 2i \end{bmatrix}.$

Therefore, $\alpha = 0, \beta = 2, \vec{p} = [1, -1], \vec{q} = [0, -2].$

(a) $\tan 2\beta t^* = \dfrac{2\vec{p} \cdot \vec{q}}{\|\vec{q}\|^2 - \|\vec{p}\|^2} = \dfrac{2(2)}{4-2} = 2.$

$2\beta t^* = \tan^{-1}(2) \approx 1.11$ radians or $1.11 + \pi = 4.25$ radians.

Thus, the parameter $\beta t^* = 0.55$ radians or 2.12 radians.

(b) For an endpoint of *one* axis of the ellipse, the value $\beta t^* = .55$ gives coordinates

$$\cos .55 \begin{bmatrix} 1 \\ -1 \end{bmatrix} - \sin .55 \begin{bmatrix} 0 \\ -2 \end{bmatrix} \approx \begin{bmatrix} .85 \\ -.85 + 1.04 \end{bmatrix} = \begin{bmatrix} .85 \\ .19 \end{bmatrix}.$$

For an endpoint of the *other* axis of the ellipse, the value $\beta t^* = 2.12$ gives coordinates

$$\cos 2.12 \begin{bmatrix} 1 \\ -1 \end{bmatrix} - \sin 2.12 \begin{bmatrix} 0 \\ -2 \end{bmatrix} \approx \begin{bmatrix} -.52 \\ +.52 + 1.70 \end{bmatrix} \approx \begin{bmatrix} -.52 \\ 2.22 \end{bmatrix}$$

(c) See figure.

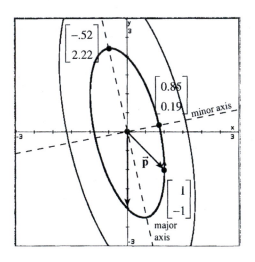

■ **3×3 System**

30. $\vec{x}' = \begin{bmatrix} -1 & 0 & 0 \\ 0 & 0 & 2 \\ 0 & -2 & 0 \end{bmatrix} \vec{x}$

(a) The characteristic equation, $\begin{vmatrix} -1-\lambda & 0 & 0 \\ 0 & -\lambda & 2 \\ 0 & -2 & -\lambda \end{vmatrix} = -(\lambda+1)(\lambda^2 + 4) = 0,$

has roots $\lambda_1 = -1, \lambda_2, \lambda_3 = \pm 2i.$

(b)   For $\lambda_1 = -1$, solving for x, y, z in the equation

$$\begin{bmatrix} -1 & 0 & 0 \\ 0 & 0 & 2 \\ 0 & -2 & 0 \end{bmatrix} \begin{bmatrix} x \\ y \\ z \end{bmatrix} = -1 \begin{bmatrix} x \\ y \\ z \end{bmatrix}$$

yields $x = \alpha$, $y = 0$, $z = 0$, α arbitrary, so the corresponding eigenvector is $\vec{x}_1 = e^{-t} \begin{bmatrix} 1 \\ 0 \\ 0 \end{bmatrix}$.

(c)   For $\lambda_2 = 2i$, solving for x, y, z in the system

$$\begin{bmatrix} -1 & 0 & 0 \\ 0 & 0 & 2 \\ 0 & -2 & 0 \end{bmatrix} \begin{bmatrix} x \\ y \\ z \end{bmatrix} = 2i \begin{bmatrix} x \\ y \\ z \end{bmatrix} \Rightarrow \vec{v}_2 = \begin{bmatrix} 0 \\ 1 \\ i \end{bmatrix}$$

Hence, we identify

$$\alpha = 0, \ \beta = 2, \ \vec{p} = [0, \ 1, \ 0], \ \vec{q} = [0, \ 0, \ 1].$$

Using complex conjugates λ_2, λ_3 and $\vec{v}_2, \vec{v}_3$, two linearly independent solutions are

$$\vec{x}_2(t) = e^{\alpha t} \cos \beta t \vec{p} - e^{\alpha t} \sin \beta t \vec{q} = \cos 2t \begin{bmatrix} 0 \\ 1 \\ 0 \end{bmatrix} - \sin 2t \begin{bmatrix} 0 \\ 0 \\ 1 \end{bmatrix} = \begin{bmatrix} 0 \\ \cos 2t \\ -\sin 2t \end{bmatrix}$$

$$\vec{x}_3(t) = e^{\alpha t} \sin \beta t \vec{p} + e^{\alpha t} \cos \beta t \vec{q} = \sin 2t \begin{bmatrix} 0 \\ 1 \\ 0 \end{bmatrix} + \cos 2t \begin{bmatrix} 0 \\ 0 \\ 1 \end{bmatrix} = \begin{bmatrix} 0 \\ \sin 2t \\ \cos 2t \end{bmatrix}.$$

(d)   The general solution, from (b) and (c), is

$$\vec{x}(t) = c_1 e^{-t} \begin{bmatrix} 1 \\ 0 \\ 0 \end{bmatrix} + c_2 \begin{bmatrix} 0 \\ \cos 2t \\ -\sin 2t \end{bmatrix} + c_3 \begin{bmatrix} 0 \\ \sin 2t \\ \cos 2t \end{bmatrix},$$

or,

$$\begin{aligned} x(t) &= c_1 e^{-t} \\ y(t) &= c_2 \cos 2t + c_3 \sin 2t \\ z(t) &= c_3 \cos 2t - c_2 \sin 2t. \end{aligned}$$

(e)   Substituting the IC:

$$\vec{x}(0)=c_1\vec{x}_1(0)+c_2\vec{x}_2(0)+c_3\vec{x}_3(0)=c_1\begin{bmatrix}1\\0\\0\end{bmatrix}+c_2\begin{bmatrix}0\\1\\0\end{bmatrix}+c_3\begin{bmatrix}0\\0\\1\end{bmatrix}=\begin{bmatrix}1\\0\\1\end{bmatrix}\Rightarrow\begin{matrix}c_1=1\\c_2=0\\c_3-1\end{matrix}.$$

The solution of the IVP is, therefore,

$$\vec{x}(t)=\vec{x}_1(t)+\vec{x}_3(t)=e^{-t}\begin{bmatrix}1\\0\\0\end{bmatrix}+\begin{bmatrix}0\\\sin 2t\\\cos 2t\end{bmatrix},\ \text{or, in coordinate form,}\ \begin{matrix}x(t)=e^{-t}\\y(t)=\sin 2t\\z(t)=\cos 2t.\end{matrix}$$

(f)   The trajectory of

$$\left(x(t),\ y(t),\ z(t)\right)$$

in 3D space is a helix (i.e., it rotates around the x-axis but approaches the yz-plane.)

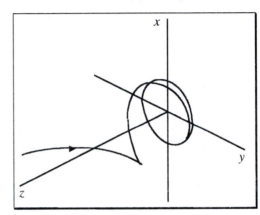

■   Threefold Solutions

33. $\vec{x}'=\begin{bmatrix}1&0&0\\2&1&-2\\3&2&1\end{bmatrix}\vec{x}$

The characteristic polynomial is given by $-\lambda^3+3\lambda^2-7\lambda+5=-(\lambda-1)(\lambda^2-2\lambda+5)$.

Hence, the eigenvalues and corresponding eigenvectors are

$$\lambda_1=1\Rightarrow\vec{v}_1=[2,\ -3,\ 2],$$
$$\lambda_2,\lambda_3=1\pm 2i\Rightarrow\vec{v}_2,\vec{v}_3=[0,\ 1,\ \mp i].$$

Therefore, from λ_2, $\vec{v}_2$ we have

$$\alpha=1,\ \beta=2,\ \vec{p}=[0,\ 1,\ 0],\ \vec{q}=[0,\ 0,\ -1].$$

Hence the general solution is

$$\vec{x}(t)=c_1e^t\begin{bmatrix}2\\-3\\2\end{bmatrix}+c_2e^t\begin{bmatrix}0\\\cos 2t\\\sin 2t\end{bmatrix}+c_3e^t\begin{bmatrix}0\\\sin 2t\\-\cos 2t\end{bmatrix}.$$

■   **Triple IVPs**

36. $\vec{\mathbf{x}}' = \begin{bmatrix} 0 & 1 & 0 \\ -1 & 0 & -1 \\ 0 & 1 & 0 \end{bmatrix} \vec{\mathbf{x}}, \ \vec{\mathbf{x}}(0) = \begin{bmatrix} 0 \\ 1 \\ 1 \end{bmatrix}$

The eigenvalues and eigenvectors of the coefficient matrix are

$$\lambda_1 = 0 \ \Rightarrow \ \vec{\mathbf{v}}_1 = [-1, \ 0, \ 1],$$

$$\lambda_2, \lambda_3 = \pm i\sqrt{2} \Rightarrow \vec{\mathbf{v}}_{2,3} = \left[1, \ \pm i\sqrt{2}, \ 1 \right],$$

therefore, three independent solutions are

$$\vec{\mathbf{x}}_1(t) = \begin{bmatrix} -1 \\ 0 \\ 1 \end{bmatrix}$$

$$\vec{\mathbf{x}}_2(t) = e^{\alpha t} \cos \beta t \vec{\mathbf{p}} - e^{\alpha t} \sin \beta t \vec{\mathbf{q}} \ = \ \cos\sqrt{2}t \begin{bmatrix} 1 \\ 0 \\ 1 \end{bmatrix} - \sqrt{2} \sin\sqrt{2}t \begin{bmatrix} 0 \\ 1 \\ 0 \end{bmatrix}$$

$$\vec{\mathbf{x}}_3(t) = e^{\alpha t} \sin \beta t \vec{\mathbf{p}} + e^{\alpha t} \cos \beta t \vec{\mathbf{q}} \ = \ \sin\sqrt{2}t \begin{bmatrix} 1 \\ 0 \\ 1 \end{bmatrix} + \sqrt{2} \cos\sqrt{2}t \begin{bmatrix} 0 \\ 1 \\ 0 \end{bmatrix}.$$

Substituting the initial conditions

$$\vec{\mathbf{x}}(0) = c_1 \begin{bmatrix} -1 \\ 0 \\ 1 \end{bmatrix} + c_2 \begin{bmatrix} 1 \\ 0 \\ 1 \end{bmatrix} + c_3 \begin{bmatrix} 0 \\ \sqrt{2} \\ 0 \end{bmatrix} = \begin{bmatrix} 0 \\ 1 \\ 1 \end{bmatrix}$$

yields $c_1 = c_2 = \dfrac{1}{2}$ and $c_3 = \dfrac{\sqrt{2}}{2}$. Hence, the solution of the IVP is

$$\vec{\mathbf{x}} = \frac{1}{2}\vec{\mathbf{x}}_1 + \frac{1}{2}\vec{\mathbf{x}}_2 + \frac{\sqrt{2}}{2}\vec{\mathbf{x}}_3$$

$$= \frac{1}{2}\begin{bmatrix} -1 \\ 0 \\ 1 \end{bmatrix} + \frac{1}{2}\begin{bmatrix} \cos\sqrt{2}t \\ -\sqrt{2}\sin\sqrt{2}t \\ \cos\sqrt{2}t \end{bmatrix} + \frac{\sqrt{2}}{2}\begin{bmatrix} \sin\sqrt{2}t \\ \sqrt{2}\cos\sqrt{2}t \\ \sin\sqrt{2}t \end{bmatrix}$$

$$= \frac{1}{2}\begin{bmatrix} -1 \\ 0 \\ 1 \end{bmatrix} + \frac{1}{2}\cos\sqrt{2}t \begin{bmatrix} 1 \\ \sqrt{2} \\ 1 \end{bmatrix} + \frac{\sqrt{2}}{2}\sin\sqrt{2}t \begin{bmatrix} 1 \\ -\sqrt{2} \\ 1 \end{bmatrix}.$$

■   **Coupled Mass-Spring System**

39.   The coupled mass-spring matrix

$$\begin{bmatrix} 0 & 1 & 0 & 0 \\ -\dfrac{k_1+k_2}{m_1} & 0 & \dfrac{k_2}{m_1} & 0 \\ 0 & 0 & 0 & 1 \\ \dfrac{k_2}{m_2} & 0 & -\dfrac{k_2+k_3}{m_2} & 0 \end{bmatrix}$$

simplifies, with $k_1 = k_2 = k_3 = m_1 = m_2 = 1$ to

$$\begin{bmatrix} 0 & 1 & 0 & 0 \\ -2 & 0 & 1 & 0 \\ 0 & 0 & 0 & 1 \\ 1 & 0 & -2 & 0 \end{bmatrix}.$$

We find purely complex eigenvalues and their corresponding eigenvectors to be

$$\lambda_{1,2} = \pm i, \qquad \vec{v}_{1,2} = \begin{bmatrix} \mp i, & 1, & \mp i, & 1 \end{bmatrix}$$

$$\lambda_{3,4} = \pm i\sqrt{3}, \qquad \vec{v}_{3,4} = \begin{bmatrix} -1, & \mp i\sqrt{3}, & 1, & \pm i\sqrt{3} \end{bmatrix}.$$

For $\lambda_{1,2}$ we have

$$\alpha = 0, \ \beta = 1 \ \text{and} \ \vec{p}_{1,2} = [0,1,0,1] \ , \ \vec{q}_{1,2} = [-1,0,-1,0]$$

For $\lambda_{3,4}$ we have

$$\alpha = 0, \ \beta = \sqrt{3} \ \text{and} \ \vec{p}_{3,4} = [-1,0,1,0] \ , \ \vec{q}_{3,4} = [0,-\sqrt{3},0,\sqrt{3}].$$

Then four linearly independent solutions are

$$\vec{x}_1(t) = e^{\alpha t}\cos\beta t\, \vec{p}_{1,2} - e^{\alpha t}\sin\beta t\, \vec{q}_{1,2} = \cos t \begin{bmatrix} 0 \\ 1 \\ 0 \\ 1 \end{bmatrix} - \sin t \begin{bmatrix} -1 \\ 0 \\ -1 \\ 0 \end{bmatrix},$$

$$\vec{x}_2(t) = e^{\alpha t}\sin\beta t\, \vec{p}_{1,2} + e^{\alpha t}\cos\beta t\, \vec{q}_{1,2} = \sin t \begin{bmatrix} 0 \\ 1 \\ 0 \\ 1 \end{bmatrix} + \cos t \begin{bmatrix} -1 \\ 0 \\ -1 \\ 0 \end{bmatrix},$$

$$\vec{x}_3(t) = e^{\alpha t} \cos \beta t \, \vec{p}_{3,4} - e^{\alpha t} \sin \beta t \, \vec{q}_{3,4} = \cos \sqrt{3}t \begin{bmatrix} -1 \\ 0 \\ 1 \\ 0 \end{bmatrix} - \sin \sqrt{3}t \begin{bmatrix} 0 \\ -\sqrt{3} \\ 0 \\ \sqrt{3} \end{bmatrix},$$

$$\vec{x}_4(t) = e^{\alpha t} \sin \beta t \, \vec{p}_{3,4} + e^{\alpha t} \cos \beta t \, \vec{q}_{3,4} = \sin \sqrt{3}t \begin{bmatrix} -1 \\ 0 \\ 1 \\ 0 \end{bmatrix} + \cos \sqrt{3}t \begin{bmatrix} 0 \\ -\sqrt{3} \\ 0 \\ \sqrt{3} \end{bmatrix}.$$

The general solution is

$$\vec{x}(t) = c_1 \vec{x}_1 + c_2 \vec{x}_2 + c_3 \vec{x}_3 + c_4 \vec{x}_4.$$

Substituting the initial conditions

$$x_1(0) = 0, \ x_2(0) = 0, \ x_3(0) = 2, \text{ and } x_4(0) = 0$$

we get

$$c_1 = 0, \ c_2 = -1, \ c_3 = 1, \text{ and } c_4 = 0.$$

Finally, because $x = x_1$, and $y = x_3$, we have the desired result

$$x(t) = \cos t - \cos \sqrt{3}t$$
$$y(t) = \cos t + \cos \sqrt{3}t.$$

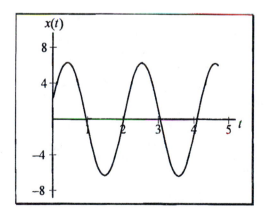
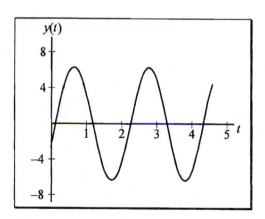

■ **Suggested Journal Entry**

42. Student Project

6.4   Stability and Linear Classification

■   **Classification Verification**

3. $\vec{x}' = \begin{bmatrix} -2 & 0 \\ 0 & -2 \end{bmatrix} \vec{x}$ (star node)

The matrix has eigenvalues –2 and –2. Because both eigenvalues are negative, the origin is an asymptotically stable equilibrium point. Also the matrix has two linearly independent eigenvectors (in fact every vector in the plane is an eigenvector), and hence, the origin is a star node.

6. $\vec{x}' = \begin{bmatrix} 0 & 1 \\ -1 & -1 \end{bmatrix} \vec{x}$ (spiral sink)

The matrix has eigenvalues $-\dfrac{1}{2} \pm i \dfrac{\sqrt{3}}{2}$. Because the real part of the eigenvalues is negative, the origin is an asymptotically stable equilibrium point. The fact that the eigenvalues are complex with negative real parts also means the origin is a spiral sink.

■   **One Zero Eigenvalue**

9.   (a)   If $\lambda_1 = 0$ and $\lambda_2 \neq 0$, then **A** is a singular matrix because $|\mathbf{A}| = |\mathbf{A} - \lambda_1 \mathbf{I}| = 0$. Hence, the rank of **A** is less than 2. But the rank of **A** is not 0 because if it were it would be the matrix of all zeros, which would have both eigenvalues 0. The rank of **A** is 1, which means the kernel of **A** consists of a one-dimensional subspace of $\mathbb{R}^2$, a line through the origin. But the kernel of **A** is simply the set of solutions of $\mathbf{A}\vec{x} = \vec{0}$, which are the equilibrium points of $\vec{x}' = \mathbf{A}\vec{x}$. We use the solution of the form

$$\vec{x}(t) = \begin{bmatrix} x \\ y \end{bmatrix} = c_1 \begin{bmatrix} a \\ b \end{bmatrix} + c_2 e^{\lambda_2 t} \begin{bmatrix} c \\ d \end{bmatrix}$$

to find the equilibrium points. We compute the derivatives and set them to zero. Setting $\dot{x} = \dot{y} = 0$, yields the equation

$$\vec{x}'(t) = c_2 \lambda_2 e^{\lambda_2 t} \begin{bmatrix} c \\ d \end{bmatrix} = \begin{bmatrix} 0 \\ 0 \end{bmatrix},$$

which implies $c_2 = 0$. The points that satisfy $\dot{x} = \dot{y} = 0$ are the points

$$\vec{x}(t) = \begin{bmatrix} x \\ y \end{bmatrix} = c_1 \begin{bmatrix} a \\ b \end{bmatrix},$$

which consists of all multiples of a given vector (i.e., a line through the origin.)

(b)   If a solution starts off the line of equilibrium points, then $c_2 \neq 0$. If $\lambda_2 > 0$, the second term

$$c_2 e^{\lambda_2 t} \begin{bmatrix} c \\ d \end{bmatrix}$$

becomes larger and larger. Hence, the solution moves farther and farther away from the line of equilibrium points. On the other hand, if $\lambda_2 < 0$, the second term becomes smaller and smaller, the solution moves towards the line.

■ Both Eigenvalues Zero

12. $\vec{x}' = \begin{bmatrix} 2 & 1 \\ -4 & -2 \end{bmatrix} \vec{x}$ 　　　　 $\lambda = 0, 0$

For $\lambda = 0$, there is only one linearly independent eigenvector: $\vec{v} = \begin{bmatrix} 1 \\ -2 \end{bmatrix}$

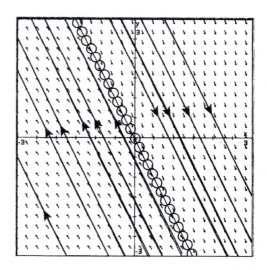

■ Zero Again

15. $\vec{x}' = \begin{bmatrix} 1 & -2 \\ 1 & -2 \end{bmatrix} \vec{x}$

(a)   The characteristic equation of this system is $\lambda^2 + \lambda = 0$, yielding $\lambda_1 = 0$ and $\lambda_2 = -1$. The corresponding eigenvectors can be seen to be $\vec{v}_1 = [2, 1]$, $\vec{v}_2 = [1, 1]$.

(b)   Setting

$$x' = y' = 0,$$

we see that all points on the line

$$x - 2y = 0$$

are equilibrium points, and thus $(0, 0)$ is not an isolated equilibrium point. Also from the differential equations,

$$x' = y' = x - 2y,$$

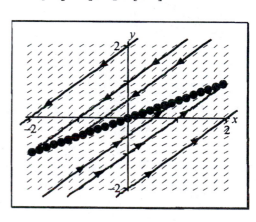

Sample trajectories of a singular system

we see that solutions move along trajectories on 45-degree lines. Above the line

$$x - 2y = 0, \; x' = y' = x - 2y < 0$$

and the movement is downward and to the left. Below the line $x - 2y = 0$, movement is upward and to the right. This outcome is shown in the phase plane. (See the figure.) Note that the solutions below the equilibrium line approach the line because the trajectories move along the 45-degree lines, but the equilibrium line goes up by less than 45 degrees, and the solutions above the equilibrium line move down towards the line.

■   **Bifurcation Point**

18.   The characteristic equation of $\vec{\mathbf{x}}' = \begin{bmatrix} 0 & 1 \\ -1 & k \end{bmatrix} \vec{\mathbf{x}}$ is $\lambda^2 - k\lambda + 1 = 0$, which has roots

$$\lambda_1 = \lambda_2 = \frac{1}{2}\left(k + \sqrt{k^2 - 4}\right).$$

When $|k| < 2$, the roots are complex and the solutions oscillate. When $|k| \geq 2$,

the solutions emanate from an unstable node. Hence, the bifurcation values are $k = \pm 2$.

■   **Interpreting the Trace-Determinant Graph**

21. $|\mathbf{A}| < 0$

Using the basic formula, the determinant of $\mathbf{A}$ is negative then $\sqrt{(Tr\mathbf{A})^2 - 4|\mathbf{A}|} > 0$ and $|Tr\mathbf{A}| > 0$, so the eigenvalues must be positive and have opposite signs. Hence, the origin is a saddle point and an unstable equilibrium.

24. $(Tr\mathbf{A})^2 - 4|\mathbf{A}| = 0$, $Tr\mathbf{A} \neq 0$

Using the basic formula, (real) nonzero eigenvalues are repeated. Hence, the origin is a degenerate or star node.

27. $Tr\mathbf{A} < 0$ and $|\mathbf{A}| > 0$

Using the basic formula, the eigenvalues are real and both negative. Hence, the origin is asymptotically stable.

6.5   Decoupling a Linear DE System

■   **Decoupling Homogeneous Linear Systems**

3. $\vec{\mathbf{x}}' = \begin{bmatrix} 0 & -1 \\ -1 & 0 \end{bmatrix} \vec{\mathbf{x}}$

The coefficient matrix has eigenvalue and eigenvectors

$$\lambda_1 = 1, \ \vec{\mathbf{v}}_1 = [-1, \ 1]$$
$$\lambda_2 = -1, \ \vec{\mathbf{v}}_2 = [1, \ 1].$$

The matrix of eigenvectors is

$$\mathbf{P} = \begin{bmatrix} -1 & 1 \\ 1 & 1 \end{bmatrix}, \text{ and } \mathbf{P}^{-1} = \frac{1}{2}\begin{bmatrix} -1 & 1 \\ 1 & 1 \end{bmatrix}.$$

Therefore,

$$\mathbf{P}^{-1}\mathbf{A}\mathbf{P} = \frac{1}{2}\begin{bmatrix} -1 & 1 \\ 1 & 1 \end{bmatrix}\begin{bmatrix} 0 & -1 \\ -1 & 0 \end{bmatrix}\begin{bmatrix} -1 & 1 \\ 1 & 1 \end{bmatrix} = \begin{bmatrix} 1 & 0 \\ 0 & -1 \end{bmatrix}.$$

Hence, transforming from $\vec{\mathbf{x}}$ to the new variable $\vec{\mathbf{w}} = \mathbf{P}^{-1}\vec{\mathbf{x}}$ yields the decoupled system $w_1' = w_1$ and $w_2' = -w_2$. Solving this decoupled system yields $w_1(t) = c_1 e^t$ and $w_2(t) = c_2 e^{-t}$. The solution of the original system is

$$\vec{\mathbf{x}}(t) = \mathbf{P}\vec{\mathbf{w}}(t) = \begin{bmatrix} -1 & 1 \\ 1 & 1 \end{bmatrix}\begin{bmatrix} c_1 e^t \\ c_2 e^{-t} \end{bmatrix} = c_1 e^t \begin{bmatrix} -1 \\ 1 \end{bmatrix} + c_2 e^{-t}\begin{bmatrix} 1 \\ 1 \end{bmatrix}.$$

6. $\vec{\mathbf{x}}' = \begin{bmatrix} 0 & 1 \\ 1 & 0 \end{bmatrix} \vec{\mathbf{x}}$

The coefficient matrix has eigenvalue and eigenvectors

$$\lambda_1 = -1, \ \vec{\mathbf{v}}_1 = [-1, \ 1]$$
$$\lambda_2 = 1, \ \vec{\mathbf{v}}_2 = [1, \ 1].$$

The matrix of eigenvectors is

$$\mathbf{P} = \begin{bmatrix} -1 & 1 \\ 1 & 1 \end{bmatrix}, \text{ and } \mathbf{P}^{-1} = \frac{1}{2}\begin{bmatrix} -1 & 1 \\ 1 & 1 \end{bmatrix}.$$

Therefore, $\mathbf{P}^{-1}\mathbf{A}\mathbf{P} = \frac{1}{2}\begin{bmatrix} -1 & 1 \\ 1 & 1 \end{bmatrix}\begin{bmatrix} 0 & 1 \\ 1 & 0 \end{bmatrix}\begin{bmatrix} -1 & 1 \\ 1 & 1 \end{bmatrix} = \begin{bmatrix} -1 & 0 \\ 0 & 1 \end{bmatrix}.$

continued on the next page

Hence, transforming from $\vec{x}$ to the new variable $\vec{w} = \mathbf{P}^{-1}\vec{x}$ yields the decoupled system $w_1' = -w_1$ and $w_2' = w_2$. Solving this decoupled system yields $w_1(t) = c_1 e^{-t}$ and $w_2(t) = c_2 e^t$. Hence, the solution of the original system is

$$\vec{x}(t) = \mathbf{P}\vec{w}(t) = \begin{bmatrix} -1 & 1 \\ 1 & 1 \end{bmatrix} \begin{bmatrix} c_1 e^{-t} \\ c_2 e^t \end{bmatrix} = c_1 e^{-t} \begin{bmatrix} -1 \\ 1 \end{bmatrix} + c_2 e^t \begin{bmatrix} 1 \\ 1 \end{bmatrix}.$$

9. $\vec{x}' = \begin{bmatrix} 1 & 0 & 0 \\ -4 & 3 & 0 \\ -4 & 2 & 1 \end{bmatrix} \vec{x}$ (See Problem 43 in Section 5.4)

The eigenvalues are $\lambda_1 = 1$, 1 and $\lambda_2 = 3$, with eigenvectors

$$\vec{v}_1 = \begin{bmatrix} 1 \\ 2 \\ 0 \end{bmatrix}, \begin{bmatrix} 0 \\ 0 \\ 1 \end{bmatrix} \text{ and } \vec{v}_2 = \begin{bmatrix} 0 \\ 1 \\ 1 \end{bmatrix},$$

so that

$$\mathbf{D} = \begin{bmatrix} 1 & 0 & 0 \\ 0 & 1 & 0 \\ 0 & 0 & 3 \end{bmatrix} \text{ and } \mathbf{P} = \begin{bmatrix} 1 & 0 & 0 \\ 2 & 0 & 1 \\ 0 & 1 & 1 \end{bmatrix}.$$

We change the variable to $\vec{w} = \mathbf{P}^{-1}\vec{x}$, to find $\vec{w}' = \mathbf{D}\vec{w}$:

$$\vec{w}'(t) = \begin{bmatrix} w_1' \\ w_2' \\ w_3' \end{bmatrix} = \begin{bmatrix} 1 & 0 & 0 \\ 0 & 1 & 0 \\ 0 & 0 & 3 \end{bmatrix} \begin{bmatrix} w_1 \\ w_2 \\ w_3 \end{bmatrix} = \begin{bmatrix} w_1 \\ w_2 \\ 3w_3 \end{bmatrix}.$$

Solving we obtain $w_1(t) = c_1 e^t$, $w_2(t) = c_2 e^t$, $w_3(t) = c_3 e^{3t}$.

Thus $\vec{x}(t) = \mathbf{P}\vec{w}(t) = \begin{bmatrix} 1 & 0 & 0 \\ 2 & 0 & 1 \\ 0 & 1 & 1 \end{bmatrix} \begin{bmatrix} c_1 e^t \\ c_2 e^t \\ c_3 e^{3t} \end{bmatrix} = \begin{bmatrix} c_1 e^t \\ 2c_1 e^t + c_3 e^{3t} \\ c_2 e^t + c_3 e^{3t} \end{bmatrix}.$

■   **Decoupling Nonhomogeneous Linear Systems**

12. $\vec{x}' = \begin{bmatrix} -3 & 1 \\ 1 & -3 \end{bmatrix} \vec{x} + \begin{bmatrix} \sin t \\ 0 \end{bmatrix}$

The eigenvalues are -2 and -4, and their two independent eigenvectors are $[1, 1]$ and $[-1, 1]$. We form the matrices

$$\mathbf{P} = \begin{bmatrix} 1 & -1 \\ 1 & 1 \end{bmatrix} \text{ and } \mathbf{P}^{-1} = \frac{1}{2} \begin{bmatrix} 1 & 1 \\ -1 & 1 \end{bmatrix}.$$

We change to the variable $\vec{w} = \mathbf{P}^{-1}\vec{x}$, to yield the decoupled system

$$\vec{w}' = \begin{bmatrix} -2 & 0 \\ 0 & -4 \end{bmatrix}\vec{w} + \frac{1}{2}\begin{bmatrix} 1 & 1 \\ -1 & 1 \end{bmatrix}\begin{bmatrix} \sin t \\ 0 \end{bmatrix}, \quad \text{or} \quad \begin{matrix} w_1' = -2w_1 + \sin t \\ w_2' = -4w_2. \end{matrix}$$

Solving these yields $w_1(t) = c_1 e^{-2t} - \dfrac{1}{5}\cos t + \dfrac{2}{5}\sin t, \quad w_2(t) = c_2 e^{-4t}.$

Thus

$$\vec{x}(t) = \mathbf{P}\vec{w}(t) = \begin{bmatrix} 1 & -1 \\ 1 & 1 \end{bmatrix}\begin{bmatrix} c_1 e^{-2t} - \dfrac{1}{5}\cos t + \dfrac{2}{5}\sin t \\ c_2 e^{-4t} \end{bmatrix}$$

$$= c_1 e^{-2t}\begin{bmatrix} 1 \\ 1 \end{bmatrix} + c_2 e^{-4t}\begin{bmatrix} -1 \\ 1 \end{bmatrix} + \frac{1}{5}\begin{bmatrix} -\cos t + 2\sin t \\ -\cos t + 2\sin t \end{bmatrix}.$$

15. $\vec{x}' = \begin{bmatrix} 1 & 4 \\ 2 & 3 \end{bmatrix}\vec{x} + \begin{bmatrix} t \\ 2t \end{bmatrix}$

$\lambda_1 = 5$, eigenvector $\begin{bmatrix} 1 \\ 1 \end{bmatrix}$, $\lambda_2 = -1$, eigenvector $\begin{bmatrix} 2 \\ -1 \end{bmatrix}$,

$$\mathbf{D} = \begin{bmatrix} 5 & 0 \\ 0 & -1 \end{bmatrix}, \quad \mathbf{P} = \begin{bmatrix} 1 & 2 \\ 1 & -1 \end{bmatrix}, \quad \mathbf{P}^{-1} = -\frac{1}{3}\begin{bmatrix} -1 & -2 \\ -1 & 1 \end{bmatrix} = \frac{1}{3}\begin{bmatrix} 1 & 2 \\ 1 & -1 \end{bmatrix}.$$

We change to the variable $\vec{w} = \mathbf{P}^{-1}\vec{x}$, so that $\vec{w}' = \mathbf{D}\vec{w} + \mathbf{P}^{-1}\vec{f}(t)$.

$$\vec{w}'(t) = \begin{bmatrix} 5 & 0 \\ 0 & -1 \end{bmatrix}\begin{bmatrix} w_1 \\ w_2 \end{bmatrix} + \frac{1}{3}\begin{bmatrix} 1 & 2 \\ 1 & -1 \end{bmatrix}\begin{bmatrix} t \\ 2t \end{bmatrix} = \begin{bmatrix} 5w_1 + \dfrac{5}{3}t \\ -w_2 - \dfrac{1}{3}t \end{bmatrix}.$$

Solving these linear DEs gives $\vec{w}(t) = \begin{bmatrix} e^{5t}\displaystyle\int \dfrac{5}{3}te^{-5t}\,dt \\ e^{-t}\displaystyle\int -\dfrac{1}{3}te^{t}\,dt \end{bmatrix}$ (integration by parts)

$$= \begin{bmatrix} -\dfrac{t}{3} - \dfrac{1}{15} + c_1 e^{5t} \\ -\dfrac{t}{3} + \dfrac{1}{3} + c_2 e^{-t} \end{bmatrix}.$$

Thus we obtain

$$\vec{x}(t) = \mathbf{P}\vec{w}(t) = \begin{bmatrix} 1 & 2 \\ 1 & -1 \end{bmatrix}\begin{bmatrix} -\dfrac{t}{3} - \dfrac{1}{15} + c_1 e^{5t} \\ -\dfrac{t}{3} + \dfrac{1}{3} + c_2 e^{-t} \end{bmatrix} = \begin{bmatrix} c_1 e^{5t} + 2c_2 e^{-t} - t + \dfrac{3}{5} \\ c_1 e^{5t} - c_2 e^{-t} - \dfrac{2}{5} \end{bmatrix}.$$

18. $\vec{x}' = \begin{bmatrix} 3 & -2 & 0 \\ 1 & 0 & 0 \\ -1 & 1 & 3 \end{bmatrix} \vec{x} + \begin{bmatrix} 4 \\ 6 \\ 1 \end{bmatrix}$

$\lambda_1 = 1, \ \vec{v}_1 = \begin{bmatrix} 1 \\ 1 \\ 0 \end{bmatrix}; \ \lambda_2 = 2, \ \vec{v}_2 = \begin{bmatrix} 2 \\ 1 \\ 1 \end{bmatrix}; \ \lambda_2 = 3, \ \vec{v}_3 = \begin{bmatrix} 0 \\ 0 \\ 1 \end{bmatrix}.$

$$P = \begin{bmatrix} 1 & 2 & 0 \\ 1 & 1 & 0 \\ 0 & 1 & 1 \end{bmatrix}, \ D = \begin{bmatrix} 1 & 0 & 0 \\ 0 & 2 & 0 \\ 0 & 0 & 3 \end{bmatrix}, \ P^{-1} = \begin{bmatrix} -1 & 2 & 0 \\ 1 & -1 & 0 \\ -1 & 1 & 1 \end{bmatrix}.$$

$$\vec{w}' = D\vec{w} + P^{-1}\vec{f}(t) = \begin{bmatrix} 1 & 0 & 0 \\ 0 & 2 & 0 \\ 0 & 0 & 3 \end{bmatrix}\begin{bmatrix} w_1 \\ w_2 \\ w_3 \end{bmatrix} + \begin{bmatrix} -1 & 2 & 0 \\ 1 & -1 & 0 \\ -1 & 1 & 1 \end{bmatrix}\begin{bmatrix} 4 \\ 6 \\ 1 \end{bmatrix} = \begin{bmatrix} w_1 + 8 \\ 2w_2 - 2 \\ 3w_3 + 3 \end{bmatrix} \Rightarrow \begin{array}{l} w_1 = c_1 e^t - 8 \\ w_2 = c_2 e^t + 1 \\ w_3 = c_3 e^{3t} - 1 \end{array}$$

$$\vec{x}(t) = P\vec{w}(t) = \begin{bmatrix} 1 & 2 & 0 \\ 1 & 1 & 0 \\ 0 & 1 & 1 \end{bmatrix}\begin{bmatrix} c_1 e^t - 8 \\ c_2 e^{2t} + 1 \\ c_3 e^{3t} - 1 \end{bmatrix} = \begin{bmatrix} c_1 e^t + 2c_2 e^{2t} - 6 \\ c_1 e^t + c_2 e^{2t} - 7 \\ c_2 e^{2t} + c_3 e^{3t} \end{bmatrix}.$$

■   **Working Backwards**

21.   Given eigenvalues are 1 and −1 and respective eigenvectors are $[1, \ 1]$ and $[1, \ 2]$, we form the matrices

$$P = \begin{bmatrix} 1 & 1 \\ 1 & 2 \end{bmatrix} \text{ and } P^{-1} = \begin{bmatrix} 2 & -1 \\ -1 & 1 \end{bmatrix}, \text{ and then the diagonal matrix } D = \begin{bmatrix} 1 & 0 \\ 0 & -1 \end{bmatrix},$$

whose diagonal elements are the eigenvalues. Using the relation $D = P^{-1}AP$, we premultiply by P, and postmultiply by P^{-1}, yielding

$$A = PDP^{-1} = \begin{bmatrix} 1 & 1 \\ 1 & 2 \end{bmatrix}\begin{bmatrix} 1 & 0 \\ 0 & -1 \end{bmatrix}\begin{bmatrix} 2 & -1 \\ 1 & 2 \end{bmatrix} = \begin{bmatrix} 3 & -2 \\ 4 & -3 \end{bmatrix}.$$

■   **Suggested Journal Entry**

24.   Student Project

6.6   Matrix Exponential

■   **Matrix Exponential Functions**

3. $\mathbf{A} = \begin{bmatrix} 1 & 0 \\ 1 & 0 \end{bmatrix}$ $\mathbf{A}$ has eigenvalues 0 and 1, with eigenvectors $\begin{bmatrix} 0 \\ 1 \end{bmatrix}$ and $\begin{bmatrix} 1 \\ 1 \end{bmatrix}$, respectively.

Therefore a fundamental matrix is $\mathbf{X}(t) = \begin{bmatrix} 0 & e^t \\ 1 & e^t \end{bmatrix}$, and

$$e^{\mathbf{A}t} = \mathbf{X}(t)\mathbf{X}^{-1}(0) = \begin{bmatrix} 0 & e^t \\ 1 & e^t \end{bmatrix}\begin{bmatrix} -1 & 1 \\ 1 & 0 \end{bmatrix} = \begin{bmatrix} e^t & 0 \\ e^t - 1 & 1 \end{bmatrix}$$

6. $\mathbf{A} = \begin{bmatrix} 0 & 1 & 1 \\ 0 & 0 & 1 \\ 0 & 0 & 0 \end{bmatrix}$ Note that $\mathbf{A}^3 = \mathbf{0}$, so that

$$e^{\mathbf{A}t} = \mathbf{I} + \mathbf{A}t + \frac{(\mathbf{A}t)^2}{2} = \begin{bmatrix} 1 & 0 & 0 \\ 0 & 1 & 0 \\ 0 & 0 & 1 \end{bmatrix} + t\begin{bmatrix} 0 & 1 & 1 \\ 0 & 0 & 1 \\ 0 & 0 & 0 \end{bmatrix} + \frac{t^2}{2}\begin{bmatrix} 0 & 0 & 1 \\ 0 & 0 & 0 \\ 0 & 0 & 0 \end{bmatrix} = \begin{bmatrix} 1 & t & t+t^2/2 \\ 0 & 1 & t \\ 0 & 0 & 1 \end{bmatrix}.$$

■   **DE Solutions using Matrix Exponentials**

9. $\begin{aligned} x' &= x + y \\ y' &= y \end{aligned}$ Note that $\mathbf{A} = \begin{bmatrix} 1 & 1 \\ 0 & 1 \end{bmatrix}$ is not diagonalizable.

So we must use the definition of matrix exponential.

We find that $\mathbf{A} = \begin{bmatrix} 1 & 1 \\ 0 & 1 \end{bmatrix}$, $\mathbf{A}^2 = \begin{bmatrix} 1 & 2 \\ 0 & 1 \end{bmatrix}$, $\mathbf{A}^3 = \begin{bmatrix} 1 & 3 \\ 0 & 1 \end{bmatrix}$, ..., $\mathbf{A}^n = \begin{bmatrix} 1 & n \\ 0 & 1 \end{bmatrix}$, ...,

and so $e^{\mathbf{A}t} = \mathbf{I} + t\begin{bmatrix} 1 & 1 \\ 0 & 1 \end{bmatrix} + \frac{t^2}{2!}\begin{bmatrix} 1 & 2 \\ 0 & 1 \end{bmatrix} + \frac{t^3}{3!}\begin{bmatrix} 1 & 3 \\ 0 & 1 \end{bmatrix} + \dots + \frac{t^k}{k!}\begin{bmatrix} 1 & k \\ 0 & 1 \end{bmatrix} + \dots = \sum_{k=0}^{\infty}\frac{t^k}{k!}\begin{bmatrix} 1 & k \\ 0 & 1 \end{bmatrix}$

$$= \begin{bmatrix} \displaystyle\sum_{k=0}^{\infty}\frac{t^k}{k!} & \displaystyle\sum_{k=1}^{\infty}\frac{t^k}{(k-1)!} \\ 0 & \displaystyle\sum_{k=0}^{\infty}\frac{t^k}{k!} \end{bmatrix} = \begin{bmatrix} e^t & te^t \\ 0 & e^t \end{bmatrix}.$$

Note: we have used the fact that $te^t = t\displaystyle\sum_{k=0}^{\infty}\frac{t^k}{k!} = \sum_{k=0}^{\infty}\frac{t^{k+1}}{k!} = 0 + \sum_{k=1}^{\infty}\frac{t^k}{(k-1)!}$.

Hence, $\mathbf{x}(t) = \begin{bmatrix} e^t & te^t \\ 0 & e^t \end{bmatrix}\begin{bmatrix} c_1 \\ c_2 \end{bmatrix}$.

12. $\vec{x}' = \begin{bmatrix} 2 & 0 \\ 0 & 3 \end{bmatrix} \begin{bmatrix} x_1 \\ x_2 \end{bmatrix} + \begin{bmatrix} 0 \\ 6 \end{bmatrix}$ Because **A** is diagonal, we have $e^{\mathbf{A}t} = \begin{bmatrix} e^{2t} & 0 \\ 0 & e^{3t} \end{bmatrix}$, so

$$\vec{x}(t) = \begin{bmatrix} e^{2t} & 0 \\ 0 & e^{3t} \end{bmatrix} \begin{bmatrix} c_1 \\ c_2 \end{bmatrix} + \begin{bmatrix} e^{2t} & 0 \\ 0 & e^{3t} \end{bmatrix} \int_0^t \begin{bmatrix} e^{-2s} & 0 \\ 0 & e^{-3s} \end{bmatrix} \begin{bmatrix} 0 \\ 6 \end{bmatrix} ds$$

$$= \begin{bmatrix} e^{2t} & 0 \\ 0 & e^{3t} \end{bmatrix} \begin{bmatrix} c_1 \\ c_2 \end{bmatrix} + \begin{bmatrix} e^{2t} & 0 \\ 0 & e^{3t} \end{bmatrix} \int_0^t \begin{bmatrix} 0 \\ 6e^{-3s} \end{bmatrix} ds$$

$$= \begin{bmatrix} e^{2t} & 0 \\ 0 & e^{3t} \end{bmatrix} \begin{bmatrix} c_1 \\ c_2 \end{bmatrix} + \begin{bmatrix} e^{2t} & 0 \\ 0 & e^{3t} \end{bmatrix} \begin{bmatrix} 0 \\ -2e^{-3t} + 2 \end{bmatrix}$$

■   **Products of Matrix Exponentials**

15. $\mathbf{A} = \begin{bmatrix} 0 & -1 \\ 0 & 0 \end{bmatrix}$ and $\mathbf{B} = \begin{bmatrix} 0 & 0 \\ 1 & 0 \end{bmatrix}$

(a)   Note that $\mathbf{A}^2 = \mathbf{0} = \mathbf{B}^2$, so that $e^{\mathbf{A}t} = \mathbf{I} + \mathbf{A}t = \begin{bmatrix} 1 & t \\ 0 & 1 \end{bmatrix}$ and $e^{\mathbf{B}t} = \mathbf{I} + \mathbf{B}t = \begin{bmatrix} 1 & 0 \\ t & 1 \end{bmatrix}$.

(b)   To find $e^{(\mathbf{A}+\mathbf{B})t}$, we note that $\mathbf{A} + \mathbf{B} = \begin{bmatrix} 0 & -1 \\ 1 & 0 \end{bmatrix}$ has eigenvalues $\lambda = \pm i$,

and a fundamental matrix $\mathbf{X}(t) = \begin{bmatrix} \cos t & -\sin t \\ \sin t & \cos t \end{bmatrix}$.

Note that $\mathbf{X}(0) = \mathbf{I} = \mathbf{X}(0)^{-1}$.   Then $e^{(\mathbf{A}+\mathbf{B})t} = \mathbf{X}(t)\mathbf{I} = \begin{bmatrix} \cos t & -\sin t \\ \sin t & \cos t \end{bmatrix}$

(c)   No, because $e^{(\mathbf{A}+\mathbf{B})t} = \begin{bmatrix} \cos t & -\sin t \\ \sin t & \cos t \end{bmatrix}$

$$\neq\ e^{\mathbf{A}t}e^{\mathbf{B}t} = \begin{bmatrix} 1 & t \\ 0 & 1 \end{bmatrix} \begin{bmatrix} 1 & 0 \\ t & 1 \end{bmatrix} = \begin{bmatrix} 1+t^2 & t \\ t & 1 \end{bmatrix}.$$

■   **Nilpotent Example**

18.   (a) $\mathbf{A} = \begin{bmatrix} 1 & 1 & -1 \\ 1 & 0 & -1 \\ 1 & 1 & -1 \end{bmatrix}$, $\mathbf{A}^2 = \begin{bmatrix} 1 & 0 & -1 \\ 0 & 0 & 0 \\ 1 & 0 & -1 \end{bmatrix}$, $\mathbf{A}^3 = \begin{bmatrix} 0 & 0 & 0 \\ 0 & 0 & 0 \\ 0 & 0 & 0 \end{bmatrix}$.

(b)   Since $\mathbf{A}^3 = \mathbf{0}$, we have

$$e^{\mathbf{A}t} = \mathbf{I}t + t\mathbf{A} + \frac{t^2}{2!}\mathbf{A}^2$$

$$= \begin{bmatrix} 1 & 0 & 0 \\ 0 & 1 & 0 \\ 0 & 0 & 1 \end{bmatrix} + t\begin{bmatrix} 1 & 1 & -1 \\ 1 & 0 & -1 \\ 1 & 1 & -1 \end{bmatrix} + \frac{t^2}{2!}\begin{bmatrix} 1 & 0 & -1 \\ 0 & 0 & 0 \\ 1 & 0 & -1 \end{bmatrix}$$

$$= \begin{bmatrix} 1+t^2+\dfrac{t^2}{2!} & t & -t-\dfrac{t^2}{2!} \\[2mm] t & 1 & -t \\[2mm] t+\dfrac{t^2}{2!} & t & 1-t-\dfrac{t^2}{2!} \end{bmatrix}.$$

The general solution of $\vec{\mathbf{x}}' = \mathbf{A}\vec{\mathbf{x}}$ is $e^{\mathbf{A}t}\vec{\mathbf{c}}$, or $\vec{\mathbf{x}}(t) = \begin{bmatrix} 1+t^2+\dfrac{t^2}{2!} & t & -t-\dfrac{t^2}{2!} \\[2mm] t & 1 & -t \\[2mm] t+\dfrac{t^2}{2!} & t & 1-t-\dfrac{t^2}{2!} \end{bmatrix}\begin{bmatrix} c_1 \\ c_2 \\ c_3 \end{bmatrix}$.

■   **Fundamental Matrices**

21.   For $\mathbf{A} = \begin{bmatrix} 1 & 2 \\ 0 & 1 \end{bmatrix}$, $\lambda = 1, 1$ but there is only one linearly independent eigenvector $\begin{bmatrix} 1 \\ 0 \end{bmatrix}$.

A generalized eigenvector is $\begin{bmatrix} 0 \\ 1 \end{bmatrix}$.

A fundamental matrix $\mathbf{X}(t) = \begin{bmatrix} e^t & te^t \\ 0 & e^t \end{bmatrix} = e^{\mathbf{A}t}$, because $\mathbf{X}(0) = \mathbf{I} = \mathbf{X}^{-1}(0)$.

Because $\mathbf{A}$ is not diagonalizable, the second method $e^{\mathbf{A}t} = \mathbf{P}e^{\mathbf{D}t}\mathbf{P}^{-1}$ is not applicable.

■   **Computer Lab**

24. $\mathbf{A} = \begin{bmatrix} 0 & 0 & 0 & 1 \\ 0 & 0 & 1 & 0 \\ 0 & 1 & 0 & 0 \\ 1 & 0 & 0 & 0 \end{bmatrix}$; $e^{\mathbf{A}t} = \begin{bmatrix} \cosh t & 0 & 0 & \sinh t \\ 0 & \cosh t & \sinh t & 0 \\ 0 & \sinh t & \cosh t & 0 \\ \sinh t & 0 & 0 & \cosh t \end{bmatrix}$.

■   **Computer DE Solutions**

27. $A = \begin{bmatrix} 1 & 5 \\ -2 & -1 \end{bmatrix}$ has characteristic equation $\lambda^2 + 9 = 0$, so $\lambda_1 = \pm 3i$; $\vec{v} = \begin{bmatrix} -5 \\ 1 \mp 3i \end{bmatrix}$.

$$\vec{x}_{\text{Re}}(t) = \cos 3t \begin{bmatrix} -5 \\ 1 \end{bmatrix} - \sin 3t \begin{bmatrix} 0 \\ -3 \end{bmatrix}$$

$$\vec{x}_{\text{Im}}(t) = \sin 3t \begin{bmatrix} -5 \\ 1 \end{bmatrix} + \cos 3t \begin{bmatrix} 0 \\ -3 \end{bmatrix}$$

$$X(t) = \begin{bmatrix} \vec{x}_{\text{Re}} & \vec{x}_{\text{Im}} \end{bmatrix} = \begin{bmatrix} -5\cos 3t & -5\sin t \\ \cos 3t + 3\sin 3t & \sin 3t - 3\cos t \end{bmatrix}$$

$$X(0) = \begin{bmatrix} -5 & 0 \\ 1 & -3 \end{bmatrix} \qquad X^{-1}(0) = \frac{1}{15}\begin{bmatrix} -3 & 0 \\ -1 & -5 \end{bmatrix}$$

$$e^{At} = \frac{1}{15}\begin{bmatrix} -5\cos 3t & -5\sin 3t \\ \cos 3t + 3\sin 3t & \sin 3t - 3\cos 3t \end{bmatrix}\begin{bmatrix} -3 & 0 \\ -1 & -5 \end{bmatrix}$$

$$= \frac{1}{15}\begin{bmatrix} 15\cos 3t + 5\sin 3t & 25\sin 3t \\ -10\sin 3t & -5\sin 3t + 15\cos 3t \end{bmatrix}$$

$$\vec{x}(t) = e^{At}\vec{x}_0 = e^{At}\begin{bmatrix} 1 \\ 1 \end{bmatrix} = \begin{bmatrix} \cos 3t + 2\sin 3t \\ \cos 3t - \sin 3t \end{bmatrix}$$

30. $A = \begin{bmatrix} 6 & 3 & -2 \\ -4 & -1 & 2 \\ 13 & 9 & -3 \end{bmatrix}, \quad \vec{x}(0) = \begin{bmatrix} 1 \\ 0 \\ 0 \end{bmatrix}$

$$e^{At} = \begin{bmatrix} -3e^{-t} + 5e^{t} - e^{2t} & -e^{2t} + 3e^{t} - 2e^{-t} & -e^{t} + e^{-t} \\ -5e^{t} + 3e^{-t} + 2e^{2t} & 2e^{-t} + 2e^{2t} - 3e^{t} & e^{t} - e^{-t} \\ 5e^{t} - 6e^{-t} + e^{2t} & e^{2t} + 3e^{t} - 4e^{-t} & 2e^{-t} - e^{t} \end{bmatrix} \Rightarrow \vec{x}(t) = e^{At}\vec{x}(0) = \begin{bmatrix} 5e^{t} - e^{2t} - 3e^{-t} \\ -5e^{t} + 2e^{2t} + 3e^{-t} \\ 5e^{t} + e^{2t} - 6e^{-t} \end{bmatrix}.$$

6.7   Theory of Linear DE Systems

■   **Nonhomogeneous Illustration**

3.   As seen in Section 6.2 of the text, the general solution of the homogeneous linear system

$$\vec{x}' = A\vec{x} = \begin{bmatrix} 1 & 1 \\ 4 & 1 \end{bmatrix} \vec{x}$$

is

$$\vec{x}_h(t) = c_1 e^{3t} \begin{bmatrix} 1 \\ 2 \end{bmatrix} + c_2 e^{-t} \begin{bmatrix} 1 \\ -2 \end{bmatrix}.$$

Hence, by the principle of superposition, the nonhomogenous system

$$\vec{x}' = A\vec{x} + \begin{bmatrix} t - 2 + e^t \\ 4t - 1 - 4e^t \end{bmatrix}$$

has the general solution of

$$\vec{x}(t) = \vec{x}_h(t) + \vec{x}_p(t) = c_1 e^{3t} \begin{bmatrix} 1 \\ 2 \end{bmatrix} + c_2 e^{-t} \begin{bmatrix} 1 \\ -2 \end{bmatrix} + \begin{bmatrix} e^t - t \\ 1 - e^t \end{bmatrix}.$$

■   **Systematic Prediction**

6. $$\vec{x}' = \begin{bmatrix} 1 & 4 \\ 1 & 1 \end{bmatrix} \vec{x} + \begin{bmatrix} 0 \\ e^t \end{bmatrix}.$$

The homogeneous solution is $\vec{x}_h(t) = c_1 e^{3t} \begin{bmatrix} 2 \\ 1 \end{bmatrix} + c_2 e^{-t} \begin{bmatrix} 2 \\ -1 \end{bmatrix}.$

We seek a particular solution of the form $\vec{x}_p(t) = e^t \begin{bmatrix} A \\ B \end{bmatrix}.$

Substituting this expression into the nonhomogeneous system yields equations in A and B, that give $A = -1$, and $B = 0$. Hence, the general solution of the nonhomogeneous system is

$$\vec{x}(t) = \vec{x}_h(t) + \vec{x}_p(t) = c_1 e^{3t} \begin{bmatrix} 2 \\ 1 \end{bmatrix} + c_2 e^{-t} \begin{bmatrix} 2 \\ -1 \end{bmatrix} + e^t \begin{bmatrix} -1 \\ 0 \end{bmatrix}.$$

■   **Nonhomogeneous 2 × 2 Linear Systems**

Note to Instructors and Students: In the second printing of the text, the title of these problems was changed from **Variation of Parameters** *to the current title in order to allow students to select the method.*

9. $\vec{x}' = \begin{bmatrix} 1 & 1 \\ 4 & 1 \end{bmatrix} \vec{x} + \begin{bmatrix} -3 \\ -9 \end{bmatrix}$

The eigenvalues and vectors of the coefficient matrix are

$$\lambda_1 = 3, \quad \vec{v}_1 = \begin{bmatrix} 1 \\ 2 \end{bmatrix}; \quad \lambda_2 = -1, \quad \vec{v}_2 = \begin{bmatrix} 1 \\ -2 \end{bmatrix}.$$

Thus, the homogeneous solution is given by

$$\vec{x}_h(t) = c_1 e^{3t} \begin{bmatrix} 1 \\ 2 \end{bmatrix} + c_2 e^{-t} \begin{bmatrix} 1 \\ -2 \end{bmatrix}.$$

We seek a particular solution of the form $\vec{x}_p(t) = \begin{bmatrix} k_1 \\ k_2 \end{bmatrix}$. Substituting this into the system yields

$$\begin{bmatrix} 1 & 1 \\ 4 & 1 \end{bmatrix} \begin{bmatrix} k_1 \\ k_2 \end{bmatrix} = \begin{bmatrix} 3 \\ 9 \end{bmatrix},$$

which gives $k_1 = 2$ and $k_2 = 1$. Hence, we have the particular solution $\vec{x}_p(t) = \begin{bmatrix} 2 \\ 1 \end{bmatrix}$,

and the general solution $\vec{x}(t) = \vec{x}_h(t) + \vec{x}_p(t) = c_1 e^{3t} \begin{bmatrix} 1 \\ 2 \end{bmatrix} + c_2 e^{-t} \begin{bmatrix} 1 \\ -2 \end{bmatrix} + \begin{bmatrix} 2 \\ 1 \end{bmatrix}.$

12. $\vec{x}' = \begin{bmatrix} 1 & 1 \\ 4 & 1 \end{bmatrix} \vec{x} + \begin{bmatrix} 2e^{3t} \\ 0 \end{bmatrix}$

The eigenvalues and eigenvectors of the coefficient matrix are

$$\lambda_1 = 3, \quad \vec{v}_1 = \begin{bmatrix} 1 \\ 2 \end{bmatrix}; \quad \lambda_2 = -1, \quad \vec{v}_2 = \begin{bmatrix} 1 \\ -2 \end{bmatrix}.$$

The fundamental matrix and its inverse are

$$\mathbf{X}(t) = \begin{bmatrix} e^{3t} & e^{-t} \\ 2e^{3t} & -2e^{-t} \end{bmatrix}; \quad \mathbf{X}^{-1}(t) = \frac{e^{-2t}}{-4} \begin{bmatrix} -2e^{-t} & -e^{-t} \\ -2e^{3t} & e^{3t} \end{bmatrix} = \frac{1}{4} \begin{bmatrix} 2e^{-3t} & e^{-3t} \\ 2e^{t} & -e^{t} \end{bmatrix}.$$

We first find a particular solution by computing

$$\mathbf{X}^{-1}(t)\vec{f}(t) = \frac{1}{4} \begin{bmatrix} 2e^{-3t} & e^{-3t} \\ 2e^{t} & -e^{t} \end{bmatrix} \begin{bmatrix} 2e^{3t} \\ 0 \end{bmatrix} = \begin{bmatrix} 1 \\ e^{4t} \end{bmatrix}.$$

Integrating yields

$$\int \mathbf{X}^{-1}(t)\vec{\mathbf{f}}(t)\,dt = \begin{bmatrix} t \\ \dfrac{1}{4}e^{4t} \end{bmatrix}.$$

Finally, multiplying by $\mathbf{X}(t)$ gives a particular solution

$$\vec{\mathbf{x}}_p = \mathbf{X}(t)\int \mathbf{X}^{-1}(t)\vec{\mathbf{f}}(t)\,dt = \begin{bmatrix} e^{3t} & e^{-t} \\ 2e^{3t} & -2e^{-t} \end{bmatrix}\begin{bmatrix} t \\ \dfrac{1}{4}e^{4t} \end{bmatrix} = \begin{bmatrix} t+\dfrac{1}{4} \\ 2t-\dfrac{1}{2} \end{bmatrix}e^{3t}.$$

Thus the general solution of the system is

$$\vec{\mathbf{x}}(t) = \mathbf{X}(t)\vec{\mathbf{c}} + \vec{\mathbf{x}}_p(t) = c_1 e^{3t}\begin{bmatrix} 1 \\ 2 \end{bmatrix} + c_2 e^{-t}\begin{bmatrix} 1 \\ -2 \end{bmatrix} + \begin{bmatrix} t+\dfrac{1}{4} \\ 2t-\dfrac{1}{2} \end{bmatrix}e^{3t}.$$

15. $\quad \vec{\mathbf{x}}' = \begin{bmatrix} 4 & -2 \\ 8 & -4 \end{bmatrix}\vec{\mathbf{x}} + \begin{bmatrix} t^{-3} \\ -t^{-2} \end{bmatrix}$

The characteristic equation is $\lambda^2 = 0$, which yields eigenvalues $\lambda_1 = \lambda_2 = 0$, but only a single eigenvector $\vec{\mathbf{v}}_1 = \begin{bmatrix} 1 \\ 2 \end{bmatrix}$, which yields one solution for the homogeneous system:

$$\vec{\mathbf{x}}_1 = e^{0t}\begin{bmatrix} 1 \\ 2 \end{bmatrix} = \begin{bmatrix} 1 \\ 2 \end{bmatrix}.$$

We seek a second linearly independent solution $\vec{\mathbf{x}}_2$ of the form

$$\vec{\mathbf{x}}_2 = te^{0t}\begin{bmatrix} 1 \\ 2 \end{bmatrix} + e^{0t}\vec{\mathbf{v}}_2 = t\begin{bmatrix} 1 \\ 2 \end{bmatrix} + \vec{\mathbf{v}}_2,$$

where $\vec{\mathbf{v}}_2 = \begin{bmatrix} a \\ b \end{bmatrix}$ is the generalized eigenvector that satisfies $(\mathbf{A} - 0\mathbf{I})\vec{\mathbf{v}}_2 = \begin{bmatrix} 1 \\ 2 \end{bmatrix}$.

Solving

$$\begin{bmatrix} 4 & -2 \\ 8 & -4 \end{bmatrix}\begin{bmatrix} a \\ b \end{bmatrix} = \begin{bmatrix} 1 \\ 2 \end{bmatrix}, \text{ or } 4a - 2b = 1,\ 8a - 4b = 2,$$

$$\vec{\mathbf{v}}_2 = \begin{bmatrix} a \\ b \end{bmatrix} = \begin{bmatrix} a \\ \dfrac{4a-1}{2} \end{bmatrix} = \begin{bmatrix} 1 \\ 2 \end{bmatrix}a + \begin{bmatrix} 0 \\ -\dfrac{1}{2} \end{bmatrix}, \text{ and } \vec{\mathbf{x}}_2 = t\begin{bmatrix} 1 \\ 2 \end{bmatrix} + \begin{bmatrix} 1 \\ 2 \end{bmatrix}a + \begin{bmatrix} 0 \\ -\dfrac{1}{2} \end{bmatrix}.$$

continued on the next page

For convenience, we choose $a = 0$, and write the general solution of the homogeneous system as

$$\vec{x}_h = c_1 \vec{x}_1 + c_2 \vec{x}_2 = c_1 \begin{bmatrix} 1 \\ 2 \end{bmatrix} + c_2 \left(t \begin{bmatrix} 1 \\ 2 \end{bmatrix} + \begin{bmatrix} 0 \\ -\frac{1}{2} \end{bmatrix} \right).$$

Hence,

$$\mathbf{X}(t) = \begin{bmatrix} 1 & t \\ 2 & 2t - \frac{1}{2} \end{bmatrix}, \text{ and } \mathbf{X}^{-1}(t) = \begin{bmatrix} 1 - 4t & 2t \\ 4 & -2 \end{bmatrix}.$$

We compute

$$\mathbf{X}^{-1}(t) \vec{\mathbf{f}}(t) = \begin{bmatrix} 1 - 4t & 2t \\ 4 & -2 \end{bmatrix} \begin{bmatrix} t^{-3} \\ -t^{-2} \end{bmatrix} = \begin{bmatrix} \dfrac{-2t^2 - 4t + 1}{t^3} \\ \dfrac{2t + 4}{t^3} \end{bmatrix},$$

and integrate,

$$\int \mathbf{X}^{-1}(t) \vec{\mathbf{f}}(t) \, dt = \begin{bmatrix} -\dfrac{1}{2t^2} + \dfrac{4}{t} - 2\ln t \\ -\dfrac{2}{t^2} - \dfrac{2}{t} \end{bmatrix},$$

to get the particular solution

$$\vec{\mathbf{x}}_p = \mathbf{X}(t) \int \mathbf{X}^{-1}(t) \vec{\mathbf{f}}(t) \, dt = \begin{bmatrix} 1 & t \\ 2 & 2t - \frac{1}{2} \end{bmatrix} \begin{bmatrix} -\dfrac{1}{2t^2} + \dfrac{4}{t} - 2\ln t \\ -\dfrac{2}{t^2} - \dfrac{2}{t} \end{bmatrix} = \begin{bmatrix} \dfrac{-1 + 4t - 4t^2 \ln t - 4t^2}{2t^2} \\ \dfrac{5t - 4t^2 \ln t - 4t^2}{t^2} \end{bmatrix}$$

$$= \frac{1}{2t^2} \begin{bmatrix} -1 + 4t - 4t^2 (\ln t + 1) \\ 10t + 8t^2 (-\ln t - 1) \end{bmatrix}.$$

Finally, we have the general solution,

$$\vec{\mathbf{x}}(t) = c_1 \vec{\mathbf{x}}_1 + c_2 \vec{\mathbf{x}}_2 + \vec{\mathbf{x}}_p = c_1 \begin{bmatrix} 1 \\ 2 \end{bmatrix} + c_2 \begin{bmatrix} t \\ 2t - \frac{1}{2} \end{bmatrix} + \frac{1}{2t^2} \begin{bmatrix} -1 + 4t - 4t^2 (\ln t + 1) \\ 10t - 8t^2 (\ln t + 1) \end{bmatrix}.$$

■ Two-Loop Circuit

18. $R_{AB} = 2$ ohms, $R_{EF} = 1$ ohm, $L_{KL} = 1$ henry, $L_{DG} = 5$ henries.

From Kirchoff's Laws we obtain

$$\text{Loop 1: } I_1' + 2I_1 + (I_1 - I_2) = 60$$

$$\text{Loop 2: } 5I_2' - (I_1 - I_2) = 0$$

Hence the IVP is

$$\begin{bmatrix} I_1' \\ I_2' \end{bmatrix} = \begin{bmatrix} -3 & 1 \\ 1/5 & -1/5 \end{bmatrix} \begin{bmatrix} I_1 \\ I_2 \end{bmatrix} + \begin{bmatrix} 60 \\ 0 \end{bmatrix}, \begin{bmatrix} I_1(0) \\ I_2(0) \end{bmatrix} = \begin{bmatrix} 0 \\ 0 \end{bmatrix}.$$

and

$$\lambda^2 + \frac{16}{5}\lambda + \frac{2}{5} = 0 \implies \lambda = -3.070, \; -0.1303.$$

Solving (with Maple) gives

$$\begin{bmatrix} I_1(t) \\ I_2(t) \end{bmatrix} = 19e^{-3.07t} \begin{bmatrix} -1 \\ 0.07 \end{bmatrix} + 33.33e^{-0.13t} \begin{bmatrix} -.33 \\ -.94 \end{bmatrix} + \begin{bmatrix} 30 \\ 30 \end{bmatrix}$$

Check:

$$\vec{I}'(0) = \begin{bmatrix} -19 + 33.3(-0.33) + 30 \\ 19(0.07) + 33.3(-0.94) + 30 \end{bmatrix} \approx \begin{bmatrix} 0 \\ 0 \end{bmatrix}.$$

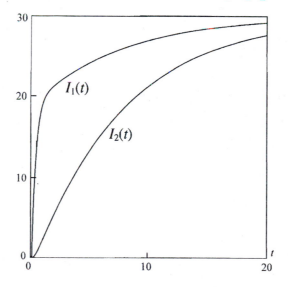

When the current reaches steady-state, the voltage drop across the 2 henry inductor is zero, so the 1 ohm resistor is shorted out. Loop 1 then has only 2 ohms of resistance.

Thus the current $I_1 = I_2 = 30$ amps.

CHAPTER 7

Nonlinear Systems of Differential Equations

7.1   Nonlinear Systems

■ Review of Classifications

3. $x_1' = \kappa x_2$

$x_2' = -\sin x_1$

Dependent variables: x_1, x_2

Parameter: κ

Autonomous nonlinear $(\sin x_1)$ system

■ Verification Review

6. $x' = x$

$y' = y$

Substituting $x = y = e^t$ into the two differential equations, we get

$$e^t = e^t$$
$$e^t = e^t$$

9. Use direct substitution, as in Problems 6 & 7.

■ A Habit to Acquire

For the phase portraits of Problems 10-13 we focus on the *slope information* contained in the DEs, using the following general principles:

- Setting $x' = 0$ gives the v-nullcline of vertical slopes

- Setting $y' = 0$ gives the v-nullcline of vertical slopes

- The equilibria are located where an h-nullcline intersects a v-nullcline, i.e., where $x' = 0$ and $y' = 0$ simultaneously

- In the regions between nullclines the DEs tell whether trajectories move left or right (sign of x'), up or down (sign of y')

- The direction picture that results shows the stability of the equilibria

Note: if the trajectories circle around an equilibrium, further argument is necessary to distinguish between a center and a spiral (which could be either stable or unstable).

Note: For computer-drawn trajectories, the Runge-Kutta method will be far more accurate than Euler's method at answering these questions.

Note: Recall from Section 6.5 that an equilibrium with trajectories that head toward it in one direction and others that head away in *another* direction is a *saddle*. Unique trajectories (*separatrices*) head to or from a saddle and *separate* the behaviors.

12. $x' = x^2 - y + 2$ v-nullclines: $y = x^2 + 2$

    $y' = y + 2x$ h-nullclines: $y = -2x$

Equilibria: none, because h- and v-nullclines do not intersect.

The nullclines/direction figure indicates that trajectories come from the left and move to the right; those with large enough y-values to cross the parabola will head back to the left and move upward toward the left half of the parabola, otherwise they move toward the right forever. The trajectories and vector field confirm all of the above information; see figures.

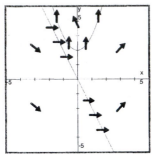

Nullclines and directions

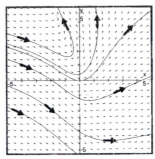

Typical trajectories

■   **Phase Portraits from Nullclines**

For Problems 14-19, note the general procedures listed with Problems 10-13. Note that these procedures when combined with computer pictures of trajectories and vector fields should give redundant information. That is, if ever any of these do not agree, you can know there is an error. Furthermore, it should not matter in what order you apply these procedures. (E.g., if the nullclines are difficult to plot, as in Problem 16, you might start with a computer phase portrait and vector field, then use the slope marks to locate and sketch approximately the nullclines.) *Focus on looking for (and checking for) consistency.*

15. $x' = y - \ln|x|$ v-nullclines: $y = \ln|x|$ or $x = \pm e^y$

    $y' = x - \ln|y|$ h-nullclines: $x = \ln|y|$ or $y = \pm e^x$

Equilibria: approximately at (-1.31, 0.27), (-0.57, -0.57), (0.27, 1.31)

The equilibria in the 2nd and 4th quadrants are saddle points (hence unstable); the equilibrium in the 3rd quadrant is an unstable node. The trajectories and vector field confirm all of the above information; see figures.

Questions arise, however. Note that in the computer pictures some trajectories cross the axes while others stop there. Should they cross or not? Technically the DEs are not defined when $x = 0$ or $y = 0$; why should they appear to have the same slopes on either side of an axis? Does symmetry help find answers? Learn to be on the lookout for issues like this that require additional analysis; even if you don't have answers, it is important to list any unresolved questions.

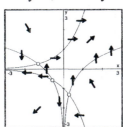
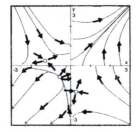
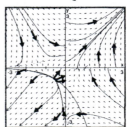

18. $x' = y - x^2 + 1$ v-nullclines: $y = x^2 - 1$

$y' = y + x^2 - 1$ h-nullclines: $y = -x^2 + 1$

Equilibria: (-1,0), 1,0)

The equilibrium at (-1, 0) is an unstable saddle. The equilibrium at (1,0) is an unstable spiral (the direction field is not symmetric about either x or y direction, and the direction arrows spiral us outward). The trajectories and vector field confirm all of the above information; see figures.

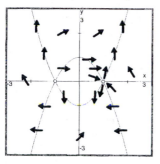

Nullclines and directions   Typical trajectories

■ Equilibria for Second-Order DEs

For second order DEs (e.g., Problems 20-25) we find a lovely shortcut to determining directions: Our first order system begins with introducing a new variable y for the first derivative, e.g., $x' = y$; this immediately determines that trajectories move

- to the right in the upper half plane,

- to the left in the lower half plane, and

- vertically when they cross the x-axis.

21. $\theta'' + (g/L)\sin\theta = 0$

(a),(b) Letting $y = \theta'$ we obtain the first order system

$\theta' = y$ v-nullclines: $y = 0$ (θ-axis)

$y' = -\left(\dfrac{g}{L}\right)\sin\theta$ h-nullclines: $\theta = \pi/n, \ n = 0, \pm 1, \pm 2...$

Equilibria: $(n/\pi, \theta), \ n = 0, \pm 1, \pm 2...$

(c),(d) The figures (for $g = L = 1$) show that the equilibrium points
                $(0,0), (\pm 2\pi, 0), (\pm 4\pi, 0)...$

are center points (hence stable), because trajectories near them circle around and form closed loops (by symmetry of the slope marks about the x-axis); equilibrium points
                $(\pm\pi, 0), (\pm 3\pi, 0), (\pm 5\pi, 0)...$

are saddles (hence unstable). The constant solutions $x(t) = 0, \pm 2\pi, 4\pi, \ ...$ of the second-order DE are stable. The constant solutions $x(t) = \pi, 3\pi, 5\pi ...$ are unstable.

continued on the next page

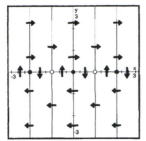

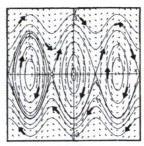

The periodic solutions are not limit cycles because none attract nearby solutions.

Nullclines and directions   Typical trajectories

24. $x'' + |x|x' + x = 0$

(a),(b) Letting $y = x'$ we obtain the first order system

$$x' = y$$ v-nullclines: $y = 0$ (x-axis)

$$y' = -|x|x' - x$$ h-nullclines: $x + |x|y = 0$;

i.e., the y-axis as well as $y = \begin{cases} 1, x < 0 \\ -1, x > 0 \end{cases}$

Equilibrium: (0,0)

(c),(d) The figures show that trajectories spiral into the origin (note that the slope marks are not symmetric about either axis). Hence the origin is a stable spiral equilibrium point, and $x(t) = 0$ is a stable solution of the second-order DE. There are no periodic solutions.

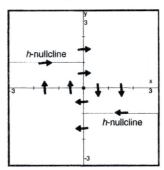

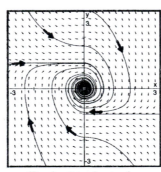

Nullclines and directions   Typical trajectories

■ Finding Equations for Trajectories

27. $x' = y, \; y' = x$

We write $\dfrac{dy}{dx} = \dfrac{y'}{x'} = \dfrac{x}{y}$

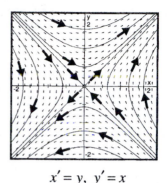

Separating variables yields $y\,dy = x\,dx$. Hence,

$$\frac{1}{2}y^2 = \frac{1}{2}x^2 + c$$

or $x^2 - y^2 = 2c$

$$x' = y, \; y' = x$$

This is a family of hyperbolas in the phase plane, which can be seen in the figure. The direction that trajectories follow can be determined by looking at the original system. From equations $x' = y$ and $y' = x$ we see that trajectories in the first and third quadrants move away from the origin, but in the second and fourth quadrants, trajectories move towards the origin.

30. $x' = 1, \; y' = x + y$

We write $\dfrac{dy}{dx} = \dfrac{y'}{x'} = x + y$,

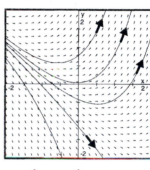

which we can solve easily as a linear equation, yielding

$$y = ce^x - x - 1$$

$$x' = y, \; y' = x + y$$

See figure for this family of curves in the phase plane. We can determine the direction in which solutions move along the trajectories because $x' > 0$ means x is always increasing.

■ Nonlinear Systems from Applications

33. $\dot{x} = y$

$\dot{y} = -x - \text{sgn}(y)$

Solving $x' = y' = 0$, we find the single equilibrium point $(0,0)$. We draw several trajectories showing that when the trajectory crosses the x-axis between -1 and +1 , the trajectory simply stops and turns around, only to turn around again and again, giving rise to a "chattering" motion.

Note: The figure suggests there exist equilibria at $(\pm 1, 0)$, but algebra shows they are not, because $y' = 1 \neq 0$ at those points.

We can interpret the solution physically as a vibrating spring represented by the single equation

$$x'' + \text{sgn}(x') + x = 0,$$

continued on the next page

where the friction always opposes the direction of motion of the spring with *constant* magnitude 1. When the displacement x is small, the friction force is stronger than the spring force $-x$, with the net result that when $-1 \le x \le 1$, trajectories simply chatter back and forth across the x-axis.

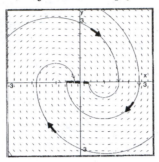
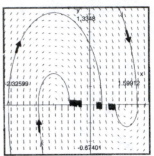

Trajectories of Coulomb damping   Zoom on chattering effect

Note: We saw this chattering in a phase portrait only when the approximation crosses the x-axis, which it does for Euler's method with $h = 0.5$. Our plots by Euler at $h = 0.1$ or by Runge-Kutta at $h = 0.5$ did not show the chattering phenomenon.

■   **Polar Limit Cycles**

Phase portraits for Problem 36 can be most easily sketched by hand. However, if you wish to make a computer drawing with an xy DE solver, note that

$$x'(r\cos\theta)' = r'\cos\theta - r\theta'\sin\theta$$
$$y'(r\sin\theta)' = r'\sin\theta + r\theta'\cos\theta$$

Rewrite the right-hand sides in terms of x and y using

$$\cos\theta = x/r, \ \sin\theta = y/r, \ r = \sqrt{x^2 + y^2}$$

and the given expressions for r' and θ'.

36. $r' = r(a - r)$

$\theta' = 1$

Because $r' = 0$ when $r = 0$ or when $r = a$, we have an equilibrium at the origin and a closed trajectory at $r = a$ in the xy phase-plane.

The equation $\theta' = 1$ tells us that trajectories rotate around the origin at a constant angular velocity (1 radian per unit time) in the counterclockwise direction, regardless of r-value, so the origin is the only equilibrium.

We also note that $r' > 0$ for $0 < r < a$, and that $r' < 0$ for $r > a$, so xy-trajectories both inside and outside the circle $r = a$ approach it asymptotically. Hence $r = a$ is a stable limit cycle.

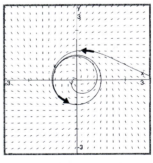

$r' = r(2 - r), \ \theta' = 1$

■   **Testing Existence and Uniqueness**

39. $x' = 1 + x = f(x, y)$

$y' = (1 + x)\sqrt{y} = g(x, y)$

(a)   The domain of the system of differential equations is the upper-half xy-plane; that is, all points (x, y) for which $y \geq 0$. Also because f is continuous everywhere, and g is continuous for $y \geq 0$, there exists a solution passing through each point (x_0, y_0) for which $y_0 \geq 0$ (the existence theorem does not tell how far the solution can be extended).

We now compute for partial derivatives

$$\frac{\partial f}{\partial x} = 1 \qquad \frac{\partial f}{\partial y} = 0$$

$$\frac{\partial g}{\partial x} = \sqrt{y} \qquad \frac{\partial g}{\partial y} = \frac{1 + x}{2\sqrt{y}}$$

Which are continuous for $y > 0$, and so there exists a unique solution passing through each point (x_0, y_0) for which $y_0 > 0$.

(b)   The direction field as shown in the figure seems to confirm the analysis so far, and looks suspiciously nonunique along y = 0 as indeed we have seen in Section 1.5, Example 3.

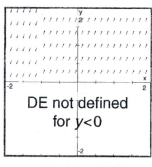

DE not defined
for y<0

Direction field

However, this is definitely not the whole story for the system given. Although the system reduces to $y' = \sqrt{y}$ when t is eliminated, and although the extended theorem assures us there will be a unique solution whenever $y > 0$, the fact that we have a system of *two* equations introduces the need to also look at them separately.

If you think about left/right or up/down directions, or if you use phase-plane software, you will notice that the arrowheads on the slope marks on the far left of our window have to point in the opposite direction from those on the right.

Furthermore, if you seek equilibria by setting $x' = 0$ and $y' = 0$, you will see that equilibria cover the entire half-line $x = -1$, separating the trajectories heading to the upper right from those heading to the lower left. Notice that this does *not* contradict the statements above re existence and uniqueness, but rather emphasizes the wisdom of going as far as you can with analyzing a phase portrait. In particular, it shows that attention to direction arrows is essential to proper mastery of a system.

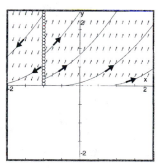

Phase portrait of
$x' = 1 + x, \; y' = (1 + x)\sqrt{y}$

■   **Computer Lab: Phase Plane Analysis**

Computer phase portraits, as in Problems 42-47, must clearly indicate the *directions* of the trajectories, with arrows visible either on the vector field or on the trajectories. For the tiny figures in this manual we have chosen the latter as being the most easily readable.

42. $x' = x(x - y)$

$y' = y(1 - y)$

(a),(b)   See the figure for the vector field with sample trajectories (and the v-nullcline).

(c),(d)   The equilibria are at $(0,0)$, $(0,1)$, $(1,1)$. The figure shows that $(0,1)$ is stable and attracting; $(1,1)$ is a saddle (unstable); $(0,0)$ is unstable in an interesting way (a saddle on the left and a node on the right!).

(e),(f)   The long-term behavior of this system depends on the initial conditions. Most trajectories starting in the upper half plane head toward $(0,1)$, while some are deflected by the saddle on the right and move toward $y = 1$ then toward infinity. Trajectories starting in the lower half plane move downward toward negative infinity.

The phase portrait does not detect any periodic solutions, which would appear as closed loop trajectories.

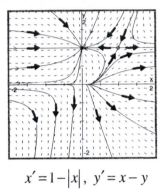

$x' = 1 - |x|, \ y' = x - y$

45. $x' = x(2 - x - y)$

$y' = -y$

(a),(b)   See the figure for the vector field, plus the v-nullcline, with sample trajectories.

(c),(d)   The equilibria are at $(0,0)$ and $(2,0)$. The figure shows that $(0,0)$ is unstable (a saddle) and $(2,0)$ is stable.

(e)   The long-term behavior of this system depends on the initial conditions. For $x > 0$, trajectories move toward the stable equilibrium at $(2,0)$. For $x < 0$, trajectories approach the x-axis and go off to $-\infty$.

(f)   The phase portrait does not detect any periodic solutions, which would appear as closed loop trajectories.

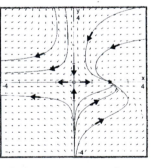

$x' = x(2 - x - y), \ y' = -y$

■   **Computer Lab: Graphing in Two Dimensions**

48.   Student Lab Projects with IDE

■   **Suggested Journal Entry**

51.   Student Project

7.2   Linearization

Review of Classifications

At a given nonlinear equilibrium of a system $x' = f(x, y)$, $y' = g(x, y)$, we can use the Jacobian matrix

$$\mathbf{J} = \begin{bmatrix} f_x(x, y) & f_y(x, y) \\ g_x(x, y) & g_y(x, y) \end{bmatrix}$$

to quickly and algebraically calculate the stability, by finding either

- the eigenvalues of $\mathbf{J}$ (Reference: Table 7.2.1 in the text)

or

- the location of the trace and determinant of $\mathbf{J}$ in the linear classification diagram. (Reference: Figure 7.2.7 in the text)

Occasionally for a nonlinear equilibrium further analysis is still necessary.

- If the linearization is a center, the nonlinear equilibrium could be a center or a spiral of either stability.

- Sometimes an equilibrium is a combination, due to a linearization that is a borderline case (i.e., degenerate). E.g., you could find a nonlinear equilibrium to be a saddle on one side and a node on the other.

- Sometimes there is a whole line or curve of equilibria.

For each of Problems 1-19 we show various ways of reaching the conclusions – a phase portrait of the nonlinear system, the linearizations about each equilibrium, each with its Jacobian and stability analysis, and sometimes small zooms of these linearizations. *Focus on the elements that work best for you, in whatever order.*

■   **Original Equilibrium**

3.
$$x' = x + y + 2xy$$
$$y' = -2x + y + y^3$$

$$\mathbf{J}(x, y) = \begin{bmatrix} 1 + 2y & 1 + 2x \\ -2 & 1 + 3y^2 \end{bmatrix}$$

At $(0,0)$, $x' = y' = 0$, so the origin is an equilibrium point. The phase portrait, linearization, and Jacobian calculations are shown.

Calculations are shown on the next page.

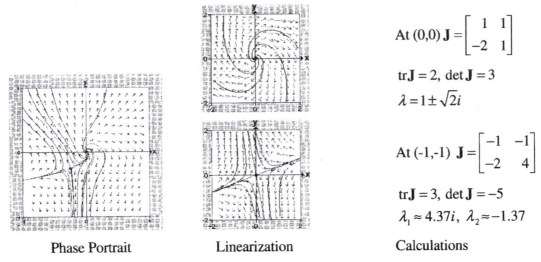

| Phase Portrait | Linearization | Calculations |

At $(0,0)$ $\mathbf{J} = \begin{bmatrix} 1 & 1 \\ -2 & 1 \end{bmatrix}$

$\mathrm{tr}\mathbf{J} = 2$, $\det\mathbf{J} = 3$

$\lambda = 1 \pm \sqrt{2}i$

At $(-1,-1)$ $\mathbf{J} = \begin{bmatrix} -1 & -1 \\ -2 & 4 \end{bmatrix}$

$\mathrm{tr}\mathbf{J} = 3$, $\det\mathbf{J} = -5$

$\lambda_1 \approx 4.37i$, $\lambda_2 \approx -1.37$

We conclude that $(0,0)$ is an *unstable spiral* (complex eigenvalues with positive real part). We also know that $(-1, -1)$ is a *saddle* (real eigenvalues of opposite sign).

6. $x' = \sin y^2$
         $y' = -\sin x + y$
         $\mathbf{J}(x, y) = \begin{bmatrix} 0 & \cos y \\ -\cos x & 1 \end{bmatrix}$

At $(0,0)$ $x' = y' = 0$, so the origin is an equilibrium point. There are other equilibria at $(\pm n\pi, 0)$ for any integer n. We will investigate those that appear in the phase portrait shown.

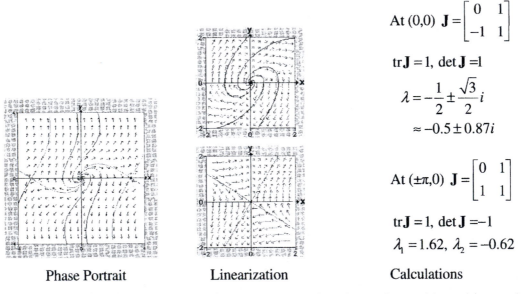

| Phase Portrait | Linearization | Calculations |

At $(0,0)$ $\mathbf{J} = \begin{bmatrix} 0 & 1 \\ -1 & 1 \end{bmatrix}$

$\mathrm{tr}\mathbf{J} = 1$, $\det\mathbf{J} = 1$

$\lambda = -\dfrac{1}{2} \pm \dfrac{\sqrt{3}}{2}i$

$\approx -0.5 \pm 0.87i$

At $(\pm\pi, 0)$ $\mathbf{J} = \begin{bmatrix} 0 & 1 \\ 1 & 1 \end{bmatrix}$

$\mathrm{tr}\mathbf{J} = 1$, $\det\mathbf{J} = -1$

$\lambda_1 = 1.62$, $\lambda_2 = -0.62$

We conclude that (0.0) is an *unstable spiral* (complex eigenvalues with positive real part). The equilibria at $(\pm\pi, 0)$ are both *saddles* (real eigenvalues of opposite sign), while those at $(0, \pm\pi)$ are both *unstable spirals* (complex eigenvalues with positive real part). We leave it to the reader to make similar calculations and conclusions for other equilibria.

■ Unusual Equilibria

9.

$$x' = 4x - x^3 - xy^2$$
$$y' = 4y - x^2 y - y^3$$

$$\mathbf{J}(x,y) = \begin{bmatrix} 4 - 3x^2 - y^2 & -2xy \\ -2xy & 4 - x^2 - 3y^2 \end{bmatrix}$$

We find equilibria at (0,0) and (2/3,2/5). The phase portrait, linearizations, and Jacobian calculations are shown.

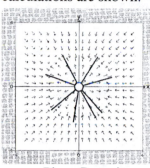

Phase Portrait

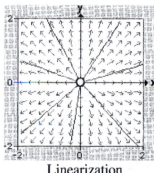

Linearization

At (0,0)

$$\mathbf{J} = \begin{bmatrix} 4 & 0 \\ 0 & 4 \end{bmatrix}$$

$tr \mathbf{J} = 8$, det $\mathbf{J} = 16$

double eigenvalue $\lambda = 4$

Calculations

We conclude that the origin is an *unstable node* (real eigenvalues, both positive), and could easily be fooled into thinking that was the end of the story. *But*, a sharper eye either to the algebra or to computer trajectories shows there is more – a whole circle of equilibria where $x^2 + y^2 = 4$.

Because each point on the circle has different coordinates, it is difficult to examine the stability of these equilibrium points using the Jacobian. However if from the nonlinear system we write

$$\frac{dy}{dx} = \frac{y'}{x'} = \frac{y}{x'}$$

we can see that the trajectories in the phase plane are simply straight lines $y = cx$. Analysis of the signs of x' and y' shows by the quadratic factor that movement is always toward the circle $x^2 + y^2 = 2$. Hence, any trajectory beginning at an initial condition outside the circle will move towards the circle. The points on the circle are stable equilibria. In the final figures we show sample linearizations for two such equilibrium points.

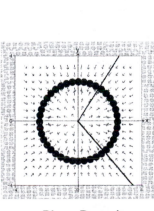

Phase Portrait

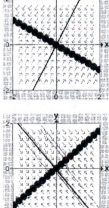

Linearization

At (0.953,1.759)

$$\mathbf{J} \approx \begin{bmatrix} -1.8 & -3.35 \\ -3.35 & -6.2 \end{bmatrix}$$

$tr\mathbf{J} = -8$, det $\mathbf{J} = 0$

$\lambda_1 = 0$, $\lambda_2 = -8$

At (1.275,-1.541)

$$\mathbf{J} \approx \begin{bmatrix} -3.25 & 3.93 \\ 3.93 & -4.75 \end{bmatrix}$$

$tr\mathbf{J} = -8$, det $\mathbf{J} = 0$

$\lambda_1 = 0$, $\lambda_2 = -8$

Calculations

$$x' = 4x - x^3 - xy^2, y' = 4y - x^2 y - y^3$$

■   **Uncertainty**

12. $x'' + x' + x + x^3 = 0$ can be written as a system

$x' = y$

$y' = -y - x + x^3$

$$J = \begin{bmatrix} 0 & 1 \\ -1 + 3x^2 & -1 \end{bmatrix}$$

Setting $x' = y' = 0$ we find a single equilibria at (0,0), (1,0) and (-1,0) which we analyze as follows.

At (0,0)

$$J = \begin{bmatrix} 0 & 1 \\ -1 & -1 \end{bmatrix}$$

$\text{tr}\,J = -1$, $\det J = 1$

$\lambda = (-1 \pm \sqrt{3}i)/2$

Stable spiral

At (±1,0)

$$J = \begin{bmatrix} 0 & 1 \\ 2 & -1 \end{bmatrix}$$

$\text{tr}\,J = -1$, $\det J = -2$

$\lambda_1 = -2$, $\lambda_2 = 1$

Saddles

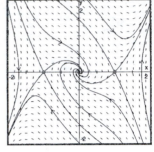

$x' = y, \quad y' = -y - x + x^3$

■   **Predator-Prey Equations**

15. $x' = (a - by)x = f(x, y)$

$y' = (cx - d)y = g(x, y)$

(a)   The Jacobian of this system is

$$J(x, y) = \begin{bmatrix} f_x & f_y \\ g_x & g_y \end{bmatrix} = \begin{bmatrix} a - by & -bx \\ cy & cx - d \end{bmatrix}$$

so at the equilibrium point $\left(\dfrac{d}{c}, \dfrac{a}{b} \right)$

$$J = \begin{bmatrix} 0 & -\dfrac{bd}{c} \\ \dfrac{ac}{b} & 0 \end{bmatrix}$$, with eigenvalues $\pm i\sqrt{ad}$. Hence,

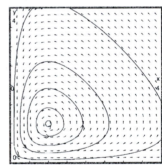

Trajectories of
the predator-prey system

the equilibium point $\left(\dfrac{d}{c}, \dfrac{a}{b} \right)$ could be either a center or a spiral of unknown stability.

The phase plane portrait for this system when $a = b = c = d = 1$ shows the equilibrium point (1,1) as a center. (See the answer for Problem 9 in section 2.6.)

■   Damped Mass-Spring Systems

18. $\ddot{x} + \dot{x} - \dot{x}^3 + x = 0$

Writing the equation as a system yields

$$\dot{x} = y$$
$$\dot{y} = -x - y + y^3$$

Setting $\dot{x} = \dot{y} = 0$ yields the equilibrium point $(0,0)$. The linearized equations about this point are

$$\begin{bmatrix} \dot{x} \\ \dot{y} \end{bmatrix} = \begin{bmatrix} 0 & 1 \\ -1 & -1 \end{bmatrix} \begin{bmatrix} x \\ y \end{bmatrix},$$

which has eigenvalues

$$\lambda_1, \lambda_2 = -\frac{1}{2} \pm i\frac{\sqrt{3}}{2}.$$

Hence, the origin is an asymptotically stable spiral point of the nonlinear system. Intuition suggests the zero solution $x = \dot{x} = 0$ is stable because in a neighborhood of zero the positive damping term $\dot{x}$ is larger than the negative damping term $-\dot{x}^3$.

The phase portrait for this nonlinear system shows a periodic solution (limit cycle) that is unstable.

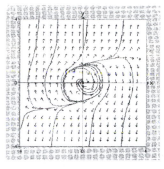

Trajectories for
$\dot{x} = y, \ \dot{y} = -x - y + y^3$

Liapunov Functions

21. $x' = y - 2x^3$ with $L(x,y) = 2x^2 + y^2$

$\quad y' = -2x - 3y^5$

Solving $x' = y' = 0$, we see that $(0,0)$ is an equilibrium, and analysis of the linearization tells us it could be either a center or a spiral, of either stability. The phase portrait does not readily answer the question, because the trajectories go ever more slowly as they approach the origin; a legitimate question is whether there is a limit cycle surrounding the origin. Liapunov's direct method gives a quick answer.

The given function $L(x,y)$ is clearly positive definite. Furthermore, calculation shows that

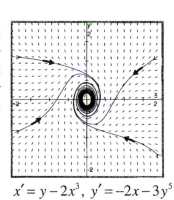

$x' = y - 2x^3, \ y' = -2x - 3y^5$

$$\begin{aligned}
\frac{dL}{dt} = L_x x' + L_y y' &= 4xx' + 2yy' \\
&= 4x(y - 2x^3) + 2y(-2x - 3y^5) \\
&= -(8x^4 + 6y^6)
\end{aligned}$$

is negative definite. Hence Liapunov's result tells us that the origin is asymptotically stable and there is no limit cycle.

■   **Computer Lab: Trajectories**

24. $x'' + x \sin x = 0$

Letting $y = \dot{x}$, the equation can be written as the first-order system

$$x' = y = f(x, y)$$
$$y' = -x \sin x = g(x, y)$$

The phase plane trajectories of this system can be studied by looking at the direction field of

$$\frac{dy}{dx} = \frac{y'}{x'} = -\frac{x \sin x}{y}$$

in the xy plane. See figure.

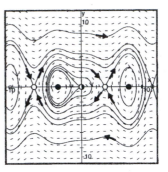

The nonlinear system has equilibria at the points $(\pm n\pi, 0)$. We examine the stability of the nonlinear system at the three equilibria $(\pm \pi, 0)$ and $(0,0)$. The Jacobian of the system is

$$J = \begin{bmatrix} f_x & f_y \\ g_x & g_y \end{bmatrix} = \begin{bmatrix} 0 & 1 \\ -\sin x - x \cos x & 0 \end{bmatrix}$$

and so the Jacobians at the three equilibrium points are

$$\mathbf{J}(0,0) = \begin{bmatrix} 0 & 1 \\ 0 & 0 \end{bmatrix} \qquad \mathbf{J}(-\pi, 0) = \begin{bmatrix} 0 & 1 \\ -\pi & 0 \end{bmatrix} \qquad \mathbf{J}(\pi, 0) = \begin{bmatrix} 0 & 1 \\ \pi & 0 \end{bmatrix}$$

At $(0,0)$ the Jacobian is singular indicating that the linear system ($x' = y$, $y' = 0$) does not have an isolated equilibria at the origin. Hence, the nonlinear system cannot be linearized about the origin and shows nothing about the stability of the nonlinear system at $(0,0)$. See figure for phase drawing that indicates $(0,0)$ is an unstable equilibrium point.

At $(-\pi,0)$ the Jacobian of the linearized system has eigenvalues $\lambda = \pm i\sqrt{\pi}$; hence, the nonlinear system has either a center or stable or unstable spiral point at $(-\pi, 0)$. The figure shows it to be a center.

At $(\pi,0)$ the Jacobian of the linearized system has eigenvalues $\lambda = \pm\sqrt{\pi}$, and so the nonlinear system has an unstable saddle at $(\pi,0)$, as shown in the figure.

In fact, as we see from the figure, all along the x-axis the equilibria at multiples of π alternate between saddles and centers.

27. $x'' + x - 0.25x^2 = 0$

Letting $y = x'$, the equation can be written as the first-order system

$$x' = y$$
$$y' = -x + 0.25x^2$$

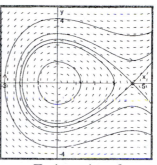

having equilibrium points $(0,0)$ and $(4,0)$ whose trajectories in the phase plane can be studied by looking at the direction field of

$$\frac{dy}{dx} = \frac{y'}{x'} = -\frac{-x + 0.25x^2}{y}$$

Trajectories of
the nonlinear system

in the xy plane. Note that the trajectories cross the x-axis in a vertical manner due to the y in the denominator of $\dfrac{dy}{dx}$.

The linearized equation at equilibrium $(0,0)$ are

$$x' = y$$
$$y' = -x$$

whose trajectories are the circles drawn in Problem 25. The linearization at $(4,0)$ can also be determined and classified.

■   **Suggested Journal Entry II**

30.   Student Project

7.3 Numerical Solutions

■ Spreadsheet Calculation

3. $x' = y$ $x(0) = 1$

$y' = -x - y^3$ $y(0) = 1$

We approximate the solution of the IVP on the interval $0 \leq t \leq 1$ using Euler's method with step sizes $h=0.1$ and $h=0.05$. The results are summarized in the following table and figure.

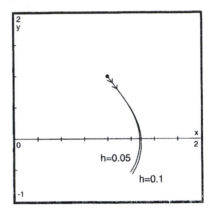

Euler's Method with h=0.1, 0.05				
	(h=0.1)		(h=0.05)	
t	x	y	xdot	ydot
0	1	1	1	1
0.1	1.1	0.8	1.0950	0.8111
0.2	1.18	0.6388	1.1720	0.6534
0.3	1.2439	0.4947	1.2337	0.5108
0.4	1.2934	0.3582	1.2814	0.3752
0.5	1.3292	0.2243	1.3156	0.2420
0.6	1.3516	0.0903	1.3365	0.1089
0.7	1.3606	-0.045	1.3440	-0.0251
0.8	1.3561	-0.181	1.3381	-0.1594
0.9	1.338	-0.3161	1.3189	-0.2921
1	1.3064	-0.4467	1.2864	-0.4197

Comparison of Euler numerical approximations

The Euler approximation using the smaller step size is consistently lower in x and higher in y than that using the larger step size. We expect the exact solution to follow this pattern.

■ Changing Views

An xy phase portrait does not indicate the speed at which a point moves while tracing a trajectory. We can observe more effects of using Euler's method with different step sizes by looking at tx and ty graphs. In all cases a tighter curvature occurs with smaller step size.

A *scaled* vector field shows *speed*, helping to give a rough idea of how these three graphs relate. Note that in some cases we should have used larger bounds for x and y.

6. $x' = x + y$ $x(0) = 0$

$y' = x + y$ $y(0) = 0.1$

For $0 \leq t \leq 2$, the figures show the requested views and include the scaled vector field, for a sample IC. The phase portrait is the same for both step sizes; we cannot see that it gives larger x and y more quickly for smaller h.

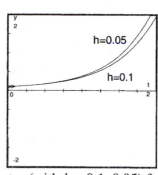
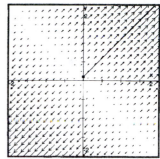

Euler approximates (with $h = 0.1, 0.05$) for $x' = x + y$, $y' = x + y$

(b)   For this system we find speed $= \sqrt{(x')^2 + (y')^2} = \sqrt{2}\,|x + y|$, which increases with distance from the line $x + y = 0$ and causes tx and ty curves to turn upward.

9. $x' = y$ $x(0) = 0$

$y' = -x - x^3$ $y(0) = 1$

For $0 \le t \le 2\pi$, the figures show the requested views and include the scaled vector field.

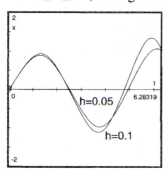
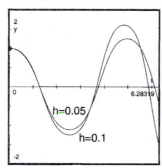
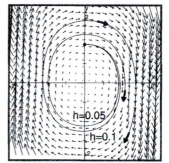

Euler approximates for $x' = y$, $y' = -x - x^3$

(b)   For this system we find speed $= \sqrt{(x')^2 + (y')^2} = \sqrt{y^2 + x^2 + 2x^4 + x^6}$. Compare both the speed and the trajectories with those of Problem 7, which uses the same initial condition. Here the slopes become steeper as $|x|$ increase (causing "cycles" to be vertically elongated) and faster as well (causing a cycle to be completed for $t < 2\pi$). Note that the smaller stepsize, the longer the period between "cycles". As in Problem 7, the exact solution would produce closed cycles.

■   **Changing Parameters**

12. $x' = y$, $y' = -x + \varepsilon(1 - x^2)y$ The phase portraits shown illuminate the role of ε.

The only equilibrium is at $(0,0)$, for all values of ε. Critical values of ε are $\varepsilon = 0$, where the origin's stability changes, and $\varepsilon = \pm 2$, where the origin is between a node and a spiral.

For $\varepsilon < 0$, the origin is stable, but we see in the phase portraits there is an unstable limit cycle that sends far away trajectories off toward infinity. At $\varepsilon = 0$ (not shown) van der Pol's equation is simply the harmonic oscillator, with circular clockwise trajectories centered at the origin.

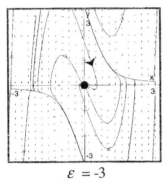
$\varepsilon = -3$

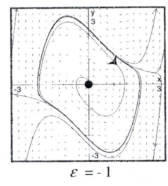
$\varepsilon = -1$

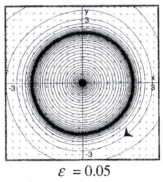
$\varepsilon = 0.05$

For $\varepsilon > 0$, the origin is unstable, but we find a stable limit cycle, drawing in trajectories from far away as well as from the origin. This case is of the most interest, because the long term bahavior will be periodic. Note: As ε increases, the limit cycles become increasingly irregular in shape.

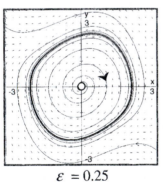
$\varepsilon = 0.25$

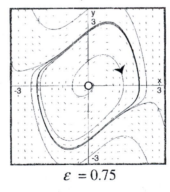
$\varepsilon = 0.75$

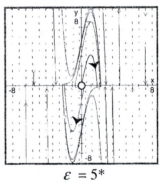
$\varepsilon = 5*$

When $\varepsilon > 1$, the van der Pol system experiences "relaxation oscillations", where energy is slowly stored and then suddenly released almost instantaneously. Note in the phase portrait for $\varepsilon = 5$ that trajectories head straight for the x-axis, only at the last moment they turn almost at right angles, indicating a sudden change of motion. The final figures shows $x(t)$ and $y(t)$ for $\varepsilon = 5$, to show these relaxation oscillations. With an initial condition close to the origin, we note that the solution reaches periodicity almost instantaneously.

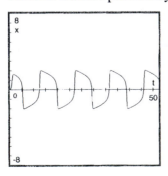

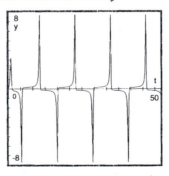

$x(t)$ and $y(t)$ for $\varepsilon = 5$

Note the change of scale for $\varepsilon = 5$. Axes now extend to 8 in all directions in order to catch the height of the limit cycle.

■   **Euler for 3x3 Systems**

For Problem 15 we extended the simple spreadsheet with Euler's method used in Problems 1-4, by inserting columns for z and $z$dot. We graph x, y, and z as functions of t, and summarize the numerical results in a table.

Where the graphs show very different results for step sizes $h = 0.05$ and 0.1 you know Euler is not giving a good approximation. If you needed numerical accuracy, you would want to use a more accurate method, such as Runge-Kutta (See Problems 16-22), which would require a more elaborate spreadsheet or DE software that can handle a system of three DEs. We have done this and provided the results from Runge-Kutta (RK) for comparison with Euler (E).

15.

$$x' = x + y \qquad x(0) = 2$$

$$y' = -x + tz \qquad y(0) = 1$$

$$z' = z + x^2 \qquad z(0) = 1$$

Method	h	$x(1)$	$y(1)$	$z(1)$
Euler	0.1	5.9022	1.7499	20.8353
Euler	0.05	6.1817	2.4982	23.0535
Runge-Kutta	0.1	6.4202	2.9697	25.5269
Runge-Kutta	0.05	6.4993	3.2136	25.7431

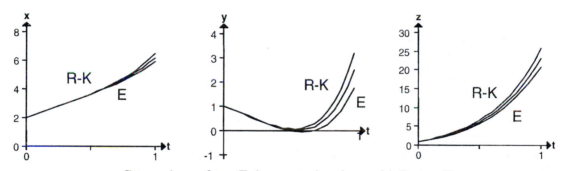

Comparison of two Euler approximations with Runge-Kutta

■   **Bug Race**

18.   We solve the three IVPs in the phase plane, with $x(0)=0$, $y(0)=1$.

A: $x' = y$
     $y' = -x$

B: $x' = y$
     $y' = -x + x^3$

C: $x' = y$
     $y' = -x - x^3$

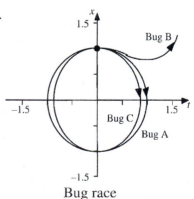

Bug race

continued on next page

Bug A returns to the starting point when $t = 2\pi$, because we know that the solution of that system is $x(t) = \sin t$.

From the phase portrait, we see that Bug B does not have a chance; her DE takes her to infinity. Bug C however has an inside track, always advancing faster up or down than A. Bug C returns to the starting point first, therefore C wins the race.

■ Runge-Kutta Method

In Problems 19-24, we compare the numerical approximation of $x(t)$ using Euler's method (from Problems 1-6) and the Runge-Kutta method, each at step sizes $h = 0.05, 0.1$. The tables show that the Runge-Kutta method refines the approximation beyond (but in the same direction) as using Euler's method with a smaller step size. Note also that in most cases Runge-Kutta does such a good job that its approximation does not change appreciably with step size, and is very close to the exact solution.

We graph $x(t)$ for both methods, using $h = 0.1$. Several graphs are extended past $t = 1$ until we can predict long-term behavior, to show how sometimes approximations alternately diverge and converge.

21. $x' = y$ $x(0) = 1$

$y' = -x - y^3$ $y(0) = 1$

Method	h	$x(1)$	$y(1)$
Euler	0.1	1.3064	- 0.4467
Euler	0.05	1.2864	- 0.4197
Runge-Kutta	0.1	1.2667	- 0.3939
Runge-Kutta	0.05	1.2667	- 0.3939

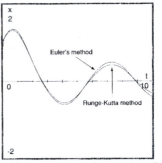

$x(t)$ approximations, $h = 0.1$

This system is the same as that of Problem 3. where the xy graph confirms that for t between 0 and 1, $x(t)$ rises and then falls as t increases. Extending the t axis shows that $x(t)$ will exhibit damped oscillation, which is appropriate because the system is close to a damped harmonic oscillator for small y.

We observe that here the maximal separations become more pronounced as t becomes larger, and that here both approximations tend to cross the t-axis at the same time. Because these equations are even closer to a damped harmonic oscillator for small y than those of Problem 17 these observations seem reasonable.

24. $\quad x' = x + y \qquad\qquad\qquad x(0) = 1$

$\qquad\quad y' = x + y \qquad\qquad\qquad y(0) = 1$

Method	h	$x(1)$	$y(1)$
Euler	0.1	6.1917	6.1917
Euler	0.05	6.7275	6.7275
Runge-Kutta	0.1	7.3889	7.3889
Runge-Kutta	0.05	7.3890	7.3890

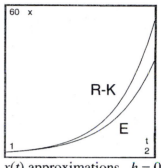

$x(t)$ approximations, $h = 0.1$

Both table and graph emphasize that Runge-Kutta approximations shoot up sooner than Euler approximations when slopes are becoming steeper. In Problem 6 we chose as initial condition, (0, 0.1), to be nonsymmetric and more general than (1,1). But as you can predict from the DEs, the Runge-Kutta approximations will show similar steeper slopes than Euler.

■ Hopf Bifurcation

27. Student Lab Project with IDE

7.4   Chaos, Strange Attractors & Period Doubling

■   Long-Term Behavior

3.   For the Lorenz equations (Problem 1) with two sets of initial conditions as in Problem 2, we let the time series of $x(t)$ run to $t = 100$. Here we show the ty and tz graphs as well, and observe that all the time series show divergence before $t = 10$. On the other hand, with longer time series the xyz phase portraits show fewer gaps and tend to look more similar than in Problem 2.

(a) $x(0) = -6.0$, $y(0) = 12$, $z(0) = 12$

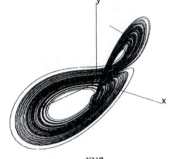

xyz

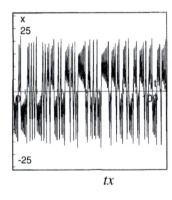

tx

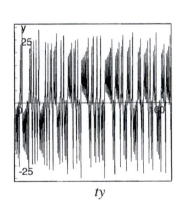

ty

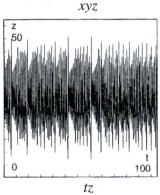

tz

(b) $x(0) = -6.01$, $y(0) = 12$, $z(0) = 12$

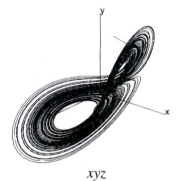

xyz

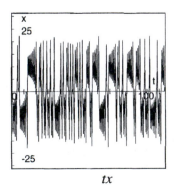

tx

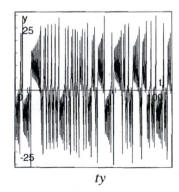

ty

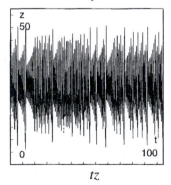

tz

■ Roessler Views

6.
$$\dot{x} = f(x,\, y,\, z) = -y - z$$
$$\dot{y} = g(x,\, y,\, z) = x + 0.2y$$
$$\dot{z} = h(x,\, y,\, z) = -5.7z + xz + 0.2$$

> For good approximations in a sensitive chaotic system we use the Runge-Kutta method with $h = 0.05$ to create the phase plane and time series graphs in the figures.

(a)   The three different 2-D phase plane projections of the Roessler system are shown below, with IC $(0,0,0)$ and $0 \le t \le 100$. The xy trajectory is attracted to a somewhat periodic looping. The xz and yz trajectories show that z never goes negative, at least for this IC.

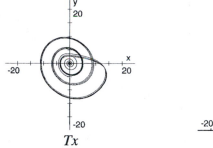

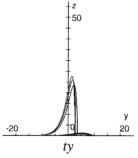

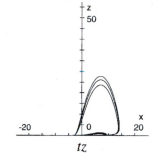

$$Tx \qquad\qquad ty \qquad\qquad tz$$

(b)   Because there is cyclic motion in the xy-plane, we expect motion in the tx- and ty-planes to oscillate; in fact the time series shown, for $0 \le t \le 100$, appear to move into an approximate 3-cycle after $t = 50$. We also observe that $z(t)$ peaks more randomly, spends a lot of time at $z = 0$, and never goes negative, at least for this IC $(0,0,0)$.

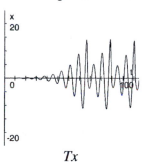

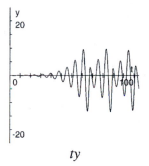

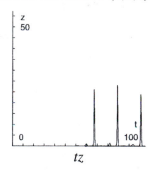

$$Tx \qquad\qquad ty \qquad\qquad tz$$

CAUTION: We must *not* assume too much from these graphs, which *seem* to settle into a pattern. If we let t run on to 250, we see some surprises (below). Between $t=150$ and $t=200$ we see a burst of *chaotic* motion (which maybe settles down later to a triple cycle pattern like that observed above, but now we should have some doubts. Further exploration and analysis is in order. Such "*intermittency*", alternating chaos with approximate order, can indeed occur in nonlinear dynamics. Beware of jumping to conclusions too soon!

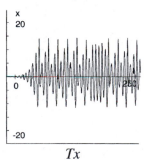

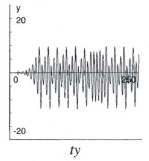

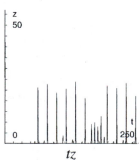

$$Tx \qquad\qquad ty \qquad\qquad tz$$

Extended time series, to $t = 250$

■   **Sensitivity**

8.

$$\dot{x} = -y - z \qquad\qquad x(0) = 1$$
$$\dot{y} = x + 0.2y \qquad\qquad y(0) = 0$$
$$\dot{z} = -5.7z + xz + 0.2 \qquad z(0) = 0$$

> To make comparisons for sensitive systems:
> - vary only one aspect at a time;
> - superimpose graphs (with different colors) or align and highlight comparisons with ruler lines.

(a)   **Sensitivity to numerical method:** the following graphs show a great sensitivity in the numerical solution x, y to the approximate method used for the *same* initial condition and stepsize. The equations are so sensitive to small differences in x, y, z that after a short time the solutions show no resemblance to each other.

The figures show the nonstability of Roessler's equations between

Euler's method (heavy curve);

Runge-Kutta (lighter curve).

Fixed stepsize $h = 0.05$

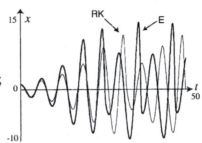
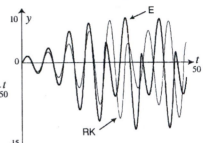

tx ty

(b)   **Sensitivity to stepsize:**

Euler approximations of Roessler's equations show nonstability with stepsizes $h = 0.1$ (heavy curve), $h = 0.05$ (lighter curve).

Note that it takes little time before tx approximations diverge.

Euler approximations (tx, $0 \le t \le 50$)

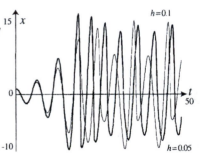

On the other hand the *Runge-Kutta* method is not as sensitive to stepsizes. For $h = 0.1$ (heavy curve), $h = 0.05$ (lighter curve),

Runge-Kutta approximations (see figure) are so close together that the difference is much smaller (barely distinguishable for much of the graph; we note also that oscillations are smaller than with Euler approximations above, because Runge-Kutta more closely approximates exact solutions.

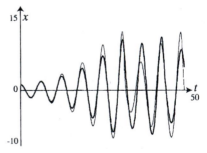

Runge-Kutta approximations (tx, $0 \le t \le 50$)

(c)   **Sensitivity to solver:** See Problem 2; the final note that gives a last figure to compare with the first figure is a good example. You'll find many more whenever you try to reproduce someone else's pictures of a sensitive or chaotic system.

(d)   We have seen how numerical approximations can differ greatly for small differences in initial conditions, stepsizes, methods, and computer numerics (Problems 2-3, 7 & 8). These differences will be most obvious in time series (tx, ty, tz) pictures; strange attractors in phase space pictures for a long run of t will look much more similar . When you notice differences that start small and get bigger, you can try to minimize them by using smaller stepsizes and more robust approximation methods (such as Runge-Kutta).

■   **Varying Lorenz**

9.   Student Lab Project with IDE

■   **Period Doubling Elsewhere**

12.   (a)   The direction field for $y' = y(1-y)$ is shown in the figure. We see that $y = 1$ is a stable equilibrium point and $y = 0$ is an unstable point. Solutions starting at positive $y(0) \neq 1$ approach 1.

NOTE: $t = nh$ $\qquad\qquad$ $y' = y(1-y)$

(b)   We show Euler's method with various stepsizes h .

When $h = 1.8$ Euler's method reaches steady state of 1.000000 after 54 iterations using Microsoft Excel and 6-place accuracy. The figure shows ever-smaller oscillations around 1 just before reaching 1.000000.

1-cycle: 1.000 $\qquad\qquad\qquad$ $h = 1.8$

When $h = 2.2$ Euler's approximation becomes periodic with repeating values 1.162844, 0.746247.

2-cycle: 11.162844, $\qquad\qquad$ $h = 2.2$
0.746247

When $h = 2.5$ we have period doubling where now the numbers repeat in blocks of four.

4-cycle: 1.1578, 0.7012, $\qquad\qquad$ $h = 2.5$
1.2250, 0.5359

When $h = 2.55$ we have another period doubling with the numbers repeating in blocks of eight (to six-place accuracy after 114 iterates).

8-cycle: 11.1313, 0.7524, 1.2274, $\qquad$ $h = 2.55$
0.5156, 1.1524, 0.7045,
1.2354, 0.4939

When $h = 2.57$ there is no periodicity (to six places) of any period in the numbers in the table, although an extended graph shows the sequence is nearly periodic.

n from 1500 to 2000 $\qquad\qquad$ $h = 2.57$

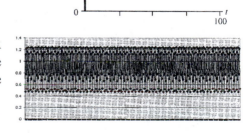

continued on the next page

When $h = 2.6$ the number show no periodicity of any kind to six places. The extended graph shows repeated highs and lows but with no sustained regularity.

n from 1500 to 2000 $h = 2.6$

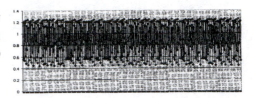

(c)   Experimenting with the spreadsheet program Microsoft Excel, we let h increase in value as we look for the h-values where the behavior changes qualitatively, we used $y(0) = 0.5$.

At $h = 1$ we observe the first bifurcation:

When $h \leq 1$, Euler approximations approach 1 from below;
When $h > 1$, Euler approximations oscillate about 1 before settling at 1.000000

At $h = 2$ there is another qualitative change in behavior:

When $h \leq 2$, Euler approximations eventually reach 1.000000.
When $h > 2$, the approximations settle into a 2-cycle oscillating above and below 1.

The next bifurcation value occurs between $h = 2.44$ and $h = 2.45$:

When $h = 2.44$ the steady state oscillates between the two values
1.196253 and 0.6233242
When $h = 2.45$ we get the repeated four-cycles of (roughly)
1.193238, 0.628326, 1.200481, and 0.610832.

We let you narrow down the subsequent bifurcations by experimenting on a spreadsheet.

The value of h where four-cycles end and eight-cycles begin is between 2.5 and 2.55.

After that the period-doubling points are closer and closer together, and it will be harder to zero in on those points. You may require more significant places on h to find them; try 10-significant-place accuracy in the spreadsheet.

(d)   A strange thing happens for $h > 3$. The Euler approximation jumps around chaotically and eventually goes to minus infinity. The larger the value of h, the sooner it goes to minus infinity. As shown, when $h = 3.04$ it goes to minus infinity around $t = 50$, but when $h = 3.01$, it doesn't go to minus infinity until around $t = 150$.

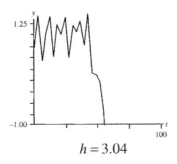

$h = 3.04$

7.5   Chaos in Forced Nonlinear Systems

> Problems 1–8 (and later Problem 15) are highly exploratory experiments such as are needed in research. We compute these curves numerically with a graphic DE solver (usually by Runge-Kutta with small step size), because an analytic solution of nonlinear DEs is (at best) awkward and difficult to obtain.
>
> Organization becomes critical in order to report results. Our solutions show some possible ways to do this. Students own experiments may lead to other ways. The goal in every case is concise communication.

■   **Solution of the Linearized Pendulum**

3.   If we linearize the pendulum equation $\ddot{\theta} + 0.1\dot{\theta} + \sin\theta = F\cos\omega t$

about $(2\pi, 0)$, we get $\ddot{\theta} + 0.1\dot{\theta} + (\theta - 2\pi) = F\cos\omega t$

(a)   The *unforced linearized* equation $\ddot{\theta} + 0.1\dot{\theta} + \theta = 2\pi$

has the homogeneous solution

$$\theta_h(t) = e^{-0.05t}\left(c_1\cos t + c_2\sin t\right),$$

and a particular solution can easily be found to be $\theta_p(t) = 2\pi$, so the general solution is

$$\theta(t) = e^{-0.05t}\left(c_1\cos t + c_2\sin t\right) + 2\pi.$$

Substituting the IC $\theta(0) = 3$ and $\dot{\theta}(0) = 1$ yields

$$c_1 = 3 - 2\pi \approx -3.28$$
$$c_2 = 1 + 0.05c_1 \approx 0.836.$$

Hence, the linearized unforced pendulum IVP solution is

$$\theta(t) = -e^{-0.05t}\left(3.28\cos t - 0.836\sin t\right) + 2\pi,$$

which approaches the steady state of 2π.

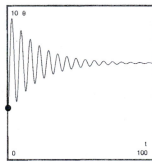

(b)   For the *forced linearized* pendulum equation $\ddot{\theta} + 0.1\dot{\theta} + \theta = \cos t + 2\pi$ we use the method of undetermined coefficients to find $\theta_p(t)$; then the general solution is

$$\theta(t) = e^{-0.05t}\left(c_1 \cos t + c_2 \sin t\right) + 2\pi + 10\sin t .$$

Substituting the IC $\theta(0) = 3$ and $\dot{\theta}(0) = 1$ yields

$$c_1 = \ 3 - 2\pi \quad \approx -3.28$$
$$c_2 = -9 + 0.05c_1 \approx -9.16.$$

Hence, the linearized forced pendulum IVP solution is

$$\theta(t) = -e^{-0.05t}\left(3.28\cos t + 9.16\sin t\right) + 2\pi + 10\sin t .$$

The first term of this algebraic solution gives a sinusoidal oscillation that disappears as $t \to 100$; we can see its diminishing effect in the $t\theta$ figure. The remaining two terms indicate the eventual steady state sinusoidal oscillation about 2π, and also confirms the surprisingly large amplitude of 10.

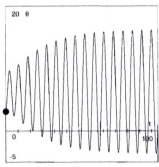

Period Doubling; Poincaré Sections in Forced Damped Pendulum

6.   Student project using IDE software.

Problem 8 is inserted as an "extra" sample response to an open-ended problem.

■   **Forced Duffing Oscillator, A Route to Chaos**

8. $\ddot{x} + 0.25\dot{x} - x + x^3 = F\cos t$ or , as a system, $\dot{x} = y$

$$\dot{y} = F\cos t - 0.25y + x - x^3$$

**E
X
T
R
A**

Problem 7(b) is a special case; here we explore some of the more hidden complexities.

To determine the motion starting near the origin, we set $F = 0$ and start at four points: $(0,\ \pm 0.25)$ and $(\pm 0.25,\ 0)$.

Two solutions approach the stable equilibrium point $(1,\ 0)$; two solutions approach $(-1,\ 0)$, as shown.

The 45-degree line $\dot{x} = -x$ separates the two basins of attraction; trajectories from initial points above the line converge to $(1,\ 0)$; those that start below the line converge to $(-1,\ 0)$.

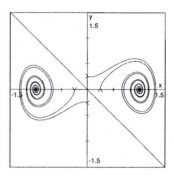

An interesting (and difficult) problem would be to determine the boundary of the basin of attraction analytically from the DE, without the phase portrait.

Four solutions, with two
basins of attraction
(separated by diagonal)

(b)   As a sample experiment, we chose the initial point $(0.25,\ 0)$ and applied forcing terms $F\cos t$ with amplitudes $F = 0.1, 0.2, 0.25$; the phase portraits are shown in the figure. Notice the increasingly chaotic (unpredictable) behavior as the amplitude increases. If you allowed the trajectories to run forever, for the larger F values, some limit cycles may appear around one of the original equilibria at $(\pm 1,\ 0)$. The decision of *which* of these points to circle around seems very sensitive to changes in either F or the initial conditions.

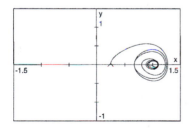
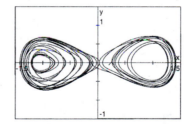
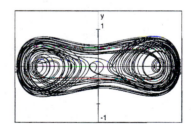

Duffing's equation with different forcing functions

■   **Period-Doubling Exploration, Forced Duffing Oscillator**

9.   Student IDE Project

■   **Forced van der Pol Equation**

12. $\ddot{x} - \varepsilon\left(1 - x^2\right)\dot{x} + x = F\cos\omega t$

The figures show phase portraits and time series for $x(0) = 1$ and $\dot{x}(0) = 1$.

(a) $\varepsilon = 0.1$, $F = 0.5$, $\omega = 1$

The solution quickly reaches a steady state that looks pretty close to a pure harmonic oscillation between 3.2 and –3.2.

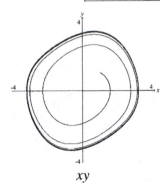

xy

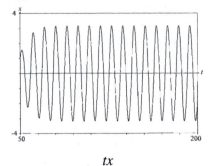

tx

(b) $\varepsilon = 1$, $F = 0.5$, $\omega = 1$

The solution quickly reaches a steady state resembling a rather warped sine or cosine curve.

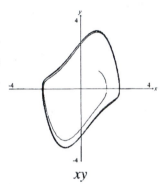

xy

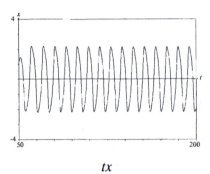

tx

(c) $\varepsilon = 1$, $F = 1$, $\omega = 0.3$

The solution gives rise to a more complicated periodic motion, with three peaks per period are displayed.

These figures start at $t = 50$ to eliminate the transient solution and isolate the

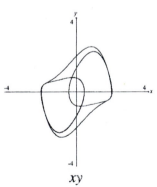

xy

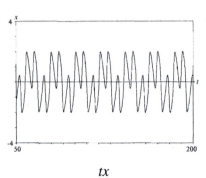

tx

(d) $\varepsilon = 1$, $F = 1$, $\omega = 0.4$

Here a different periodic solution is observed.

These figures start at $t = 50$ to eliminate the transient solution and isolate the cycle.

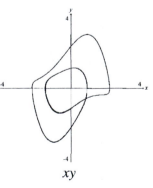

xy

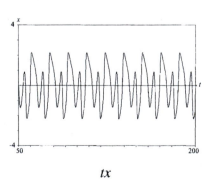

tx

■   **Stagecoach Wheels and the Poincaré Section**

For Problems 15-18 the diagrams are a schematic sketch of the strobed motion, so that you can clearly see at which angle and when the strobe will light successive points.

15.

(a) $\omega_s = \frac{1}{2}\omega_0$.

Clock appears motionless.

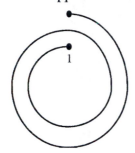

(b) $\omega_s = \frac{1}{4}\omega_0$.

Clock appears motionless.

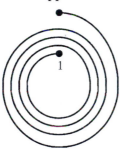

(c) $\omega_s = 2\omega_0$.

Clock appears to alternate between the bottom and top positions.

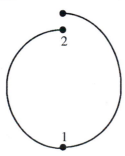

(d) $\omega_s = 4\omega_0$.

Clock appears to move clockwise in 15-second jumps.

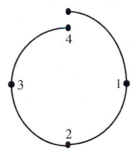

(e) $\omega_s = \frac{2}{3}\omega_0$.

Clock appears to alternate between the top and bottom positions, with 1.5 actual rotations between strobe flashes.

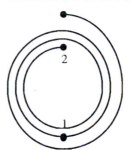

(f) $\omega_s = \frac{3}{4}\omega_0$.

Clock appears to rotate clockwise in 20-second jumps, although it is actually moving clockwise in 80-second jumps.

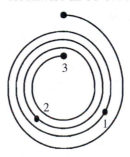

18.   (a)   The strobed motion has period 2 and moves in 30 second jumps. These means that $\omega_s = 2\omega_0$ or

$$\omega_0 = \frac{1}{2}\omega_s .$$

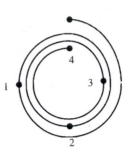

(b)   The strobed motion has period 4 and moves in 45 second jumps. These means that $\omega_s = \frac{4}{3}\omega_0$ or

$$\omega_0 = \frac{3}{4}\omega_s .$$

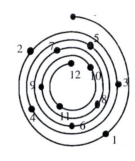

(c)   The strobed motion has period 12 and moves in 25 second jumps. These means that $\omega_s = \frac{12}{5}\omega_0$ or

$$\omega_0 = \frac{5}{12}\omega_s .$$

CHAPTER 8

Forced Equations and Systems

8.1   The Laplace Transform and Its Inverse

■ Transforms from the Definition

3. $\mathcal{L}\left\{e^{2t}\right\} = \int_0^\infty e^{2t}e^{-st}\,dt = \int_0^\infty e^{-(s-2)t}\,dt = \lim_{b\to\infty} \frac{-1}{s-2}e^{-(s-2)t}\Big|_0^b = \frac{1}{s-2}$

6. $\mathcal{L}\left\{\cos 3t\right\} = \int_0^\infty e^{-st}\cos 3t\,dt = \dfrac{s}{s^2+9}$ (integration by parts twice)

9. $\mathcal{L}\left\{f(t)\right\} = \int_0^2 (t-1)e^{-st}\,dt + \int_2^\infty e^{-st}\,dt = \int_0^2 te^{-st}\,dt + \int_2^\infty e^{-st}\,dt - \int_0^2 e^{-st}\,dt = \dfrac{1}{s^2}\left(1-e^{-2s}\right) - \dfrac{1}{s}$

■ Transforms with Tools

12. $\mathcal{L}\left\{1+e^{-t}\right\} = \mathcal{L}\{1\} + \mathcal{L}\left\{e^{-t}\right\} = \dfrac{1}{s} + \dfrac{1}{s+1} = \dfrac{2s+1}{s(s+1)}$

15. $\mathcal{L}\left\{e^{-t}\left(t+3t^2\right)\right\} = \mathcal{L}\left\{te^{-t}\right\} + 3\mathcal{L}\left\{t^2 e^{-t}\right\} = \dfrac{1}{(s+1)^2} + \dfrac{6}{(s+1)^3}$

18. $\mathcal{L}\left\{te^{-3t} + 2\sin t\right\} = \mathcal{L}\left\{te^{-3t}\right\} + 2\mathcal{L}\{\sin t\} = \dfrac{1}{(s+3)^2} + \dfrac{2}{s^2+1}$

■ Laplace Transform of Damped Sine and Cosine Functions

21.   (a) $\mathcal{L}\left\{e^{(a+ik)t}\right\} = \int_0^\infty e^{-(s-a-ik)t}\,dt = \dfrac{1}{s-a-ik} = \dfrac{s-a+ik}{(s-a)^2+k^2}$.

Breaking this into real and complex parts yields the desired result.

(b) $\mathcal{L}\left\{e^{(a+ik)t}\right\} = \mathcal{L}\left\{e^{at}\left(\cos kt + i\sin kt\right)\right\} = \mathcal{L}\left\{e^{at}\left(\cos kt\right)\right\} + i\mathcal{L}\left\{e^{at}\sin kt\right\}$.

Matching real and complex parts of the solution yields the Laplace transform formulas for $e^{at}\cos kt$ and $e^{at}\sin kt$.

■ **Power Rule**

24. (a) Given
$$L\{t^n\} = \int_0^\infty e^{-st} t^n dt,$$

we integrate by parts, letting $u = t^n$ and $dv = e^{-st} dt$, to get

$$L\{t^n\} = \lim_{b \to \infty} -\frac{t^n e^{-st}}{s} \Big|_0^b + \frac{n}{s} \int_0^\infty e^{-st} t^{n-1} dt.$$

On the right side, the left-hand term becomes 0 in the limit (for $s > 0$);

the integral terms become $\frac{n}{s} L\{t^{n-1}\}$. The result follows immediately.

(b) Performing integration by parts n times yields

$$L\{t^n\} = \frac{n!}{s^n} \int_0^\infty e^{-st} dt.$$

Integrating gives the final answer as

$$L\{t^n\} = \frac{n!}{s^{n+1}}.$$

■ **Multiplier Applications**

The multiplier rule (Problem 25) says to evaluate $L\{t f(t)\}$, we can first ignore the t and take the transform of $f(t)$, getting $F(s)$. Then to get the transform we differentiate $F(s)$ and change the sign. That is,

$$L\{tf(t)\} = -\frac{d}{ds} L\{f(t)\}.$$

---

27. $\quad L\{t \sin 3t\} = -\dfrac{d}{ds} L\{\sin 3t\} = -\dfrac{d}{ds}\left(\dfrac{3}{s^2 + 9}\right) = \dfrac{6s}{\left(s^2 + 9\right)^2}$

30. $\quad L\{-2t \sinh 2t\} = -2L\left\{t \dfrac{e^{2t} - e^{-2t}}{2}\right\} = L\{te^{-2t}\} - L\{te^{2t}\} = -\dfrac{d}{ds}\left(\dfrac{1}{s+2}\right) + \dfrac{d}{ds}\left(\dfrac{1}{s-2}\right)$

$$= \frac{1}{(s+2)^2} - \frac{1}{(s-2)^2}$$

■ Using the Shift

To find $L\{e^{at}f(t)\}$, Problem 31 says we can ignore the exponential function e^{at}, take the transform of $f(t)$, and then replace s in $F(s) = L\{f(t)\}$ by $(s-a)$. That is,

$$L\{e^{at}f(t)\} = F(s-a).$$

33. $L\{e^t \sin 2t\}$. We first compute

$$F(s) = L\{\sin 2t\} = \frac{2}{s^2+4}.$$

Then

$$L\{e^t \sin 2t\} = F(s-1) = \frac{2}{(s-1)^2+4}.$$

36. $L\{e^{-3t} \sinh t\}$. We first compute

$$F(s) = L\{\sinh t\} = \frac{1}{s^2-1}.$$

Then

$$L\{e^{-3t} \sinh t\} = F(s+3) = \frac{1}{(s+3)^2-1}.$$

■ Out of Order

39.   For any constant α, we can pick t can be large enough so that $t > \alpha$, which implies that $t^2 > \alpha t$, which, in turn, implies that $e^{t^2} > e^{\alpha t}$. Hence eventually e^{t^2} will be greater than $e^{\alpha t}$.

■ Inverse Transforms

The key for Problems 40-54 is to rewrite each function in terms of functions listed in the short Laplace transform table, on page 472 of textbook. These transforms are also included in the longer table inside the back cover of the text.

42. $L^{-1}\left\{\dfrac{5}{s^2+3}\right\} = L^{-1}\left\{\dfrac{5}{\sqrt{3}}\left(\dfrac{\sqrt{3}}{s^2+(\sqrt{3})^2}\right)\right\} = \dfrac{5}{\sqrt{3}}\sin\sqrt{3}t.$

45. $L^{-1}\left\{\dfrac{s+1}{s^2+2s+10}\right\}$

Completing the square in the denominator yields $\dfrac{s+1}{s^2+2s+10} = \dfrac{s+1}{(s+1)^2+9}.$

Hence, by the exponential law, $L^{-1}\left\{\dfrac{s+1}{s^2+2s+10}\right\} = L^{-1}\left\{\dfrac{(s+1)}{(s+1)^2+9}\right\} = e^{-t}\cos 3t.$

48. $\mathcal{L}^{-1}\left\{\dfrac{s+1}{s^2+s-2}\right\}$

Factoring the denominator and writing as a partial fraction gives

$$\frac{s+1}{s^2+s-2}=\frac{s+1}{(s+2)(s-1)}=\frac{A}{s+2}+\frac{B}{s-1}.$$

Solving for A and B yields $A=\dfrac{1}{3}$ and $B=\dfrac{2}{3}$. Hence

$$\mathcal{L}^{-1}\left\{\frac{s+1}{s^2+s-2}\right\}=\frac{1}{3}\mathcal{L}^{-1}\left\{\frac{1}{s+2}\right\}+\frac{2}{3}\mathcal{L}^{-1}\left\{\frac{1}{s-1}\right\}=\frac{1}{3}e^{-2t}+\frac{2}{3}e^{t}.$$

51. $\mathcal{L}^{-1}\left\{\dfrac{7}{s^2+4s+7}\right\}=\mathcal{L}^{-1}\left\{\dfrac{7}{\sqrt{3}}\left(\dfrac{\sqrt{3}}{(s+2)^2+\left(\sqrt{3}\right)^2}\right)\right\}=\dfrac{7}{\sqrt{3}}e^{-2t}\sin\sqrt{3}t.$

54. $\mathcal{L}^{-1}\left\{\dfrac{3}{\left(s^2+1\right)\left(s^2+4\right)}\right\}$

We use partial fractions to write

$$\frac{3}{\left(s^2+1\right)\left(s^2+4\right)}=\frac{As+B}{s^2+1}+\frac{Cs+D}{s^2+4}.$$

Multiplying the equation by the denominator on the left-hand side, we get

$$3=(A+C)s^3+(B+D)s^2+(4A+C)s+(4B+D).$$

Comparing coefficients and solving the resulting equations for A, B, C, and D yields $A=0$, $B=1$, $C=0$, and $D=-1$. Hence

$$\mathcal{L}^{-1}\left\{\frac{3}{\left(s^2+1\right)\left(s^2+4\right)}\right\}=\mathcal{L}^{-1}\left\{\frac{1}{s^2+1}-\frac{1}{2}\left(\frac{2}{s^2+4}\right)\right\}=\mathcal{L}^{-1}\left\{\frac{1}{s^2+1}\right\}-\frac{1}{2}\mathcal{L}^{-1}\left\{\frac{2}{s^2+4}\right\}=\sin t-\frac{1}{2}\sin 2t.$$

8.2   Solving DEs and IVPs with Laplace Transforms

■ **First-Order Problems**

3. $y' - y = e^t$, $y(0) = 1$

The Laplace transform yields the equation $sL\{y\} - 1 - L\{y\} = \dfrac{1}{s-1}$.

Substituting the initial condition and solving for the Laplace transform gives

$$L\{y\} = \frac{1}{(s-1)^2} + \frac{1}{s-1}.$$

Taking the inverse transform yields $y(t) = te^t + e^t$.

■ **Transformations at Work**

6. $y'' + 2y' = 4$, $y(0) = 1$, $y'(0) = -4$

Taking the transform yields the equation

$$s^2 L\{y\} - s + 4 + 2sL\{y\} - 2 = \frac{4}{s}.$$

Solving for the Laplace transform yields

$$L\{y\} = \frac{4}{s^2(s+2)} + \frac{1}{s+2} + \frac{-2}{s(s+2)}.$$

Using partial fractions we write the first term of $L\{y\}$ as

$$\frac{4}{s^2(s+2)} = \frac{A}{s} + \frac{B}{s^2} + \frac{C}{s+2} = \frac{-1}{s} + \frac{2}{s^2} + \frac{1}{s+2}.$$

Similarly, we write the last fraction of $L\{y\}$ as

$$\frac{-2}{s(s+2)} = \frac{D}{s} + \frac{E}{s+2} = \frac{-1}{s} + \frac{1}{s+2}.$$

Putting these expressions together yields the transform

$$L\{y\} = \frac{-2}{s} + \frac{2}{s^2} + \frac{3}{s+2}.$$

Taking the inverse transform yields the solution of the initial-value problem

$$y(t) = -2 + 2t + 3e^{-2t}.$$

9. $y'' + 3y' + 2y = 6$, $y(0) = 0$, $y'(0) = 2$

Taking the transform yields the equation

$$s^2 L\{y\} - s(0) - 2 + 3sL\{y\} + 2L\{y\} = \frac{6}{s}.$$

Solving for the Laplace transform yields

$$L\{y\} = \frac{6}{s(s+1)(s+2)} + \frac{2}{(s+1)(s+2)}$$

$$= \left(\frac{3}{s} - \frac{6}{s+1} + \frac{3}{s+2} \right) + \left(\frac{2}{s+1} - \frac{2}{s+2} \right).$$

$$= \left(\frac{3}{s} - \frac{4}{s+1} + \frac{1}{s+2} \right)$$

Hence, the final result is $y(t) = L^{-1}\{L\{y\}\} = 3 - 4e^{-t} + e^{-2t}.$

12. $y'' + y' + y = e^{-t}$, $y(0) = 0$, $y'(0) = 1$

Taking the Laplace transform of the equation

$$s^2 Y(s) - 1 + sY(s) + Y(s) = \frac{1}{s+1}$$

$$Y(s) = \frac{s+2}{(s+1)(s^2 + s + 1)} = \frac{1}{s+1} - \frac{s-1}{s^2 + s + 1} \quad \text{(by partial fractions)}$$

$$= \frac{1}{s+1} - \frac{s-1}{\left(s + \dfrac{1}{2} \right)^2 + \dfrac{3}{4}} \quad \text{(by completing the square)}$$

$$= 1 \frac{1}{s+1} + \sqrt{3} \, \frac{\dfrac{\sqrt{3}}{2}}{\left(s + \dfrac{1}{2} \right)^2 + \dfrac{3}{4}} - \frac{s + \dfrac{1}{2}}{\left(s + \dfrac{1}{2} \right)^2 + \dfrac{3}{4}}$$

$$y(t) = e^{-t} + \sqrt{3} e^{-(1/2)t} \sin\left(\frac{\sqrt{3}}{2} t \right) - e^{-(1/2)t} \cos\left(\frac{\sqrt{3}}{2} t \right)$$

■   **Raising the Stakes**

15. $y''' - y'' - y' + y = 6e^t$, $y(0) = y'(0) = y''(0) = 0$.

The transform yields $s^3 L\{y\} - s^2 L\{y\} - sL\{y\} + L\{y\} = \frac{6}{s-1}.$

Solving for the $L\{y\}$ yields

$$L\{y\} = \frac{6}{(s-1)(s^3 - s^2 - s + 1)}.$$

The denominator factors further, and we can then use partial fractions, so

$$\frac{6}{(s-1)^3(s+1)} = \frac{A}{s-1} + \frac{B}{(s-1)^2} + \frac{C}{(s-1)^3} + \frac{D}{s+1}$$

$$= \frac{3}{4}\frac{1}{s-1} - \frac{3}{2}\frac{1}{(s-1)^2} + \frac{3}{(s-1)^3} - \frac{3}{4}\frac{1}{s+1}.$$

Thus the solution is $y(t) = \frac{3}{4}e^t - \frac{3}{2}te^t + \frac{3}{2}t^2e^t - \frac{3}{4}e^{-t}.$

■   **Laplace Transform Using Power Series**

18.   The power series for e^t is

$$f(t) = e^t = \sum_{n=0}^{\infty} \frac{1}{n!}x^n = 1 + x + \frac{1}{2}x^2 + \ldots$$

$$F(s) = \sum_{n=0}^{\infty} \frac{1}{n!}\frac{n!}{s^{n+1}} = \sum_{n=0}^{\infty} \frac{1}{s^{n+1}} = \frac{1}{s} + \frac{1}{s^2} + \frac{1}{s^3} + \ldots$$

This is a geometric series with the first term $\frac{1}{s}$ and a common ratio of $\frac{1}{s}$. The closed form for

this series is

$$F(s) = \frac{\frac{1}{s}}{1 - \frac{1}{s}} = \frac{1}{s-1}.$$

■   **Bessel Functions with IDE**

21.   Student Lab Project

■   **Suggested Journal Entry**

24.   Student Project

8.3   The Step Function and the Delta Function

■ Stepping Out

3. $f(t) = \begin{cases} 1 & \text{if } t < 1 \\ 4t-1 & \text{if } 1 \le t < 4 \\ 1 & \text{if } t \ge 4 \end{cases}$

We write the function as

$$f(t) = 1 + \left(4t - t^2 - 1\right)\text{step}(t-1) + \left(1 - 4t + t^2\right)\text{step}(t-4).$$

■ Geometric Series

6. $f(t) = 1 + \text{step}(t-1) + \text{step}(t-2) + \text{step}(t-3) + \cdots,$

$$L\{f(t)\} = \frac{1}{s} + \frac{e^{-s}}{s} + \frac{e^{-2s}}{s} + \frac{e^{-3s}}{s} + \cdots = \frac{1}{s}\left\{1 + \left(e^{-s}\right) + \left(e^{-s}\right)^2 + \left(e^{-s}\right)^3 + \cdots\right\} = \frac{1}{s}\left(\frac{1}{1 - e^{-s}}\right).$$

9. $f(t) = 1 - 2\,\text{step}(t-1) + 2\,\text{step}(t-2) - \cdots,$

$$L\{f(t)\} = \frac{1}{s} - \frac{2e^{-s}}{s} + \frac{2e^{-2s}}{s} - \frac{2e^{-3s}}{s} - \cdots = \frac{1}{s} - \frac{2e^{-s}}{s}\left(1 - e^{-s} + \left(e^{-s}\right)^2 - \cdots\right) = \frac{1}{s} - \frac{2e^{-s}}{s}\left(\frac{1}{1 + e^{-s}}\right).$$

■ Piecewise-Continuous Functions

> In Problems 10–15 we use the **alternate delay rule**
> $$L\{f(t)\,\text{step}(t-c)\} = e^{-cs}L\{f(t+c)\}.$$

12. $f(t) = (t-1)\left[\text{step}(t-1) - \text{step}(t-3)\right] + 2\,\text{step}(t-3),$

$$L\{f(t)\} = e^{-s}L\{t\} - e^{-3s}L\{t+2\} + \frac{2e^{-3s}}{s} = \frac{e^{-s}}{s^2} - \frac{e^{-3s}}{s^2} - 2\frac{e^{-3s}}{s} + \frac{2e^{-3s}}{s} = \frac{e^{-s}}{s^2} - \frac{e^{-3s}}{s^2}.$$

15.   The two parts of the sine function can be written as

$$f(t) = \sin(\pi t)\left(1 - \text{step}(t-1)\right) - 2\sin(\pi t)\left(\text{step}(t-1) - \text{step}(t-2)\right),$$

$$L\{f(t)\} = L\{\sin \pi t\} - e^{-s}L\{\sin(\pi(t+1))\} - 2e^{-s}L\{\sin \pi(t+1)\} + 2e^{-2s}L\{\sin(\pi(t+2))\}$$

$$= \frac{\pi}{s^2 + \pi^2} + \frac{\pi}{s^2 + \pi^2}e^{-s} + \frac{2\pi}{s^2 + \pi^2}e^{-s} + \frac{2\pi}{s^2 + \pi^2}e^{-2s} = \frac{\pi}{s^2 + \pi^2}\left(1 + 3e^{-s} + 2e^{-2s}\right).$$

■ Transforming Delta

18. $L\{\delta(t) - \delta(t-1) + \delta(t-2) - \cdots\} = 1 - e^{-s} + e^{-2s} - \cdots$

■   **Laplace Step by Step**

21. $f(t) = 1 - 2\,\text{step}(t-1) + \text{step}(t-2)$

$$L\{f\} = \frac{1}{s} - \frac{2e^{-s}}{s} + \frac{e^{-2s}}{s}.$$

24. $f(t) = e^t\,\text{step}(t-3)$

The function here has not been shifted, so we must perform the shift and write the function as $e^3\,\text{step}(t-3)e^{t-3}$. Hence, the transform is

$$L\{f\} = e^3 \left(\frac{e^{-3s}}{s-1} \right) = \frac{e^{3-3s}}{s-1}.$$

■   **Inverse Transforms**

27. $\mathcal{L}^{-1}\left\{ \dfrac{e^{-s}}{s} \right\}$

Factoring out the e^{-s}, the function is simply $\dfrac{1}{s}$; the inverse transform is 1. The factor e^{-s} means there is a $\text{step}(t-1)$ multiplied in this term. Hence, the inverse transform is $\text{step}(t-1)$. Once again, the graphs of these functions are all graphs of familiar functions "delayed," and examples can be seen in the text.

30. $\mathcal{L}^{-1}\left\{ \dfrac{e^{-4s}}{s+4} \right\}$

The inverse transform of $\dfrac{1}{s+4}$ is e^{-4t}. Hence, the inverse transform is

$$\mathcal{L}^{-1}\left\{ \frac{e^{-4s}}{s+4} \right\} = e^{-4(t-4)}\,\text{step}(t-4).$$

■   **Transforming Solutions**

33. $x' = 1 - \text{step}(t-1),\ x(0) = 0$

The Laplace transform of the equation is

$$sX(s) - 0 = \frac{1}{s} - \frac{e^{-s}}{s}.$$

Solving for $X(s)$ yields $X(s) = \dfrac{1}{s^2} - \dfrac{e^{-s}}{s^2}$. Taking the inverse transform yields the solution of the initial-value problem

$$x(t) = t - (t-1)\,\text{step}(t-1).$$

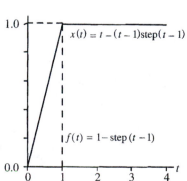

36. $x'' + x = \text{step}(t - \pi) - \text{step}(t - 2\pi)$, $x(0) = 0$, $x'(0) = 1$

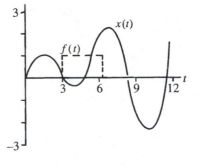

The Laplace transform of the equation is

$$s^2 X(s) - 1 + X(s) = \frac{e^{-\pi s}}{s} - \frac{e^{-2\pi s}}{s}.$$

Solving for $X(s)$ yields

$$X(s) = \frac{1}{s^2 + 1} + \frac{e^{-\pi s}}{s(s^2 + 1)} - \frac{e^{-2\pi s}}{s(s^2 + 1)}.$$

Response to one square pulse

Using a partial fraction decomposition, we write

$$\frac{1}{s(s^2 + 1)} = \frac{1}{s} - \frac{s}{s^2 + 1}.$$

Taking the inverse transform yields the solution of the initial-value problem,

$$x(t) = \sin t + (1 + \cos(t))\,\text{step}(t - \pi) - (1 - \cos(t))\,\text{step}(t - 2\pi).$$

■   Periodic Formula

39.   First, we find the Laplace transform of the single wave-form

$$f(t) = 2t\big[1 - \text{step}(t - 1)\big], \quad 0 \le t < \infty,$$

which is

$$\mathcal{L}\{\text{single wave}\} = \int_0^\infty f(t)e^{-st}\,dt.$$

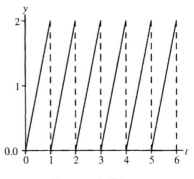

We determine the Laplace transform by applying the alternate form of the Delay Theorem instead of integrating.

Sawtooth Wave

$$\mathcal{L}\{\text{single wave}\} = 2\left(\frac{1}{s^2}\right) - e^{-s}\mathcal{L}\{2(t + 1)\} = 2\frac{1}{s^2} - e^{-s}\left(2\frac{1}{s^2} + \frac{2}{s}\right)$$

$$= 2\left[\frac{1}{s^2} - e^{-s}\left(\frac{1}{s^2} + \frac{1}{s}\right)\right].$$

The Laplace transform of the periodic wave-form of period 1 on $[0, \infty)$ is

$$\mathcal{L}\{f(t)\} = \frac{1}{1 - e^{-s}}\mathcal{L}\{\text{single wave}\}$$

$$= \frac{2}{1 - e^{-s}}\left[\frac{1}{s^2} - e^{-s}\left(\frac{1}{s^2} + \frac{1}{s}\right)\right].$$

42.   From Problem 14, we have

$$\mathcal{L}\{\text{single wave}\} = \frac{1}{s^2 + \left(\frac{\pi}{2}\right)^2}\left(\frac{\pi}{2} - se^{-s}\right) + \frac{e^{-s}}{s} - \frac{e^{-2s}}{s}$$

$$= \frac{2\pi s + \pi^2 e^{-s} - e^{-2s}\left(4s^2 + \pi^2\right)}{s\left(4s^2 + \pi^2\right)}.$$

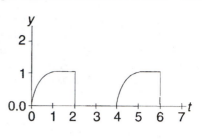

Hence, the Laplace transform of the periodic function of period 4 on $[0, \infty)$ is

$$\mathcal{L}\{f(t)\} = \frac{1}{1 - e^{-4s}}\left[\frac{1}{s^2 + \left(\frac{\pi}{2}\right)^2}\left(\frac{\pi}{2} - se^{-s}\right) + \frac{e^{-s}}{s} - \frac{e^{-2s}}{s^2}\right].$$

45.   From Problem 8, we have

$$\mathcal{L}\{\text{single wave}\} = \int_0^2 e^{-st} f(t)\,dt = \int_0^1 e^{-st}\,dt = \frac{1}{s}\left(1 - e^{-s}\right).$$

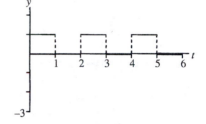

Hence, the Laplace transform of the periodic function of period 2 on $[0, \infty)$ is

$$\mathcal{L}\{f(t)\} = \frac{1}{s}\left(\frac{1 - e^{-s}}{1 - e^{-2s}}\right).$$

Square wave

■   **Solve on Impulse**

48. $x' = \delta(t)$, $x(0) = 0$

Taking the Laplace transform of both sides of the DE yields $sX(s) = 1$ or $X(s) = \dfrac{1}{s}$.

Hence, we have $x(t) = 1$.

51. $x'' + x = -\delta(t - \pi) + \delta(t - 2\pi)$, $x(0) = 0$, $x'(0) = 1$

Taking the Laplace transform of both sides of the DE yields

$$\left(s^2 + 1\right)X(s) - 1 = -e^{-\pi s} + e^{-2\pi s}.$$

Solving for $X(s)$ yields

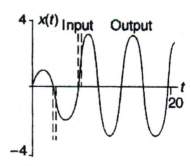

Response to two impulses

$$X(s) = \frac{1}{s^2 + 1} - e^{-\pi s}\left(\frac{1}{s^2 + 1}\right) + e^{-2\pi s}\left(\frac{1}{s^2 + 1}\right).$$

The inverse is

$$x(t) = \sin t - \sin(t - \pi)\,\text{step}(t - \pi) + \sin(t - 2\pi)\,\text{step}(t - 2\pi).$$

■   **Suggested Journal Entry**

54.   Student Project

8.4   The Convolution Integral and the Transfer Function

■ **Convolution Properties**

3.   Prove $f * 0 = 0$

$$f * 0 = \int_0^t f(t-w)(0)dw = \int_0^t 0\,dw = 0$$

■ **Calculating Convolutions**

6. $1 * t = \int_0^t 1(w)dw = \dfrac{1}{2}t^2$

9. $e^t * e^{-t} = \int_0^t e^{(t-w)}e^{-w}dw = \int_0^t e^{t-2w}dw = \left[\dfrac{-1}{2}e^{t-2w}\right]_0^t = \dfrac{-1}{2}e^{-t} + \dfrac{1}{2}e^t = \sinh t$

■ **Convoluted Solutions**

12. $x' = f(t),\ x(0) = 0$

Taking the Laplace transform of both sides of the DE yields $sX(s) = L\{f\}$.

Solving for $X(s)$ yields $X(s) = \dfrac{1}{s}L\{f\} = L\{1\}L\{f\}$.

The inverse yields the solution $y(t) = 1 * f(t) = \int_0^t f(\tau)d\tau$.

15. $x' + x = f(t),\ x(0) = 1$

Taking the Laplace transform of both sides of the DE yields

$$sX(s) - 1 + X(s) = L\{f\}.$$

Solving for $X(s)$ yields

$$X(s) = \dfrac{1}{s+1} + \left(\dfrac{1}{s+1}\right)L\{f\} = \dfrac{1}{s+1} + L\{e^{-t}\}L\{f\}.$$

The inverse yields the solution

$$x(t) = e^{-t} + e^{-t} * f(t) = e^{-t} + \int_0^t e^{-(t-\tau)}f(\tau)d\tau.$$

■   **Transfer and Impulse Response Functions**

18. $x' = f(t)$, $x(0) = 0$

Taking the transform of the DE yields

$$sX(s) = L\{f\} \text{ or } X(s) = \frac{1}{s} L\{f\}.$$

The transfer function is the coefficient of $L\{f\}$ or in this case,

$$\text{Transfer Function} = \frac{1}{s}.$$

The impulse response function $I(t)$ is the inverse transform of this function. Hence,

$$I(t) = 1.$$

The solution in terms of the transfer function is

$$x(t) = I(t) * f(t) = \int_0^t I(t-\tau) f(\tau) d\tau = \int_0^t f(\tau) d\tau.$$

21. $x'' + 4x' + 5x = f(t)$, $x(0) = x'(0) = 0$. Taking the transform of the differential equation yields

$$s^2 X(s) + 4sX(s) + 5X(s) = L\{f\} \text{ or } X(s) = \frac{L\{f\}}{s^2 + 4s + 5} = \frac{L\{f\}}{(s+2)^2 + 1}.$$

The transfer function is the coefficient of $L\{f\}$, or

$$\text{Transfer Function} = \frac{1}{(s+2)^2 + 1}.$$

The impulse response function is

$$I(t) = L\left\{ \frac{1}{(s+2)^2 + 1} \right\} = e^{-2t} \sin t.$$

The solution is $x(t) = I(t) * f(t) = \int_0^t I(t-\tau) f(\tau) d\tau = \int_0^t e^{-2(t-\tau)} \sin(t-\tau) f(\tau) d\tau.$

■   **Inverse of Convolution Theorem**

24.   Find $L^{-1}\left\{ \frac{1}{s(s+1)} \right\}$. Then $F(s) = \frac{1}{s} \frac{1}{s+1}$ so that $f(t) = 1 * e^{-t} = \int_0^t (1) e^{-w} dw = [-e^{-w}]_0^t = 1 - e^{-t}.$

27.   Find $\mathcal{L}^{-1}\left\{\dfrac{1}{(s^2+1)^2}\right\}$

$$F(s)=\frac{1}{s^2+1}\frac{1}{s^2+1}$$

$$f(t)=\sin t * \sin t = \int_0^t \sin(t-w)\sin w\,dw$$

$$=\int_0^t \frac{1}{2}(-\cos t + \cos(t-2w))\,dw \quad \text{(by Trigonometric Identity)}$$

$$=\left[\frac{-1}{2}w\cos t - \frac{1}{4}\sin(t-2w)\right]_0^t$$

$$=\frac{1}{2}\sin t - \frac{1}{2}t\cos t$$

■   **Nonzero Practice**

30. $x''+x'+x=\delta(t-2),\ x(0)=1,\ x'(0)=0.$

$$X(s)=H(s)e^{-2s}+\frac{s-1}{s^2+s+1}$$

$$=H(s)e^{-2s}+\frac{s+\dfrac{1}{2}}{\left(s+\dfrac{1}{2}\right)^2+\dfrac{3}{4}}-\sqrt{3}\frac{\dfrac{\sqrt{3}}{2}}{\left(s+\dfrac{1}{2}\right)^2+\dfrac{3}{4}}$$

$$x(t)=h(t)*\delta(t-2)+e^{-\frac{1}{2}t}\cos\left(\frac{\sqrt{3}}{2}t\right)-\sqrt{3}e^{-\frac{1}{2}t}\sin\left(\frac{\sqrt{3}}{2}t\right)$$

■   **Fractional Calculus**

33. $I_{1/2}(I_{1/2}(f))(t)=\displaystyle\int_0^t f(w)\,dw$

$$I_{1/2}(I_{1/2}(f))(t)=\frac{1}{\sqrt{\pi}}(t^{-1/2}*I_{1/2}(f))$$

Applying the convolution theorem to each side yields: ($\mathcal{L}$ indicates the Laplace transform.)

$$\mathcal{L}[I_{1/2}(I_{1/2}(f))(t)]=\frac{1}{\sqrt{\pi}}\left(\sqrt{\frac{\pi}{s}}\right)\mathcal{L}[I_{1/2}(f)]$$

$$=\frac{1}{\sqrt{\pi}}\sqrt{\frac{\pi}{s}}\sqrt{\frac{\pi}{s}}F(s)$$

$$=\frac{F(s)}{s}$$

Since Laplace transforms have unique continuous inverses, $I_{1/2}(I_{1/2}(f))(t)=\displaystyle\int_0^t f(w)\,dw$

■   **Investment and Savings**

36. $A' = .04A + 10000e^{.01t}$, $A(0) = 0$

$$A(t) = e^{.04t} * 10000e^{.01t}$$

$$A(20) = \int_0^{20} (e^{.04(20-w)})(10000e^{.01w})dw \approx 3.34712 \times 10^5 \text{ or } \$334{,}712$$

■   **Radioactive Decay Chain**

39. $A' = -.01A + e^{0.001t}$, $A(0) = 0$

$$A(t) = e^{-0.01t} * e^{0.001t}$$

$$= \int_0^t e^{-0.01(t-w)} e^{0.001w} dw$$

$$= 90.91 e^{0.001t} - 90.91 e^{-0.01t}$$

■   **Volterra Integral Equation**

42. $y(t) = t^3 + \int_0^t \sin(t-w) y(w) dw$

$$Y(s) = \frac{6}{s^4} + \frac{1}{s^2+1} Y(s)$$

$$Y(s) = \frac{6(s^2+1)}{s^6} = \frac{6}{s^4} + \frac{1}{20}\frac{5!}{s^6}$$

$$y(t) = t^3 + \frac{1}{20} t^5$$

■   **General Solution of Volterra's Equation**

45. $y(t) = g(t) + \int_0^t k(t-w) y(w) dw$

$$Y(s) = G(s) + K(s)Y(s)$$

$$Y(s) = \frac{G(s)}{1 - K(s)}$$

■   **Transfer Functions for Circuits**

48. $LI'(t) + RI(t) = V(t)$

Applying Laplace transforms yields

$$LsI(s) + RI(s) = L\{V(t)\},$$

$$I(s) = \frac{1}{Ls+R} L\{V(t)\},$$

so that $H(s) = \dfrac{1}{Ls+r}$ is the transfer function.

Then $I(t) = L^{-1}\{H(s)L\{V(t)\}\} = \dfrac{1}{L} e^{-\frac{R}{L}t} * V(t)$

■   **Using Duhamel's Principle**

51. $y'' - y = f(t), \ y(0) = y'(0) = 0$

$z'' - z = 1, \ z(0) = z'(0) = 0$

has solution

$$z(t) = e^t + e^{-t} - 2$$

so

$$y(t) = (e^t + e^{-t} - 2) * f(t)$$

8.5   Laplace Transform Solution of Linear Systems

■   **Laplace for Systems**

3. $\dot{x} = y,$ $\qquad\qquad x(0) = 1$

$\dot{y} = -2x + 3y + 12e^{4t},$ $\quad y(0) = 1$

$$\dot{\vec{x}} = \begin{bmatrix} 0 & 1 \\ -2 & 3 \end{bmatrix} \vec{x} + \begin{bmatrix} 0 \\ 12e^{4t} \end{bmatrix}$$

$$s\vec{X}(s) - \begin{bmatrix} 1 \\ 1 \end{bmatrix} = \begin{bmatrix} 0 & 1 \\ -2 & 3 \end{bmatrix} \vec{X}(s) + \begin{bmatrix} 0 \\ \dfrac{12}{s-4} \end{bmatrix}, \quad \text{so} \quad \begin{bmatrix} s & -1 \\ 2 & s-3 \end{bmatrix} \vec{X}(s) = \begin{bmatrix} 1 \\ 1 + \dfrac{12}{s-4} \end{bmatrix}.$$

$$\vec{X}(s) = \begin{bmatrix} s & -1 \\ 2 & s-3 \end{bmatrix}^{-1} \begin{bmatrix} 1 \\ 1 + \dfrac{12}{s-4} \end{bmatrix} = \frac{1}{s^2 - 3s + 2} \begin{bmatrix} s-3 & 1 \\ -2 & s \end{bmatrix} \begin{bmatrix} 1 \\ 1 + \dfrac{12}{s-4} \end{bmatrix}$$

$$= \begin{bmatrix} \dfrac{s-3}{s^2 - 3s + 2} + \dfrac{1}{s^2 - 3s + 2}\left(1 + \dfrac{12}{s-4}\right) \\[4mm] -\dfrac{2}{s^2 - 3s + 2} + \dfrac{s}{s^2 - 3s + 2}\left(1 + \dfrac{12}{s-4}\right) \end{bmatrix}$$

$$= \begin{bmatrix} \dfrac{s^2 - 6s + 20}{(s^2 - 3s + 2)(s-4)} \\[4mm] \dfrac{6s + 8 + s^2}{(s^2 - 3s + 2)(s-4)} \end{bmatrix} = \begin{bmatrix} \dfrac{5}{s-1} - \dfrac{6}{s-2} + \dfrac{2}{s-4} \\[4mm] \dfrac{5}{s-1} - \dfrac{12}{s-2} + \dfrac{8}{s-4} \end{bmatrix}$$

$$\vec{x}(t) = \begin{bmatrix} 5e^t - 6e^{2t} + 2e^{4t} \\ 5e^t - 12e^{2t} + 8e^{4t} \end{bmatrix}$$

6. $\dot{x} = -y + t,$ $x(0) = 0$

$\dot{y} = 3x + 4y - 2 - 4t,$ $y(0) = 0$

$$\dot{\vec{x}} = \begin{bmatrix} 0 & -1 \\ 3 & 4 \end{bmatrix} \vec{x} + \begin{bmatrix} t \\ -2 - 4t \end{bmatrix}$$

$$s\vec{X}(s) = \begin{bmatrix} 0 & -1 \\ 3 & 4 \end{bmatrix} \vec{X}(s) + \begin{bmatrix} \dfrac{1}{s^2} \\ \dfrac{-2}{s} - \dfrac{4}{s^2} \end{bmatrix}, \quad \text{so} \quad \begin{bmatrix} s & 1 \\ -3 & s-4 \end{bmatrix} \vec{X}(s) = \begin{bmatrix} \dfrac{1}{s^2} \\ \dfrac{-2}{s} - \dfrac{4}{s^2} \end{bmatrix}.$$

$$\vec{X}(s) = \begin{bmatrix} s & 1 \\ -3 & s-4 \end{bmatrix}^{-1} \begin{bmatrix} \dfrac{1}{s^2} \\ \dfrac{-2}{s} - \dfrac{4}{s^2} \end{bmatrix} = \frac{1}{s^2 - 4s + 3} \begin{bmatrix} s-4 & -1 \\ 3 & s \end{bmatrix} \begin{bmatrix} \dfrac{1}{s^2} \\ -\dfrac{2}{s} - \dfrac{4}{s^2} \end{bmatrix}$$

$$= \begin{bmatrix} \dfrac{3}{s(s^2 - 4s + 3)} \\ -\dfrac{-3 + 2s^2 + 4s}{(s^2 - 4s + 3)s^2} \end{bmatrix} = \begin{bmatrix} \dfrac{1}{s} - \dfrac{3}{2(s-1)} + \dfrac{1}{2(s-3)} \\ \dfrac{1}{s^2} + \dfrac{3}{2(s-1)} - \dfrac{3}{2(s-3)} \end{bmatrix}$$

$$\vec{x}(t) = \begin{bmatrix} 1 - \dfrac{3}{2}e^t + \dfrac{1}{2}e^{3t} \\ t + \dfrac{3}{2}e^t - \dfrac{3}{2}e^{3t} \end{bmatrix}$$

9. $\vec{x}' = \begin{bmatrix} 2 & 1 \\ -3 & 6 \end{bmatrix} \vec{x} + \begin{bmatrix} e^{5t} \\ e^{5t} \end{bmatrix}, \quad \vec{x}(0) = \begin{bmatrix} 0 \\ 1 \end{bmatrix}$

$$\vec{x}(t) = \mathcal{L}^{-1} \left\{ (s\mathbf{I} - \mathbf{A})^{-1} (\vec{F}(s) + \vec{x}_0) \right\}$$

$$= \mathcal{L}^{-1} \left\{ \begin{bmatrix} s-2 & -1 \\ 3 & s-6 \end{bmatrix}^{-1} \left[\begin{bmatrix} \dfrac{1}{s-5} \\ -\dfrac{1}{s-5} \end{bmatrix} + \begin{bmatrix} 0 \\ 1 \end{bmatrix} \right] \right\}$$

$$= -e^{3t} \begin{bmatrix} 1 \\ -1 \end{bmatrix} + \frac{1}{2}e^{3t} \begin{bmatrix} 1 \\ 1 \end{bmatrix} + \begin{bmatrix} 1 \\ 2 \end{bmatrix}.$$

■   General Solutions of Linear Systems

12. $\dot{x} = -x - 4y, \quad x(0) = x_0$

$\dot{y} = 1x - y, \qquad y(0) = y_0$

$$\dot{\mathbf{x}} = \begin{bmatrix} -1 & -4 \\ 1 & -1 \end{bmatrix} \vec{\mathbf{x}}$$

$$s\vec{\mathbf{X}}(s) - \begin{bmatrix} x_0 \\ y_0 \end{bmatrix} = \begin{bmatrix} -1 & -4 \\ 1 & -1 \end{bmatrix} \vec{\mathbf{X}}(s), \quad \text{so} \quad \begin{bmatrix} s+1 & 4 \\ -1 & s+1 \end{bmatrix} \vec{\mathbf{X}}(s) = \begin{bmatrix} x_0 \\ y_0 \end{bmatrix}.$$

$$\vec{\mathbf{X}}(s) = \begin{bmatrix} s+1 & 4 \\ -1 & s+1 \end{bmatrix}^{-1} \begin{bmatrix} x_0 \\ y_0 \end{bmatrix} = \frac{1}{s^2+2s+5} \begin{bmatrix} s+1 & -4 \\ 1 & s+1 \end{bmatrix} \begin{bmatrix} x_0 \\ y_0 \end{bmatrix}$$

$$= \begin{bmatrix} \dfrac{s+1}{s^2+2s+5} & -\dfrac{4}{s^2+2s+5} \\ \dfrac{1}{s^2+2s+5} & \dfrac{s+1}{s^2+2s+5} \end{bmatrix} \begin{bmatrix} x_0 \\ y_0 \end{bmatrix} = \begin{bmatrix} \dfrac{s+1}{(s+1)^2+4} & \dfrac{-4}{(s+1)^2+4} \\ \dfrac{1}{(s+1)^2+4} & \dfrac{s+1}{(s+1)^2+4} \end{bmatrix} \begin{bmatrix} x_0 \\ y_0 \end{bmatrix}$$

$$\vec{\mathbf{x}}(t) = \begin{bmatrix} e^{-t}\cos 2t & -2e^{-t}\sin 2t \\ \dfrac{1}{2}e^{-t}\sin 2t & e^{-t}\cos 2t \end{bmatrix} \begin{bmatrix} x_0 \\ y_0 \end{bmatrix}$$

■   Finding General Solutions

15. $\dot{x} = a_{11}x + a_{12}y + f_1(t), \quad x(0) = c_1$

$\dot{y} = a_{12}x + a_{22}y + f_2(t), \quad y(0) = c_2$

$$\dot{\mathbf{x}} = \mathbf{A}\vec{\mathbf{x}} + \vec{\mathbf{f}}$$

$$s\vec{\mathbf{X}}(s) - \begin{bmatrix} c_1 \\ c_2 \end{bmatrix} = \mathbf{A}\vec{\mathbf{X}}(s) + \vec{\mathbf{F}}(s), \quad \text{so} \quad (s\mathbf{I} - \mathbf{A})\vec{\mathbf{X}}(s) = \begin{bmatrix} c_1 \\ c_2 \end{bmatrix} + \vec{\mathbf{F}}(s).$$

$$\vec{\mathbf{X}}(s) = (s\mathbf{I} - \mathbf{A})^{-1} \begin{bmatrix} c_1 \\ c_2 \end{bmatrix} + (s\mathbf{I} - \mathbf{A})^{-1}\vec{\mathbf{F}}(s)$$

$$\vec{\mathbf{x}}(t) = \vec{\mathbf{x}}_h(t) + \vec{\mathbf{x}}_p(t),$$

where each term is the inverse Laplace transform of the term above. Therefore $\vec{\mathbf{x}}_h(t)$ depends only on $\mathbf{A}$, c_1, and c_2, while $\vec{\mathbf{x}}_p(t)$ depends only on $\mathbf{A}$, $f_1(t)$ and $f_2(t)$.

■   **Mass-Spring Systems**

18.   We apply $m_1 = m_2 = 2$ and $k_1 = k_2 = k_3 = 1$ to the system of DE's in Example 4 to obtain

$$2\ddot{x} = y - x$$
$$2\ddot{y} = -(y - x)$$

so that

$$2[s^2 X(s) - 1] = Y(s) - X(s) \atop 2s^2 Y(s) = -[Y(s) - X(s)], \text{ so } \begin{bmatrix} 2s^2 + 1 & -1 \\ -1 & 2s^2 + 1 \end{bmatrix} \begin{bmatrix} X(s) \\ Y(s) \end{bmatrix} = \begin{bmatrix} 2 \\ 0 \end{bmatrix}$$

$$\begin{bmatrix} X(s) \\ Y(s) \end{bmatrix} = \begin{bmatrix} 2s^2 + 1 & -1 \\ -1 & 2s^2 + 1 \end{bmatrix}^{-1} \begin{bmatrix} 2 \\ 0 \end{bmatrix} = \frac{1}{4(s^4 + s^2)} \begin{bmatrix} 2s^2 + 1 & 1 \\ 1 & 2s^2 + 1 \end{bmatrix} \begin{bmatrix} 2 \\ 0 \end{bmatrix}$$

$$= \begin{bmatrix} \dfrac{1}{2} \dfrac{2s^2 + 1}{s^2(s^2 + 1)} \\[3mm] \dfrac{1}{2s^2(s^2 + 1)} \end{bmatrix} = \begin{bmatrix} \dfrac{1}{2s^2} + \dfrac{1}{2(s^2 + 1)} \\[3mm] \dfrac{1}{2s^2} - \dfrac{1}{2(s^2 + 1)} \end{bmatrix}$$

$$\begin{bmatrix} x(t) \\ y(t) \end{bmatrix} = \begin{bmatrix} \dfrac{1}{2} t + \dfrac{1}{2} \sin t \\[3mm] \dfrac{1}{2} t - \dfrac{1}{2} \sin t \end{bmatrix}$$

Note that this solution sends the center of mass moving to the right with a constant velocity of $\dfrac{1}{2}$ in accordance with the laws of classical mechanics.

■   **Comparing Laplace**

21. $$\dot{\vec{x}} = \begin{bmatrix} -1 & 0 \\ 0 & 2 \end{bmatrix} \vec{x} + \begin{bmatrix} 1 \\ 0 \end{bmatrix}, \quad \vec{x}(0) = \begin{bmatrix} 0 \\ 0 \end{bmatrix}$$

$$s\vec{X}(s) = \begin{bmatrix} -1 & 0 \\ 0 & 2 \end{bmatrix} \vec{X}(s) + \begin{bmatrix} \dfrac{1}{s} \\ 0 \end{bmatrix}, \quad \text{so } \begin{bmatrix} s+1 & 0 \\ 0 & s-2 \end{bmatrix} \vec{X}(s) = \begin{bmatrix} \dfrac{1}{s} \\ 0 \end{bmatrix}.$$

$$\vec{X}(s) = \begin{bmatrix} s+1 & 0 \\ 0 & s-2 \end{bmatrix}^{-1} \begin{bmatrix} \dfrac{1}{s} \\ 0 \end{bmatrix} = \frac{1}{(s+1)(s-2)} \begin{bmatrix} s-2 & 0 \\ 0 & s+1 \end{bmatrix} \begin{bmatrix} \dfrac{1}{s} \\ 0 \end{bmatrix} = \begin{bmatrix} \dfrac{1}{(s+1)s} \\ 0 \end{bmatrix} = \begin{bmatrix} -\dfrac{1}{s+1} + \dfrac{1}{s} \\ 0 \end{bmatrix}$$

$$\vec{x}(t) = \begin{bmatrix} e^{-t} + 1 \\ 0 \end{bmatrix}$$

■   **Suggested Journal Entry I**

24.   Student Project

CHAPTER

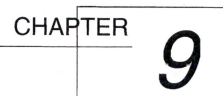

Discrete
Dynamical Systems

9.1   Iterative Equations

■   First-Order Linear Iterative Equations

For Problems 1-12 we use the fact that the solution of $y_{n+1} = ay_n + b$ is $y_n = a^n y_0 + b\left(\dfrac{a^n - 1}{a - 1}\right)$.

3. $y_{n+1} = -\dfrac{1}{2}y_n + 3$ $y_0 = 1$

(a),(c)   Because $a = -\dfrac{1}{2}$, $b = 3$, $y_0 = 1$,   **(b)**

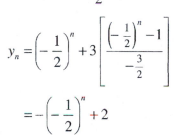

$$y_n = \left(-\frac{1}{2}\right)^n + 3\left[\frac{\left(-\frac{1}{2}\right)^n - 1}{-\frac{3}{2}}\right]$$

$$= -\left(-\frac{1}{2}\right)^n + 2$$

As n increases, the orbit approaches 2, with successive y_n alternatively above and below this value. Thus $y_n \to 2 = y_e$.

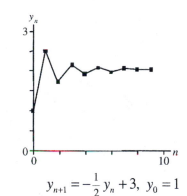

$$y_{n+1} = -\frac{1}{2}y_n + 3, \; y_0 = 1$$

6. $y_{n+1} = 2y_n + 3$ $y_0 = -1$

(a),(c)   Because $a = 2$, $b = 3$, $y_0 = -1$,   **(b)**

$$y_n = 2^n + 3\left[\frac{2^n - 1}{1}\right]$$

$$= 2^n - 3$$

As n increases, the solution grows without bound.

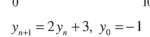

$$y_{n+1} = 2y_n + 3, \; y_0 = -1$$

9. $y_{n+1} = y_n + 3$ $y_0 = 1$

(a),(c)   Because $a = 1$, $b = 3$, $y_0 = 1$,   (b)

$$y_n = y_0 + nb$$
$$= 1 + 3n$$

As n increases, the solution grows without bound.

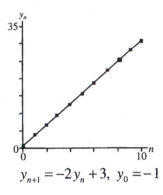

$$y_{n+1} = -2y_n + 3, \ y_0 = -1$$

12. $y_{n+1} = -y_n + 3$ $y_0 = -1$

(a),(c)   Because $a = -1$, $b = 3$, $y_0 = -1$,   (b)

$$y_n = -(-1)^n + 3\left[\frac{(-1)^n - 1}{-2}\right]$$
$$= -\frac{5}{2}(-1)^n + \frac{3}{2}$$

As n increases, the solution oscillates between -1 and 4 for all time. There is no steady state or equilibrium, but rather a cycle of period 2.

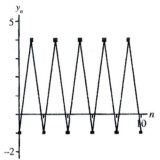

$$y_{n+1} = -y_n + 3, \ y_0 = -1$$

■   **Lab Problem: Spreadsheet Predictions**

In problems 15-21, when there is an equilibrium or steady state solution, its exact value y_e is calculated by setting $y_{n+1} = y_n$.

15. $y_{n+1} = 0.3y_n - 1$ $y_0 = -1.6$

(a)   A simple spreadsheet calculation yields the following first 10 iterates and graph.

n	y_n
0	-1.6
1	-1.48
2	-1.444
3	-1.4332
4	-1.42996
5	-1.42899
6	-1.42870
7	-1.42861
8	-1.42858
9	-1.42857
10	-1.42857

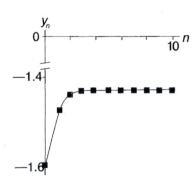

$$y_{n+1} = 0.3y_n - 1, \ y_0 = -1.6$$

(b)   The solution decreases monotonically and approaches a steady state near -1.43. Calculation gives $0.7 y_e = -1$, or $y_e = -1.428...$.

(c)   The coefficient of y_n is small, so we are not surprised to see that there is little effect as n becomes large. The constant -1 contributes to the level of the steady state. The initial value $-1.6 < y_e$ causes the orbit to approach equilibrium from below.

18. $y_{n+1} = 1.3y_n - 1$ $\qquad\qquad\qquad$ $y_0 = 0.2$

(a)   A simple spreadsheet calculation yields the following first 10 iterates and graph.

n	y_n
0	0.2
1	-0.74
2	-1.962
3	-3.5506
4	-5.61578
5	-8.30051
6	-11.7907
7	-16.3279
8	-22.2262
9	-29.8941
10	-39.8623

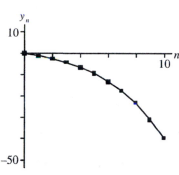

$y_{n+1} = 1.3y_n - 1$, $y_0 = 0.2$

(b)   The solution decreases monotonically and approaches infinity rather than a steady state. There *is* an equilibrium for this equation, because calculation gives $-0.3y_e = -1$, or $y_e = 3.33333...$; it is an unstable equilibrium that can only be reached by iterating backward with negative n.

(c)   The coefficient of y_n is multiplied at each step by 1.3, then -1 is subtracted. Hence once the solution becomes negative it becomes more and more negative and the orbit goes to minus infinity.

21. $y_{n+1} = -1.3y_n - 2$ $\qquad\qquad\qquad$ $y_0 = 0.2$

(a)   A simple spreadsheet calculation yields the following first 10 iterates and graph.

n	y_n
0	0.2
1	-2.26
2	0.938
3	-3.2194
4	2.18522
5	-4.84079
6	4.293022
7	-7.58093
8	7.855207
9	-12.2118
10	13.8753

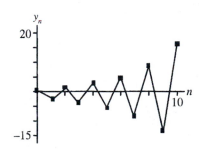

$y_{n+1} = -1.3y_n - 2$, $y_0 = 0.2$

(b)   The solution oscillates with larger and larger amplitude, not reaching a steady state (unless iteration goes backward with negative n). Equilibrium calculation $y_e = -1.3y_e - 2$ gives $y_e = -0.869...$.

(c)   Note that the constant term is now -2, so the equilibrium has moved; the long-term behavior is essentially the same.

■   **Closed-Form Sums**

24. $S_n = 1 + 3 + 3^2 + \ldots + 3^n$

(a)   This expression requires $S_0 = 1$, and we observe the general pattern

$$S_1 = S_0 + 3 = 1 + 3$$

$$S_2 = S_1 + 3^2 = 1 + 3 + 3^2$$

$$\ldots \qquad \ldots$$

$$S_n = 1 + 3 + 3^2 + 3^3 \cdots + 3^n = S_{n-1} + 3^n$$

Hence, the iterative equation for S_n is

$$S_{n+1} = S_n + 3^{n+1},$$

with $S_0 = 1$.

(b)   To show that the closed form expression $S_n = \frac{1}{2}(3^{n+1} - 1)$

satisfies the preceding iterative equation, we note that it satisfies $S_0 = 1$ and that

$$S_{n+1} - S_n = \frac{1}{2}(3^{n+2} - 1) - \frac{1}{2}(3^{n+1} - 1) = \frac{1}{2}(3 \times 3^{n+1} - 3^{n+1}) = 3^{n+1},$$

as predicted.

■   **Exceptional Case**

27.   When one eigenvalue of

$$ay_{n+2} + by_{n+1} + cy_n = 0$$

is zero, the characteristic equation is

$$a\lambda^2 + b\lambda = 0, \text{ or } \lambda(a\lambda + b) = 0,$$

and hence the solution is

$$y_n = c_1 + c_2\left(-\frac{b}{a}\right)^n.$$

For example, the equation $y_{n+2} - 2y_{n+1} = 0$ has the characteristic equation $\lambda^2 - 2\lambda = 0$ and eigenvalues 0 and 2. Hence, $y_n = c_1 + c_2 2^n$.

■   **Second-Order Linear Iterative Equations**

30. $y_{n+2} - 4y_{n+1} + 4y_n = 0$

The characteristic equation is $r^2 - 4r + 4 = 0$, which has a double root $r_1 = r_2 = 2$. Hence, the general solution is

$$y_n = c_1 2^n + c_2 n 2^n.$$

■   **Lab Problems: More Spreadsheets**

The reader should experiment with a spreadsheet and then check the results with the following solutions.

33. $y_{n+2} - y_n = 0$, $y_0 = 0$, $y_1 = 1$

(a)   The following iterates were found using Excel.

n	y_n	n	y_n
0	0	6	0
1	1	7	1
2	0	8	0
3	1	9	1
4	0	10	0
5	1		

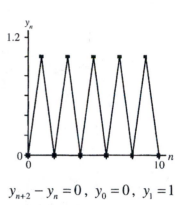

$y_{n+2} - y_n = 0$, $y_0 = 0$, $y_1 = 1$

(b)   It is clear from the results of the spreadsheet and the iterative equation itself that the solution will continue to oscillate between 0 and 1.

■   **Rabbits Again**

36.   We rewrite the Fibonacci equation as

$$y_{n+2} - y_{n+1} - y_n = 0.$$

The characteristic equation is

$$\lambda^2 - \lambda - 1 = 0,$$

which has solutions

$$\lambda_1, \lambda_2 = \frac{1}{2} \pm \frac{\sqrt{5}}{2}.$$

The general solution can then be written as

$$y_n = c_1 \left(\frac{1}{2} + \frac{\sqrt{5}}{2} \right) + c_2 \left(\frac{1}{2} - \frac{\sqrt{5}}{2} \right)^n.$$

Using the conditions $y_0 = 1$ and $y_1 = 1$ yields

$$c_1 = \frac{1}{2\sqrt{5}} \left(\sqrt{5} + 1 \right) \text{ and } c_2 = \frac{1}{2\sqrt{5}} \left(\sqrt{5} - 1 \right).$$

Thus, the solution becomes the *Binet* formula

$$y_n = \frac{1}{\sqrt{5}} \left\{ \left(\frac{1+\sqrt{5}}{2} \right)^{n+1} - \left(\frac{1-\sqrt{5}}{2} \right)^{n+1} \right\}.$$

■   **Check This with Your Banker**

39.   We use the formula for compound interest. Because interest is compounded daily, we have $r = \dfrac{0.08}{365} \approx 0.000219$. Hence, if the initial deposit is $y_0 = \$1000$, then the value in the account will be as follows.

$$1 \text{ day:}\quad y_1 = \$1000\left(1 + \frac{0.08}{365}\right)^1 = \$1000(1.000219) = \$1000.22$$

$$10 \text{ day:}\quad y_{10} = \$1000\left(1 + \frac{0.08}{365}\right)^{10} = \$1000(1.000219)^{10} = \$1002.19$$

$$1 \text{ year:}\quad y_{365} = \$1000\left(1 + \frac{0.08}{365}\right)^{365} = \$1000(1.000219)^{365} = \$1083.28$$

■   **Amazing But True**

42.   (a)   If weekly interest is $\dfrac{0.08}{52} \approx 0.00154$ (0.154%), and weekly deposits are \$25, then the amount of money y_n Wei Chen will have in the bank after n weeks satisfies the iterative IVP

$$y_{n+1} = 1.00154 y_n + 25, \quad y_0 = 0.$$

(b)   The solution of the IVP in part (a) is

$$y_n = \frac{25}{0.00154}\left[(1.00154)^n - 1\right] = \$16,237\left[(1.00154)^n - 1\right].$$

(c)   Substituting $n = 208$ (4 years) into the solution found in part (b) yields $y_{208} = \$6123.94$.

■   **Deer Population**

45.   (a)   Measuring deer in the thousands with $r = 0.10$, $d = 15$, and $y_0 = 100$, yields

$$y_{n+1} = 1.10 y_n - 15$$

with solution

$$y_n = 100(1.10)^n - \frac{15}{0.10}\left[(1.10)^n - 1\right] = -50(1.10)^n + 150 \text{ thousands of deer.}$$

(b)   In order for the population to be constant, simply harvest the new population. If initially there are 100,000 and they grow by 10% per year, harvest (hunt) 10,000 deer per year. This will keep the population at 100,000 deer.

■   **Consequence of Periodic Drug Therapy**

48.   (a)   We assume Kashkooli has no drug in his system on day zero. However, y_0 is the amount of drug in his body immediately after he takes his dose of 100 mg. Therefore, $y_0 = 100$ mg . Further, because he loses 25% per day and gains 100 mg, the iterative equation that describes the number of mg of drug he has in his body on day n immediately after taking the drug is

$$y_{n+1} = y_n - 0.25y_n + 100.$$

Thus, $y_{n+1} = 0.75y_n + 100$, $y_0 = 100$.

(b)   The iterative equation in part (a) can be written

$$y_{n+1} - 0.75y_n = 100$$

and has characteristic equation $r - 0.75 = 0$. Hence, the homogeneous solution is

$$y_n = c(0.75)^n.$$

Finding a particular solution of the nonhomogeneous equation yields $y_n = 400$, which means the general solution is

$$y_n = c(0.75)^n + 400.$$

Substituting the initial condition $y_0 = 100$ yields $c = -300$. Thus, the amount of drug he has in his body after n days is

$$y_n = 100\left[4 - 3(0.75)^n\right].$$

(c)   As n increases $(0.75)^n$ goes to zero, and hence, y_n tends to 400 mg.

(d)   The homogeneous solution $c(0.75)^n$ always tends to 0, as n increases no matter what the dosage. Thus, the particular solution is $y_n = A$ for $y_{n+1} - 0.75y_n = d$, where d is the daily dosage, which is $A = 4d$. In other words, if the limiting amount of the drug is A, then the daily dosage should be $d = 0.25A$ mg . If Kashkooli wants the limiting amount of the drug in his body to be $A = 800$ mg , then his daily dosage should be

$$d = 0.25(800) = 200 \text{ mg}.$$

■   **Very Interesting**

51.   Once you make a few sketches, you will convince yourself that the iterative equation

$$y_{n+1} = y_n + n + 1$$

is correct. Hence, we simply need to solve it. As noted in the hint, the new line divides $n+1$ regions into $2n+2$ regions, so if there were y_n regions before, there are now $y_n + n + 1$ regions.

The homogeneous equation has characteristic equation $\lambda - 1 = 0$, so the homogeneous solutions are constants $y_n = c_1$. We therefore seek a particular solution of the form $y_n = An^2 + Bn$.

Substituting this value into the nonhomogeneous equation yields

$$2An + (A + B) = n + 1.$$

Setting coefficients equal yields $A = B = \dfrac{1}{2}$. Hence, the general solution is

$$y_n = c_1 + \frac{n(n+1)}{2}.$$

Substituting the initial condition $y_0 = 1$ yields $c_1 = 1$, so

$$y_n = \frac{n(n+1)}{2} + 1 = \frac{n^2 + 2n + 2}{2} \quad \text{distinct regions.}$$

9.2   Linear Iterative Equations

■ System Classification

3.
$$\begin{bmatrix} x_{n+1} \\ y_{n+1} \end{bmatrix} = \begin{bmatrix} 0 & 1 \\ -1 & 0 \end{bmatrix} \begin{bmatrix} x_n \\ y_n \end{bmatrix}$$

Compute the eigenvalues, $\lambda_1 = i$ and $\lambda_2 = -i$, or compute $\text{Tr}\mathbf{A} = 0$ and $|\mathbf{A}| = 1$. From the trace-determinant plane in Figure 9.2.7 the point $(0, 1)$ for this matrix A lies in the center region. Hence, the equilibrium is a center, and each iterate remains at the same distance from the origin. In this case the orbit cycles around four points (see figure). Using Excel, we plot the first few points starting at the point $(x_0, y_0) = (3, 2)$.

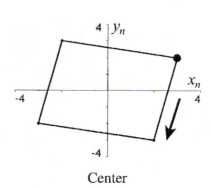

Center

6.
$$\begin{bmatrix} x_{n+1} \\ y_{n+1} \end{bmatrix} = \begin{bmatrix} 1.5 & 0 \\ 0 & -0.5 \end{bmatrix} \begin{bmatrix} x_n \\ y_n \end{bmatrix}$$

Compute the eigenvalues, $\lambda_1 = 1.5$ and $\lambda_2 = -0.5$, or compute $\text{Tr}\mathbf{A} = 1$ and $|\mathbf{A}| = -0.75$. One eigenvalue has an absolute value greater than 1, and the other is less than 1, so the equilibrium is a saddle. Because only one eigenvalue is negative, it is a flip saddle. Using Excel, we plot the first few points starting at the point $(x_0, y_0) = (3, 2)$.

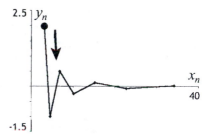

Flip saddle

9.
$$\begin{bmatrix} x_{n+1} \\ y_{n+1} \end{bmatrix} = \begin{bmatrix} 0 & 0.9 \\ -1.6 & 0 \end{bmatrix} \begin{bmatrix} x_n \\ y_n \end{bmatrix}$$

Compute the eigenvalues, $\lambda_1 = -1.2i$ and $\lambda_2 = 1.2i$, or compute $\text{Tr}\mathbf{A} = 0$ and $|\mathbf{A}| = -1.44$. We see that from the trace-determinant plane in Figure 9.2.7 the point $(0, 1.44)$ of the matrix A lies in the spiral source region. Using Excel, we plot the first few points starting at the point $(x_0, y_0) = (3, 2)$.

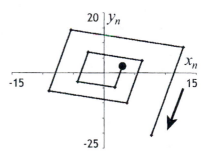

Spiral source

12.

$$\begin{bmatrix} x_{n+1} \\ y_{n+1} \end{bmatrix} = \begin{bmatrix} 0 & 1.6 \\ -0.4 & 0 \end{bmatrix} \begin{bmatrix} x_n \\ y_n \end{bmatrix}$$

Compute the eigenvalues, $\lambda_1 = 0.8i$ and $\lambda_2 = -0.8i$, or compute $\text{Tr}\mathbf{A} = 0$ and $|\mathbf{A}| = 0.64$. We see that from the trace-determinant plane in Figure 9.2.7 the point $(0, 0.64)$ for the matrix A lies in the spiral sink region. Using Excel, we plot the first few points starting at the point $(x_0, y_0) = (3, 2)$.

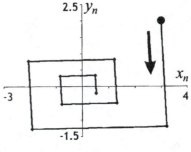

Spiral sink

■   **Classification Using Eigenvalues**

15.   (a) $\lambda_1 = -0.5$, $\lambda_2 = -0.3$

The equilibrium (at the origin) is a double-flip sink, because both eigenvalues are negative and have absolute values less than 1.

(b) $x_n = x_0 (-0.5)^n$

$y_n = y_0 (-0.3)^n$

(c)   A few iterates are shown (see figure) starting at $(2, 2)$.

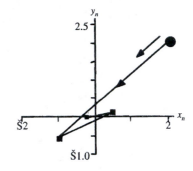

18.   (a) $\lambda_1 = -1.5$, $\lambda_2 = -3$

The equilibrium (at the origin) is a double-flip source, because both eigenvalues are negative and have absolute values greater than 1.

(b) $x_n = x_0 (-1.5)^n$

$y_n = y_0 (-3)^n$

(c)   A few iterates are shown (see figure) starting at $(2, 2)$.

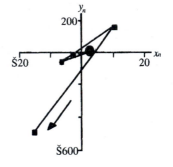

21.   (a) $\lambda_1 = -0.5$, $\lambda_2 = -2$

The equilibrium (at the origin) is a double-flip saddle, because both eigenvalues are negative and only one eigenvalue has an absolute value greater than 1.

(b) $x_n = x_0 (-0.5)^n$

$y_n = y_0 (-2)^n$

(c)   A few iterates are shown (see figure) starting at $(2, 2)$.

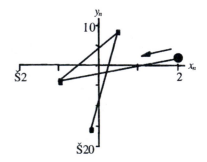

▪ Spirals or Circles?

24.

$$\vec{x}_{n+1} = \begin{bmatrix} \sqrt{2} & -\sqrt{2} \\ \sqrt{2} & \sqrt{2} \end{bmatrix} \vec{x}_n$$

We rewrite the coefficient matrix as

$$\begin{bmatrix} \sqrt{2} & -\sqrt{2} \\ \sqrt{2} & \sqrt{2} \end{bmatrix} = 2 \begin{bmatrix} \dfrac{\sqrt{2}}{2} & -\dfrac{\sqrt{2}}{2} \\ \dfrac{\sqrt{2}}{2} & \dfrac{\sqrt{2}}{2} \end{bmatrix},$$

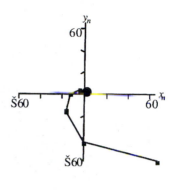

Outward spiral, counterclockwise

which is a scalar two times a rotation matrix with $\theta = \dfrac{\pi}{4}$. Hence, the action of the iterative

system is a counterclockwise rotation of $\dfrac{\pi}{4}$ at each iteration plus an expansion away from the

origin by a factor of 2. See figure for a few points of the orbit starting at $(1, 1)$.

▪ Wolf Extinction

27. $w_0 = 300, \qquad m_0 = 0$

If there are no moose present and the wolf population is initially w_0, w_0 will obey the IVP $w_{n+1} = 0.72w_n$ and $w_0 = 300$, whose solution is

$$w_n = 300(0.72)^n.$$

Without a replacement food supply, the wolf population dies out within 15 years (see figure).

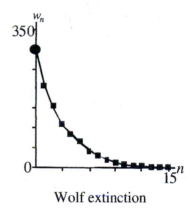

Wolf extinction

■   **System Analysis**

30. $A = \begin{bmatrix} -0.04 & 2.28 \\ 0.38 & 0.34 \end{bmatrix}$

(a)   The eigenvalues are computed as $\lambda_1 = -0.8$ and $\lambda_2 = 1.1$ with corresponding eigenvectors

$$\vec{v}_1 = [-3, 1] \text{ and } \vec{v}_2 = [2, 1].$$

(b) $\text{Tr}A = 0.3$, $|A| = -0.88$

(c)   The line with negative slope is the eigenvector corresponding to $\lambda_1 = -0.8$ and the line with positive slope is the eigenvector corresponding to $\lambda_2 = 1.1$.

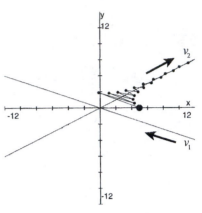

Single-flip saddle solution

The figure shows the solution starting from $x_0 = 5$, $y_0 = 0$, which gets closer and closer to the eigenvector $\vec{v}_2$. Although iterates flip back and forth around $\vec{v}_2$ (due to the negative eigenvalue) for small n, they will just move out along the eigenvector going away from the origin.

(d) $\vec{x}_n = c_1 (-0.8)^n \begin{bmatrix} -3 \\ 1 \end{bmatrix} + c_2 (1.1)^n \begin{bmatrix} 2 \\ 1 \end{bmatrix}$

The first term in the solution goes to zero as a result of the factor $(-0.8)^n$. Hence, the orbit gets closer and closer to $\vec{v}_2$ and oscillates around this eigenvector with smaller and smaller amplitudes as a result of the negative value of λ_1. Meanwhile the factor $(1.1)^n$ causes the orbit to move steadily away from the origin.

■   **Conversion Job**

33.   Using the equation numbers in the text, the two equations are

$$x_{n+1} = ax_n + by_n \tag{11}$$

$$y_{n+1} = cx_n + dy_n. \tag{12}$$

Rewrite Equation (12) as

$$y_{n+2} = cx_{n+1} + dy_{n+1} \tag{13}$$

and then work with equations (11) and (12) to get all the terms expressed in y's. E.g., multiply Equation (11) by c, to get

$$cx_{n+1} = cax_n + cby_n. \tag{14}$$

Now solve, respectively, Equation (12) for cx_n and Equation (13) for cx_{n+1}, we get

$$cx_n = y_{n+1} - dy_n$$
$$cx_{n+1} = y_{n+2} - dy_{n+1}.$$

Substituting these values in Equation (14) yields

$$y_{n+2} - dy_{n+1} = ay_{n+1} - ady_n + cby_n$$

or

$$y_{n+2} - (a + d) y_{n+1} + (ad - bc) y_n = 0.$$

■ Saddle Regions

36.   (a)   The characteristic equation of the coefficient matrix of

$$x_{n+1} = ax_n + by_n$$
$$y_{n+1} = cx_n + dy_n$$

is

$$\begin{vmatrix} a - \lambda & b \\ c & d - \lambda \end{vmatrix} = \lambda^2 - \mathrm{Tr}\mathbf{A} \ \lambda + |\mathbf{A}| = 0.$$

This is a quadratic equation in λ, so we have

$$\mathrm{Tr}\mathbf{A} = \lambda_1 + \lambda_2 \text{ and } |\mathbf{A}| = \lambda_1 \lambda_2$$

where λ_1 and λ_2 are the eigenvalues of the coefficient matrix.

Hence, because the region S is defined by

$$-\mathrm{Tr}\mathbf{A} - 1 < |\mathbf{A}| < \mathrm{Tr}\mathbf{A} - 1,$$

this inequality provides the relationship between the eigenvalues as

$$-(\lambda_1 + \lambda_2) - 1 < \lambda_1 \lambda_2 < \lambda_1 + \lambda_2 - 1$$

or equivalently

$$\lambda_1 \lambda_2 - \lambda_1 - \lambda_2 + 1 < 0$$

and $\lambda_1 \lambda_2 + \lambda_1 + \lambda_2 + 1 > 0.$

(b)   We can factor each of the left-hand sides of the inequalities in part (a) getting

$$\lambda_1 \lambda_2 - \lambda_1 - \lambda_2 + 1 = (\lambda_1 - 1)(\lambda_2 - 1)$$

and $\lambda_1 \lambda_2 + \lambda_1 + \lambda_2 + 1 = (\lambda_1 + 1)(\lambda_2 + 1).$

continued on the next page

Thus, the inequalities become

$$(\lambda_1 - 1)(\lambda_2 - 1) < 0$$

and $(\lambda_1 + 1)(\lambda_2 + 1) > 0.$

But it is easy to see that *both* of the previous inequalities hold if *either* of the following sets of inequalities hold:

$$-1 < \lambda_1 < 1 \text{ and } \lambda_2 > 1$$

or $-1 < \lambda_2 < 1 \text{ and } \lambda_1 > 1.$

The details of this simple verification are left for the student.

(c)   As we obtained in part (b) one eigenvalue is positive and greater than 1, and the other eigenvalue has an absolute value lower then 1. One can see that the origin is a saddle point.

■   **Suggested Journal Entry**

39.   Student Project

9.3 Linear Iterative Equations Chaos Again

■ **Attractors and Repellers**

3. $x_{n+1} = x_n^3$

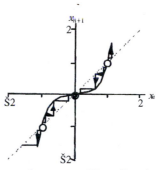

Set $x = x^3$, which yields $x^3 - x = 0$, so we have three fixed points

$$x_e = -1,\ 0,\ 1.$$

The cobweb diagram shows that 0 is attracting and that 1 and −1 are repelling.

One attracting and two repelling fixed points

■ **Stability of Fixed Points**

6.

(a) $f(x) = \dfrac{3x+2}{x+2}$

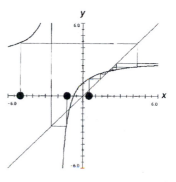

The slope of $f(x)$ at the left-hand equilibrium is unstable, and the right-hand equilibrium is stable.

Note: the text Figure 9.3.16(a) shows only the part of $f(x)$ to the right of $x = -2$, which is an asymptote for $f(x)$.

If we try a cobweb from an x_0 just to the left of −2, we discover we must graph $f(x)$ to left of −2 as well.

(b) $f(x) = 3x(1-x)$

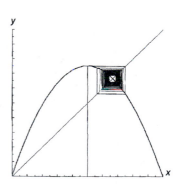

Eyeballing the slopes of $f(x)$ at the equilibria shows that the origin is repelling and unstable, but at the right-hand fixed point $x_e = 2/3$ the slope appears close to −1, so we take the derivative to check it out.
$$f'(x) = 3 - 6x,\ \text{so}\ f'(2/3) = -1.$$

Note that a straight line with slope −1 produces a cyclic orbit of period 2; however the curvature of $f(x) = 3x(1-x)$ spoils the possibility of a cycle. In fact, if the orbit is extended you will see it actually converges, very very slowly, to $x_e = 2/3$.

■   **Analyzing the Data**

There are many ways to show a solution sequence for an iterative equation. In problems 7-12 we chose a time series with a "typical" seed, usually $x_0 = 0.5$. A list of values, or sequences from other seeds (not at an equilibrium) should give the same information. Excel is an excellent source of all these options.

9. $x_{n+1} = 3.2 x_n (1 - x_n)$

(a)   Starting at $x_0 = 0.5$, the function iterates in a cycle, shown in gray.

Starting at $x_0 = 0.2$, the function iterates toward the same cycle, as shown in black.

(b)   The equilibrium point(s) x_e of this iteration are the root(s) of

$$x = f(x) = 3.2x(1-x)$$

or

$$3.2x^2 - 2.2x = 0,$$

which yields $x_e = 0$ and $x_e = \dfrac{2.2}{3.2} \approx 0.6875..$

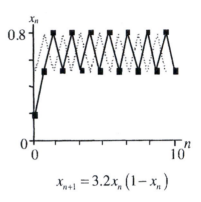

$$x_{n+1} = 3.2 x_n (1 - x_n)$$

(c) $\left| f'(0) \right| = 3.2 > 1$, and

$$\left| f'\left(\frac{2.2}{3.2}\right) \right| = \left| 3.2 - 6.4\left(\frac{2.2}{3.2}\right) \right| = \left| 3.2 - 4.4 \right| = 1.2 > 1.$$

Hence both fixed points are unstable or repelling, with an attracting *cycle* passing between them. See Problem 30.

12. $x_{n+1} = \cos x_n$

(a)   Starting at $x_0 = 0.5$, the function iterates in an oscillatory fashion to an equilibrium $x_e \approx 0.75$.

(b)   The equilibrium point(s) x_e of this iteration are the root(s) of $x = f(x) = \cos x$, or

$$x_e = 0.739.$$

(c)   Because $f'(x) = -\sin(x)$ yields

$$\left| f'(0.739) \right| = \left| -\sin(0.739) \right| = 0.674 < 1,$$

0.739 is an asymptotically stable equilibrium, as the graph shows.

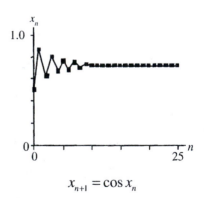

$$x_{n+1} = \cos x_n$$

15. $x_{n+1} = x_n^2 - 2$

(a)   Starting at $x_0 = 0.5$, the function seems to show chaotic iterative behavior, and extending the orbit does not change that view.

(b)   The equilibrium point(s) x_e of this iteration are the root(s) of

$$x = f(x) = x^2 - 2,$$

or

$$x^2 - x - 2 = 0,$$

which yields $x_e = -1$ and $x_e = 2$.

(c)   Because $f'(x) = 2x$ yields

$$|f'(-1)| = |2(-1)| = 2 > 1,$$

$$|f'(2)| = |2(2)| = 4 > 1,$$

both fixed points are unstable, with no evidence of anything attracting between. This gives rise to the chaos we see in the time series.

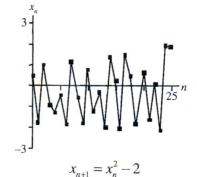

$x_{n+1} = x_n^2 - 2$

■   **Repeat Pete's Repeat**

18. $x_{n+1} = \sin x_n$

Starting at $x_0 = 0.5$, the solution appears to be converging ever so slowly to 0. We see that, for a linear iterative equation, an equilibrium or fixed point x_e of $x_{n+1} = f(x_n)$ is asymptotically stable if $|f'(x_e)| < 1$ and unstable if $|f'(x_e)| > 1$. The only equilibrium point x_e of this iteration is the root of $x = f(x) = \sin x$ or $x_e = 0$. Because $f'(x) = \cos x$ yields

$$|f'(x_e)| = |\cos(0)| = 1,$$

so the derivative test is inconclusive.

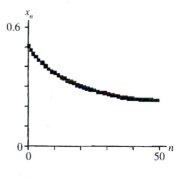

time series $x_{n+1} = x_n^2 - 0.1$

■   **The Bernoulli Mapping**

21. $x_{n+1} = \begin{cases} 2x_n & 0 \le x_n < 0.5 \\ 2x_n - 1 & 0.5 \le x_n \le 1 \end{cases}$

The cobweb diagram is shown with the orbit for $x_0 = 0.4$. The orbit seems to act like "bouncing ball" in the diagram and doesn't tend to go anywhere, but "bounces" all over the diagram indefinitely. This is precisely the behavior of chaotic motion.

Orbits will be chaotic for any x_0 other than the three fixed points 0, 0.5 or 1 (0.5 is not really a fixed point, but a single iteration moves it to 0 to stay).

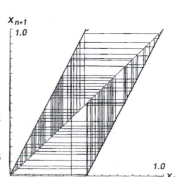

Cobweb diagram for
Bernoulli mapping

■   **Orbit Diagram of the Tent Mapping**

24.   It is an easy matter to write a computer program to find the

$$x_{n+1} = \begin{cases} 2rx_n & 0 \le x_n < 0.5 \\ 2r(1-x_n) & 0.5 \le x_n \le 1 \end{cases}$$

for the parameter $0.4 \le r \le 1$.

The sample program given below produces the figure shown.

Orbit diagram for tent mapping

BASIC Program to Compute Orbit Diagrams

10	REM ORBIT DIAGRAM	140	LET X = 0.5
20	REM N = # ITERATIONS FOR EACH R	150	LET R = MINR + (I - 1)*D
30	REM MINR = MINIMUM R	160	FOR J = 1 TO N
40	REM MAXR = MAXIMUM R	170	IF X > 0.5 THEN GO TO 200
50	REM RSTEPS = # OF R VALUES	180	X = 2*R*X
60	SCREEN 2	190	GOTO 210
70	WINDOW (0, 0.4) - (1, 1)	200	X = 2*R*(1 - X)
80	LET N = 3150	210	IF J < 3000 THEN GOTO 230
90	LET MINR = 0.4	220	PLOT (R, X)
100	LET MAXR = 1.0	230	NEXT J
110	LET RSTEPS = 101	240	NEXT I
120	LET D = (MAXR - MINR)/(RSTEPS - 1)	250	END
130	FOR I = 1 TO RSTEPS		

It is easy to change this program to draw the orbit diagram for any other iteration.

To change the range of the parameter r, simply change statements 90 and 100. (Note: For some iterations the parameter is not called r. It is suggested that you simply call it R and not change the program to a new name.)

To change the iteration function simply replace the statements 170, 180, 190, and 200 by the new iteration function. (Note: Most iteration functions will only take one line; the tent mapping is a conditional function that requires more than one line.)

Line 140 gives the initial condition, which can be changed. (For the tent mapping, the diagram comes more quickly if $x_0 = 0$.)

In line 210, J < # transient points to skip in the plotting. This number is adjustable. (For some functions, including the tent mapping, the number of transients to skip must be large to avoid extraneous patterns.)

Sequential Analysis

27.   (a)   Because x_{n+1} is a monotone decreasing sequence bounded below by 0 we know that it converges to its greatest lower bound L. Also the sequence x_{n+1} is always decreasing because if $0 \leq r \leq 1$, then $r(1 - x_n) < 1$. Multiplying by x_n yields

$$rx_n (1 - x_n) < x_n$$

or, in other words, $x_{n+1} < x_n$.

(b)   We have

$$x_{n+1} = rx_n (1 - x_n) \geq 0, \quad n = 1, 2, \ldots$$

for $0 \leq x_0 \leq 1$ and $0 \leq r \leq 1$, hence the greatest lower bound L of the sequence is greater than or equal to zero also.

(c)   The greatest lower bound L of the monotone decreasing sequence $\{x_{n+1}\}$ is 0 inasmuch as it will eventually lie below any positive number. Hence $L = 0$.

Finding Two-Cycle Values

30.   If $f(x) = rx(1 - x)$, then

$$f(f(x)) = rf(x)[1 - f(x)] = r[rx(1 - x)][1 - rx(1 - x)].$$

If we set this value to x we get

$$3.2[3.2x(1 - x)][1 - 3.2x(1 - x)] = x.$$

This fourth-degree equation was solved with Maple, yielding the four (approximate) roots 0, 0.513, 0.687, and 0.799. The four roots can been seen in the figure as the intersection of the curves

$$y = f(f(x)) \quad \text{and} \quad y = x.$$

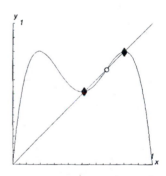

Graph of $y = f(f(x))$

Of the four fixed points for the second iterative function $y = f(f(x))$, 0 and 0.687 are unstable and 0.513 and 0.799 are stable. The first two are the fixed points (repelling) of $f(x)$. The latter two are cyclic points (attracting) of period 2 for $f(x)$.

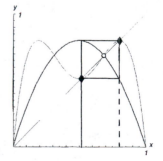

■   **Windows in the Orbit Diagram**

33.   (a)   The first 15 computed values of the iteration $x_{n+1} = r x_n (1 - x_n)$ with $r = 3.84$ are shown in the table. Note the three-cycle of approximately 0.959, 0.148, and 0.485, which fits with the cobweb diagram shown.

n	x_n	n	x_n
0	0.5	8	0.14801
1	0.96	9	0.484237
2	0.147456	10	0.959046
3	0.482737	11	0.150823
4	0.958856	12	0.491811
5	0.151494	13	0.959742
6	0.493607	14	0.148366
7	0.959843	15	0.485196

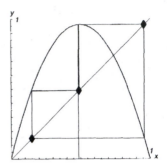

Cobweb diagram showing three-cycle $(r = 3.84)$

(b)   (i)   When $r = 3.628$ the solution approaches a six-cycle with values 0.306, 0.771, 0.641, 0.834, 0.501, as shown in the cobweb diagram.

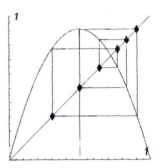

Cobweb diagram showing six-cycle $(r = 3.628)$

When $r = 3.85$ the solution approaches a six-cycle with values 0.155, 0.506, 0.962, 0.139, 0.462, and 0.956. as shown in the cobweb diagram.

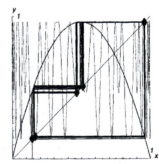

Cobweb diagram showing six-cycle $(r = 3.85)$

(ii)   The cobweb diagram for $r = 3.85$ illustrates the period doubling phenoenon tearing apart from the 3-cycle for $r = 3.84$. Compare with figure in part (a), the cobweb diagram for $r = 3.628$ shows an entirely different six-cycle behavior.

Suppose we start an orbit at the lower left most cycle point, and label the points along the diagonal in the order they are visited. For $r = 3.628$, the visitation labels read from the left along the left line.

Continued from previous page.

(c)   When $r = 3.74$ the solution approaches a five-cycle with values 0.935, 0.227, 0.657, 0.842, and 0.496, as shown in the cobweb diagram.

Another 5-cycle occurs for $r = 3.9057$. See the solution for Problem 32 to compare the cobwebs for these two cycles.

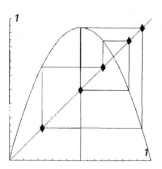

Cobweb diagram showing
five-cycle $(r = 3.74)$

■   **Class Project: The Real Bifurcation Diagram**

36.   This problem is left as a project for a group of students. The larger the group, the better the result.

CHAPTER 10   Control Theory

10.1   Feedback Controls

■   **A Matter of Control**

In Problems 1–6, note the steady-state error when proportional control is present. Note, too, the additional damping when derivative control is present.

3. $2\dot{x}+3x=u$, $x(0)=1$. The uncontrolled system is represented by

$$2\dot{x}+3x=0,\ \ x(0)=1$$

and the uncontrolled response is $x(t)=e^{-3t/2}$.

(a)   **Proportional feedback:** The proportional feedback is $u=2(1-x)$ Thus, the controlled response would satisfy the system

$$2\dot{x}+3x=2(1-x),\ \ x(0)=1,$$

or $2\dot{x}+5x=2$, which has the solution

$$x(t)=\frac{3}{5}e^{-5t/2}+\frac{2}{5}.$$

This response approaches $\frac{2}{5}$.

(b)   **Derivative feedback:** The derivative feedback is $u=-3\dot{x}$. Thus, the controlled response would satisfy the system

$$2\dot{x}+3x=-3\dot{x},\ \ x(0)=1,$$

or $5\dot{x}+3x=0$, which has the solution

$$x(t)=e^{-3t/5}.$$

This response approaches zero, but more slowly than the uncontrolled response.

(c)   **Derivative + proportional:** The derivative plus proportional feedback is $u = 2(1-x) - 3\dot{x}$.

Thus the controlled response would satisfy the system

$$2\dot{x} + 3x = 2(1-x) - 3\dot{x}, \quad x(0) = 1,$$

or $5\dot{x} + 5x = 2$, which has the solution

$$x(t) = \frac{3}{5}e^{-t} + \frac{2}{5}.$$

This response approaches $\frac{2}{5}$ but more slowly than with proportional feedback alone.

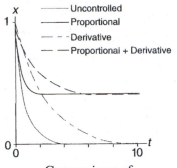

Comparison of
different controls

All responses are shown in the figure.

6. $\ddot{x} + 3\dot{x} + 2x = u$. The uncontrolled system is represented by

$$\ddot{x} + 3\dot{x} + 2x = 0, \quad x(0) = 1, \quad \dot{x}(0) = 0.$$

The uncontrolled response for the system is

$$x(t) = 2e^{-t} - e^{-2t}.$$

(a)   **Proportional feedback:** The proportional feedback is $u = 2(1-x)$. Thus, the controlled response would satisfy the system

$$\ddot{x} + 3\dot{x} + 2x = 2(1-x), \quad x(0) = 1, \quad \dot{x}(0) = 0$$

or $\ddot{x} + 3\dot{x} + 4x = 2$, which has the solution

$$x(t) = e^{-3t/2}\left[\frac{1}{2}\cos\left(\frac{\sqrt{7}}{2}t\right) + \frac{3}{14}\sqrt{7}\sin\left(\frac{\sqrt{7}}{2}t\right)\right] + \frac{1}{2}.$$

This response approaches $\frac{1}{2}$, with damped oscillation.

(b)   **Derivative feedback:** The derivative feedback is $u = -3\dot{x}$. Thus, the controlled response would satisfy the system

$$\ddot{x} + 3\dot{x} + 2x = -3\dot{x}, \quad x(0) = 1, \quad \dot{x}(0) = 0$$

or $\ddot{x} + 6\dot{x} + 2x = 0$, which has the general solution

$$x(t) = \left(\frac{1}{2} + \frac{3}{14}\sqrt{7}\right)e^{(-3+\sqrt{7})t} + \left(\frac{1}{2} - \frac{3}{14}\sqrt{7}\right)e^{(-3-\sqrt{7})t}.$$

This response approaches zero, but more slowly than the uncontrolled response, and without oscillation.

(c)   **Derivative + proportional:** The derivative plus proportional feedback is $u = 2(1-x) - 3\dot{x}$. Thus, the controlled system would satisfy the system

$$\ddot{x} + 3\dot{x} + 2x = 2(1-x) - 3\dot{x}$$

or $\ddot{x} + 6\dot{x} + 4x = 2$, $x(0) = 1$, $\dot{x}(0) = 0$, which has the solution

$$x(t) = \left(\frac{1}{4} + \frac{3}{20}\sqrt{5}\right)e^{(-3+\sqrt{5})t} + \left(\frac{1}{4} - \frac{3}{20}\sqrt{5}\right)e^{(-3-\sqrt{5})t} + \frac{1}{2}.$$

This response approaches $\frac{1}{2}$, but without oscillation.

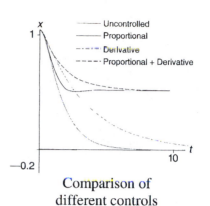

Comparison of different controls

See figure for plot comparison.

■   **Controlling an Unstable System**

9.   (a)   The controlled system is $\ddot{x} - 4\dot{x} + 3x = -k_d\dot{x}$, or

$$\ddot{x} + (k_d - 4)\dot{x} + 3x = 0,$$

which has the characteristic equation

$$\lambda^2 + (k_d - 4)\lambda + 3 = 0$$

with eigenvalues

$$\lambda_1, \lambda_2 = \frac{-(k_d - 4) \pm \sqrt{(k_d - 4)^2 - 12}}{2}.$$

The eigenvalues are complex with positive real part for $k_d < 4$ and, hence, are unstable. However, if $k_d > 4$, the eigenvalues are real and negative and, hence, the solutions have stable equilibria.

(b)   The controlled system is $\ddot{x} - 4\dot{x} + 3x = k_p(x_1 - x) - k_d\dot{x}$, or

$$\ddot{x} + (k_d - 4)\dot{x} + (k_p + 3)x = k_p x_1$$

whose homogeneous equation has characteristic equation

$$\lambda^2 + (k_d - 4)\lambda + (k_p + 3) = 0.$$

Hence the eigenvalues of this controlled system are

$$\lambda_1, \lambda_2 = \frac{-(k_d - 4) \pm \sqrt{(k_d - 4)^2 - 4(k_p + 3)}}{2},$$

which yields a double eigenvalue at -2 when

$$(k_d - 4)^2 - 4(k_p + 3) = 0 \text{ and } (k_d - 4) = 4.$$

Hence, $k_d = 8$; $k_p = 1$; offset error $= 0.75 x_1$.

The resulting controlled equation is

$$\ddot{x} - 4\dot{x} + 3x = (x_1 - x) - 8\dot{x}$$

or

$$\ddot{x} + 4\dot{x} + 4x = x_1,$$

which has the general solution

$$x(t) = c_1 e^{-2t} + c_2 t e^{-2t} + \frac{x_1}{4}.$$

■   **Integral Control**

12.   (a)   The uncontrolled system is represented by

$$\dot{x} + 4x = 0,$$

and the uncontrolled response is $x(t) = ce^{-4t}$ with a steady state response of 0.

(b)   Taking the Laplace transform of the controlled system

$$\dot{x} + 4x = 3\int_0^t \left[1 - x(w)\right] dw$$

yields

$$sX(s) + 4X(s) = \frac{3}{s^2} - \frac{3}{s}X(s)$$

or

$$\left(s + 4 + \frac{3}{s}\right)X(s) = \frac{3}{s^2}.$$

Solving for $X(s)$ gives

$$X(s) = \frac{3}{s(s^2 + 4s + 3)}.$$

Taking the inverse transform gives

$$x(t) = 1 + \frac{e^{-3t}}{2} - \frac{3e^{-t}}{2}.$$

(c)   If the final value theorem is applied to the equation found in part (b), we have

$$\lim_{t \to \infty}(x(t)) = \lim_{t \to \infty}\left(1 + \frac{e^{-3t}}{2} - \frac{3e^{-t}}{2}\right) = 1.$$

As t goes to infinity the limit approaches the set point $(x_1 = 1)$ without any offset error.

■ Out of Control

15. **(a)** Differentiating

$$\ddot{x} + 9\dot{x} + 20x = 18x_1 + 4\int_0^t \left(x_1 - x(w)\right)dw$$

we get

$$\dddot{x} + 9\ddot{x} + 20\dot{x} = 4(x_1 - x)$$

or

$$\dddot{x} + 9\ddot{x} + 20\dot{x} + 4x = 4x_1 \, .$$

(b) Using the result from part (a) and that x_p is a constant, it can be easily shown that

$$x_p = x_1 \, .$$

(c) Using Maple or another computer algebra system we find that the roots of the characteristic polynomial

$$\lambda^3 + 9\lambda^2 + 20\lambda + 4 = 0$$

are approximately –0.2215, –3.2893, and –5.4893. They are all negative; hence, the homogenous solutions go to zero, and all solutions approach the particular solution ,
$$x(t) \equiv x_1 \, .$$

■ Frequency Viewpoint of Feedback

18. $\ddot{x} + \dot{x} + x = f(t)$, $x(0) = 0$, $\dot{x}(0) = 0$. We see from Problem 17 that the transfer function was

$$G(s) = \frac{X(s)}{F(s)} = \frac{1}{s^2 + s + 1} \, .$$

Hence, $X(s) = G(s)F(s)$, which was to be proved.

21. **(a)** $u(x) = -k_p x$. Taking the Laplace transform of this equation we get

$$U(s) = -k_p X(s) \, .$$

Hence, the transfer function is

$$\frac{U(s)}{X(s)} = -k_p \, .$$

(b) $u(x) = -k_d \dot{x}$. Taking the Laplace transform of this equation we get

$$U(s) = -k_p s X(s) \, .$$

Hence, the transfer function is

$$\frac{U(s)}{X(s)} = -k_d s \, .$$

(c) $u(x) = -k_i \int x(w) dw$. Taking the Laplace transform of this equation we get

$$U(s) = -k_i \frac{1}{s} X(s).$$

Hence, the transfer function is

$$\frac{U(s)}{X(s)} = -\frac{k_i}{s}.$$

10.2 Introduction to Optimal control

■ **Tracking the Hotrod**

3. $(x_0,\ \dot{x}_0)=(3,\ 1)$

Because the initial point lies *above* the switching curve, the control is $u=-1$ until it hits the switching curve $x=\dfrac{1}{2}\dot{x}^2$. Then it is $u(t)=+1$ until it reaches the origin. Because the parabola

$$x=-\frac{1}{2}\dot{x}^2+c$$

that passes through $(3,\ 1)$ is $x=-\dfrac{1}{2}\dot{x}^2+\dfrac{7}{2}$, we follow this parabola until $x=\dfrac{1}{2}\dot{x}^2$.

We find the intersection of these curves by setting

$$-\frac{1}{2}\dot{x}^2+\frac{7}{2}=\frac{1}{2}\dot{x}^2$$

which yields $\dot{x}^2=\dfrac{7}{2}$, or $\dot{x}=-\sqrt{\dfrac{7}{2}}$.

Substituting back into $x=\dfrac{1}{2}\dot{x}^2$ we get $x=\dfrac{7}{4}$.

Hence, the switching point where control changes from -1 to +1

is $\left(\dfrac{7}{4},\ -\sqrt{\dfrac{7}{2}}\right)$. This path is shown in the figure.

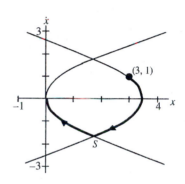

Hotrod trajectories from $(3,\ 1)$

■ **Conversion and Identification**

6. $\ddot{x}+2\dot{x}+x=u$

If $x_1=x$ and $x_2=\dot{x}$, the system is

$$\begin{bmatrix} \dot{x}_1 \\ \dot{x}_2 \end{bmatrix}=\begin{bmatrix} 0 & 1 \\ -1 & -2 \end{bmatrix}\begin{bmatrix} x_1 \\ x_2 \end{bmatrix}+\begin{bmatrix} 0 \\ u(t) \end{bmatrix}.$$

■ **More Identity Problems**

9. $$\begin{bmatrix} \dot{x} \\ \dot{y} \end{bmatrix}=\begin{bmatrix} 0 & 1 \\ -1 & 0 \end{bmatrix}\begin{bmatrix} x \\ y \end{bmatrix}+\begin{bmatrix} 0 \\ u \end{bmatrix}$$

12. $$\begin{bmatrix} \dot{x} \\ \dot{y} \end{bmatrix}=\begin{bmatrix} -1 & 2 \\ 1 & -1 \end{bmatrix}\begin{bmatrix} x \\ y \end{bmatrix}+\begin{bmatrix} 0 \\ u \end{bmatrix}$$

■ Optimal Control

15. **(a)** Using the optimal control $u*(t) = -2e^{-t}$ the controlled system is $\dot{x} - x = -2e^{-t}$ whose solution is given by $x(t) = ce^t + e^{-t}$. Applying initial condition $x(0) = 1$ gives $c = 0$. Hence, the solution of the controlled system is $x(t) = e^{-t}$. The value of the objective is

$$J = \frac{1}{2}\int_0^\infty \left[x^2(t) + u^2(t)\right]dt = \frac{1}{2}\int_0^\infty \left(e^{-2t} + 4e^{-2t}\right)dt = \frac{5}{2}\int_0^\infty e^{-2t}dt = -\frac{5}{4}e^{-2t}\Big|_0^\infty = 1.25.$$

(b) Using the control $u(t) = -5e^{-4t}$ the controlled system is $\dot{x} - x = -5e^{-4t}$ whose solution, in turn is given by $x(t) = ce^t + e^{-4t}$. Applying initial condition $x(0) = 1$ gives $c = 0$. Hence, the solution of the controlled system is $x(t) = e^{-4t}$. The value of the objective is

$$J = \frac{1}{2}\int_0^\infty \left[x^2(t) + u^2(t)\right]dt = \frac{1}{2}\int_0^\infty \left(e^{-8t} + 25e^{-8t}\right)dt = \frac{26}{2}\int_0^\infty e^{-8t}dt = -\frac{13}{8}e^{-8t}\Big|_0^\infty = 1.625.$$

This value ($J = 1.625$) is larger than J obtained in (a) for optimal control $u*(t) = -2e^{-t}$.

■ Satellite System

18. $$\ddot{r} = r(t)\dot{\theta}^2(t) - \frac{k}{r^2(t)} + u_1(t)$$

$$\ddot{\theta} = -\frac{2\dot{\theta}(t)\dot{r}(t)}{r(t)} + \frac{1}{r(t)}u_2(t)$$

Setting

$$x_1 = r - r_0$$
$$x_2 = \dot{r}$$
$$x_3 = r_0(\theta - \omega t)$$
$$x_4 = r_0(\dot{\theta} - \omega).$$

The previous equations (after some minor algebra) become

$$\dot{x}_1 = x_2$$

$$\dot{x}_2 = (x_1 + r_0)\left(\omega + \frac{x_4}{r_0}\right)^2 - \frac{r_0^3\omega^2}{(x_1 + r_0)^2} + u_1(t)$$

$$\dot{x}_3 = x_4$$

$$\dot{x}_4 = \frac{r_0}{x_1 + r_0}\left\{-2x_2\left(\omega + \frac{x_4}{r_0}\right) + u_2(t)\right\}.$$

■ Satellite Eigenvalues

21. If we expand by minors across the top row, we find the characteristic polynomial to be

$$\lambda^4 + \omega^2\lambda^2 = \lambda^2\left(\lambda^2 + \omega^2\right) = 0.$$

Hence, the eigenvalues are clearly 0, 0, and $\pm i\omega$.

10.3 Pontryagin Maximum Principle

■ Controlling a Pure Integrator

3. (a) $H(u) = -\dfrac{1}{2}u^2 + pu$ (b) $\dot{p} = -\dfrac{\partial H}{\partial x} = 0$, $p(t) = c_1$ (c_1 a constant)

(c) We now have the Hamiltonian

$$H(u) = -\frac{1}{2}u^2 + c_1 u.$$

The Pontryagin theorem states that for *each value* of t, pick the value of $u(t)$ that maximizes $H(u)$. In other words, think of t as a parameter insofar as maximizing $H(u)$ is concerned and think of $H = H(u)$ as a function of u. But $H(u)$ is a downward turning parabola with a maximum point at $u = c_1$. In other words, for all t pick $u(t) \equiv k$. Of course, the constant is not known yet, but knowing the control is a constant says something.

Calling the control $u(t) = k$, we solve the state equation $\dot{x} = k$, yielding $x(t) = kt + c_1$. Substituting this solution into the initial and final conditions, $x(0) = 1$ and $x(2) = 0$ we

get $k = -\dfrac{1}{2}$ and $c_1 = 1$. Hence, the optimal control is

$$u^*(t) = -\frac{1}{2}$$

and the optimal path is

$$x^*(t) = -\frac{1}{2}t + 1.$$

(d) $J(u) = \dfrac{1}{2}\displaystyle\int_0^2 u^2(t)\,dt = \dfrac{1}{2}\displaystyle\int_0^2 \dfrac{1}{4}\,dt = \dfrac{1}{4}.$

■ Stopping a Vibrating String

6. (a) $\vec{x}_0 = \begin{bmatrix} 0 \\ -3 \end{bmatrix}.$

From the trajectories in Figure 10.3.5 we see that the control function starts with 1 and has two switches. In other words, the control sequence is 1, –1, and 1.

(b) Physically, the spring is at equilibrium position, but moving to the left with velocity 3. We begin by pushing to the right until the spring is just past its maximum compression; then we push to the left. We continue this until the spring is just past its maximum extension, where it is then pushed to the right against the motion of the spring until it makes a soft landing at $(0, 0)$.

■   **Time Optimality and Bang-Bang Control**

9.   (a)   The objective function is $J(u) = \int_0^{t_f} dt = t_f$, and so $f_0 = 1$.

Hence, the Hamiltonian is $H(u) = -f_0 + p_1 \dot{x}_1 + p_2 \dot{x}_2 + \cdots + p_n \dot{x}_n$, which in matrix form is simply

$$H(u) = -1 + \vec{p} \cdot (\mathbf{A}\vec{x} + \vec{b}u) = -1 + \vec{p} \cdot \mathbf{A}\vec{x} + \vec{p} \cdot \vec{b}u.$$

(b)   Use simple matrix algebra resulting from factoring out the scalar function $u(t)$.

(c)   The Pontryagin theorem states that for every t in the interval of interest the optimal control maximizes the Hamiltonian $H(u)$. In other words, insofar as maximizing the Hamiltonian is concerned, think of t as a *constant*, and think of maximizing $H(u)$ as a function of a real variable u. After finding $u = u(t)$, think of t as a variable again and $u(t)$ as a function of t. So, how do we maximize $H(u)$ in this general problem? The values of the control $u(t)$ are assumed to lie in the interval $[-1, 1]$. So if we want to maximize $H(u)$ for each fixed t, we should pick $u(t) = 1$ when its (scalar) coefficient $\vec{p} \cdot \vec{b} > 0$ and $u(t) = -1$ when its coefficient $\vec{p} \cdot \vec{b} < 0$. This function $\sigma(t) = \vec{p} \cdot \vec{b}$ (the dot product of two vectors) is a scalar function of t called the switching function of the system. The goal is to find the adjoint variables $\vec{p}$.

■   **Suggested Journal Entry II**

12.   Student Project

APPENDIX

■   **Complex Exponential Numbers**

3.   (a)   Using Euler's formula, we write

$$e^{2\pi i} = \cos 2\pi + i \sin 2\pi = 1 + i(0) = 1.$$

(b)   Using Euler's formula, we write

$$e^{i\pi/2} = \cos\frac{\pi}{2} + i \sin\frac{\pi}{2} = 0 + i(1) = i.$$

(c)   Using Euler's formula, we write

$$e^{-i\pi} = \cos(-\pi) + i \sin(-\pi) = \cos\pi - i \sin\pi = -1.$$

(d)   Using the property $e^{a+b} = e^a e^b$ and using Euler's formula, we write

$$e^{(2+\pi i/4)} = e^2 e^{\pi i/4} = e^2\left(\cos\frac{\pi}{4} + i\sin\frac{\pi}{4}\right) = e^2\left(\frac{\sqrt{2}}{2} + i\frac{\sqrt{2}}{2}\right) = e^2\frac{\sqrt{2}}{2} + ie^2\frac{\sqrt{2}}{2}.$$

■   **Complex Verification II**

6.   By direct substitution we have

$$\left(\frac{1+i}{\sqrt{2}}\right)^4 = \frac{1}{4}(1+i)^4 = \frac{1}{4}(1+i)^2(1+i)^2 = \frac{1}{4}(1+2i-1)(1+2i-1) = \frac{1}{4}(4i^2) = -1.$$

■   **Roots of Unity**

9.   The m roots of $z^m = 1$ (called the roots of unity) are the m values

$$z_k = 1^{1/m} \left(\cos\left(\frac{2\pi k}{m} \right) + i \sin\left(\frac{2\pi k}{m} \right) \right), \quad k = 0, 1 \cdots m-1.$$

Note that for $z = 1$ yields polar angle $\theta = 0$ for the previous formula.

(a) $z^2 = 1$ has two roots

$$z_k = \cos\left(\frac{2\pi k}{2} \right) + i \sin\left(\frac{2\pi k}{2} \right) = \cos(\pi k) + i \sin(\pi k),$$

for $k = 0$, 1 or $z = \pm 1$.

(b) $z^3 = 1$ has three roots

$$z_k = \cos\left(\frac{2\pi k}{3} \right) + i \sin\left(\frac{2\pi k}{3} \right),$$

for $k = 0$, 1, 2 or

$$z_1 = +1$$

$$z_2 = \cos\left(\frac{2\pi}{3} \right) + i \sin\left(\frac{2\pi}{3} \right) = -\frac{1}{2} + \frac{\sqrt{3}}{2} i$$

$$z_3 = \cos\left(\frac{4\pi}{3} \right) + i \sin\left(\frac{4\pi}{3} \right) = -\frac{1}{2} - \frac{\sqrt{3}}{2} i.$$

(c) $z^4 = 1$ has four roots

$$z_k = \cos\left(\frac{2\pi k}{4} \right) + i \sin\left(\frac{2\pi k}{4} \right),$$

for $k = 0$, 1, 2, 3 or $z = 1, i, -1, -i$.

■   **Complex Exponential Functions**

12.   We use the properties of exponentials and Euler's formula to write

(a) $e^{4\pi i t} = \cos(4\pi t) + i \sin(4\pi t)$ (b) $e^{(-1+2i)t} = e^{-t} e^{2it} = e^{-t} (\cos 2t + i \sin 2t)$

LT   Linear Transformations

■   **Isomorphism Subtleties**

3. $M_{12}(\mathbb{R}) = \{[a \ \ b] : a, b \in \mathbb{R}\}$ so the elements of $M_{12}(\mathbb{R})$ are not elements of

$$M_{22}(\mathbb{R}) = \left\{ \begin{bmatrix} a & b \\ c & d \end{bmatrix} : a, b, c, d \in \mathbb{R} \right\}.$$

Thus $M_{12}(\mathbb{R})$ is not a subspace. However $T : M_{12}(\mathbb{R}) \to \left\{ \begin{bmatrix} a & b \\ 0 & 0 \end{bmatrix} : a, b \in \mathbb{R} \right\}$ is an

isomorphism to a subspace of $M_{22}(\mathbb{R})$.

■   **Isomorphisms and Bases**

6.   From Problem 32 in Section 5.1, we know that the composition of linear transformations is linear. We use the functions $L : \mathbb{U} \to \mathbb{V}$ and $T : \mathbb{V} \to \mathbb{W}$ to prove the following general results about the composition of functions $T \circ L : \mathbb{U} \to \mathbb{W}$.

The composition of injective functions is injective.

Suppose

$$T \circ L(\vec{u}_1) = T \circ L(\vec{u}_2)$$

for some $\vec{u}_1$ and $\vec{u}_2$ in $\mathbb{U}$. Then

$$T\big(L(\vec{u}_1)\big) = T\big(L(\vec{u}_2)\big),$$

so that $L(\vec{u}_1) = L(\vec{u}_2)$ since T is injective. Also, $\vec{u}_1 = \vec{u}_2$ since L is injective. Therefore $T \circ L$ is injective.

The composition of surjective functions is surjective.

Suppose

$$\vec{w} \in \mathbb{W}.$$

Because T is surjective, $T(\vec{v}) = \vec{w}$ for some $\vec{v}$ in $\mathbb{V}$. Also, because L is surjective, $L(\vec{u}) = \vec{v}$ for some in $\vec{u}$. Therefore, $T \circ L(\vec{u}) = T\big(L(\vec{u})\big) = T(\vec{v}) = \vec{w}$.

We have proved that the composition of surjective functions is surjective.

We know that $L : \mathbb{U} \to \mathbb{V}$ and $T : \mathbb{V} \to \mathbb{W}$ are isomorphisms if and only if they are injective and surjective linear transformations, so that $T \circ L$ must also have those properties.

■   **Associated Matrices**

9. $T\begin{bmatrix} a & b \\ c & d \end{bmatrix} = \begin{bmatrix} 2a & c+b \\ c+b & 2d \end{bmatrix}$,

$$\mathbf{M}_B = \left[\left(T\begin{bmatrix} 1 & 0 \\ 0 & 0 \end{bmatrix} \right)_c \middle| \left(T\begin{bmatrix} 0 & 1 \\ 0 & 0 \end{bmatrix} \right)_c \middle| \left(T\begin{bmatrix} 0 & 0 \\ 1 & 0 \end{bmatrix} \right)_c \middle| \left(T\begin{bmatrix} 0 & 0 \\ 0 & 1 \end{bmatrix} \right)_c \right]$$

$$= \left[\begin{bmatrix} 2 \\ 0 \\ 0 \\ 0 \end{bmatrix} \begin{bmatrix} 0 \\ 1 \\ 1 \\ 0 \end{bmatrix} \begin{bmatrix} 0 \\ 1 \\ 1 \\ 0 \end{bmatrix} \begin{bmatrix} 0 \\ 0 \\ 0 \\ 2 \end{bmatrix} \right] = \begin{bmatrix} 2 & 0 & 0 & 0 \\ 0 & 1 & 1 & 0 \\ 0 & 1 & 1 & 0 \\ 0 & 0 & 0 & 2 \end{bmatrix}.$$

■   **Changing Bases**

12. $\mathbb{V} = \mathbb{M}_{21}(\mathbb{R})$, $B = \left\{ \begin{bmatrix} 1 \\ 1 \end{bmatrix}, \begin{bmatrix} 3 \\ 0 \end{bmatrix} \right\}$, $C = \left\{ \begin{bmatrix} -1 \\ 1 \end{bmatrix}, \begin{bmatrix} 0 \\ 2 \end{bmatrix} \right\}$,

$$id\left(\begin{bmatrix} 1 \\ 1 \end{bmatrix} \right) = \begin{bmatrix} 1 \\ 1 \end{bmatrix} = \left(c_1 \begin{bmatrix} -1 \\ 1 \end{bmatrix} + c_2 \begin{bmatrix} 0 \\ 2 \end{bmatrix} \right)_c = \begin{bmatrix} c_1 \\ c_2 \end{bmatrix}$$

$$id\left(\begin{bmatrix} 3 \\ 0 \end{bmatrix} \right) = \begin{bmatrix} 3 \\ 0 \end{bmatrix} = \left(d_1 \begin{bmatrix} -1 \\ 1 \end{bmatrix} + d_2 \begin{bmatrix} 0 \\ 2 \end{bmatrix} \right)_c = \begin{bmatrix} d_1 \\ d_2 \end{bmatrix}.$$

To find c_1 and c_2:

$$\begin{bmatrix} 1 \\ 1 \end{bmatrix} = c_1 \begin{bmatrix} -1 \\ 1 \end{bmatrix} + c_2 \begin{bmatrix} 0 \\ 2 \end{bmatrix},$$

$1 = -c_1 + 0c_2$ so $c_1 = -1$ and $1 = 1c_1 + 2c_2$ so $c_2 = 1$. To find d_1 and d_2:

$$\begin{bmatrix} 3 \\ 0 \end{bmatrix} = d_1 \begin{bmatrix} -1 \\ 1 \end{bmatrix} + d_2 \begin{bmatrix} 0 \\ 2 \end{bmatrix},$$

$3 = -1d_1 + 0d_2$ so $d_1 = -3$ and $0 = 1d_1 + 2d_2$ so $d_2 = \dfrac{3}{2}$;

$$\mathbf{M}_B = \begin{bmatrix} -1 & -3 \\ 1 & \dfrac{3}{2} \end{bmatrix}.$$

■   **Associated Matrix Again**

15. $T : \mathbb{P}_2 \to \mathbb{R}^3$ where $T\left(at^2 + bt + c\right) = \left(a - b,\ a,\ 2c\right)$ and $B = \left\{t^2,\ t,\ 1\right\}$ as in Problem 8, but the basis for $\mathbb{R}^3$ is $D = \left\{\vec{e}_1,\ \vec{e}_1 - \vec{e}_2,\ 5\vec{e}_3 + \vec{e}_1\right\}$.

$$\left[T\left(t^2\right)\right]_D = \left[1 - 0,\ 1,\ 0\right]_D = \left[a_1\begin{bmatrix}1\\0\\0\end{bmatrix} + a_2\begin{bmatrix}1\\-1\\0\end{bmatrix} + a_3\begin{bmatrix}1\\0\\5\end{bmatrix}\right]_D = \begin{bmatrix}a_1\\a_2\\a_3\end{bmatrix},$$

$1 = 1a_1 + 1a_2 + 1a_3$, $1 = 0a_1 - 1a_2 + 0a_3$, and $0 = 0a_1 + 0a_2 + 5a_3$ which yields the results

$$\begin{bmatrix}a_1\\a_2\\a_3\end{bmatrix} = \begin{bmatrix}2\\-1\\0\end{bmatrix}.$$

$$\left[T\left(t\right)\right]_D = \left[-1, 0, 0\right]_D = \left[b_1\begin{bmatrix}1\\0\\0\end{bmatrix} + b_2\begin{bmatrix}1\\-1\\0\end{bmatrix} + b_3\begin{bmatrix}1\\0\\5\end{bmatrix}\right]_D = \begin{bmatrix}b_1\\b_2\\b_3\end{bmatrix},$$

$-1 = 1b_1 + 1b_2 + 1b_3$, $0 = 0b_1 - 1b_2 + 0b_3$, and $0 = 0b_1 + 0b_2 + 5b_3$, which yields the results

$$\begin{bmatrix}b_1\\b_2\\b_3\end{bmatrix} = \begin{bmatrix}-1\\0\\0\end{bmatrix}.$$

$$\left[T\left(1\right)\right]_D = \left[0, 0, 2\right]_D = \left[c_1\begin{bmatrix}1\\0\\0\end{bmatrix} + c_2\begin{bmatrix}1\\-1\\0\end{bmatrix} + c_3\begin{bmatrix}1\\0\\5\end{bmatrix}\right]_D = \begin{bmatrix}c_1\\c_2\\c_3\end{bmatrix},$$

$0 = 1c_1 + 1c_2 + 1c_3$, $0 = 0c_1 - 1c_2 + 0c_3$, and $2 = 0c_1 + 0c_2 + 5c_3$, which yields the results

$$\begin{bmatrix}c_1\\c_2\\c_3\end{bmatrix} = \begin{bmatrix}-\dfrac{2}{5}\\0\\\dfrac{2}{5}\end{bmatrix}$$

and

$$\mathbf{M}_B^* = \begin{bmatrix}2 & -1 & -\dfrac{2}{5}\\-1 & 0 & 0\\0 & 0 & \dfrac{2}{5}\end{bmatrix}.$$

PF   Partial Fractions

■   **Practice Makes Perfect**

3. $\dfrac{x}{(x+1)(x+2)}$. We write this fraction in the form

$$\frac{x}{(x+1)(x+2)} = \frac{A}{x+1} + \frac{B}{x+2}.$$

Clearing of fractions yields

$$x = A(x+2) + B(x+1).$$

Collecting terms yields

$$(A+B-1)x + (2A+B) = 0.$$

Equating coefficients we find the equations

$$A + B = 1$$
$$2A + B = 0$$

which has the solutions $A = -1$ and $B = 2$. Hence, we have

$$\frac{x}{(x+1)(x+2)} = \frac{-1}{x+1} + \frac{2}{x+2}.$$

6. $\dfrac{3}{\left(x^2+1\right)\left(x^2+4\right)}$. We write this fraction in the form

$$\frac{3}{\left(x^2+1\right)\left(x^2+4\right)} = \frac{Ax+B}{x^2+1} + \frac{Cx+D}{x^2+4}.$$

Clearing of fractions yields

$$3 = (A+C)x^3 + (B+D)x^2 + (4A+C)x + 4B + D.$$

Equating coefficients and solving we find $A = 0$, $B = 1$, $C = 0$, and $D = -1$. Hence, we have the partial fraction decomposition

$$\frac{3}{\left(x^2+1\right)\left(x^2+4\right)} = \frac{1}{x^2+1} - \frac{1}{x^2+4}.$$

9. $\dfrac{x^2 + 9x + 2}{(x-1)^2(x+3)}$. We write this fraction in the form

$$\frac{x^2 + 9x + 2}{(x-1)^2(x+3)} = \frac{A}{x-1} + \frac{B}{(x-1)^2} + \frac{C}{x+3}.$$

Clearing of fractions yields

$$x^2 + 9x + 2 = A(x-1)(x+3) + B(x+3) + C(x-1)^2.$$

Equating coefficients we find

$$(A + C - 1)x^2 + (2A + B - 2C - 9)x + (-3A + 3B + C - 2) = 0.$$

Setting the coefficients equal yields

$$A + C - 1 = 0$$
$$2A + B - 2C = 9$$
$$-3A + 3B + C = 2.$$

Solving these equations we find $A = 2$, $B = 3$, and $C = -1$. Hence, we have the partial fraction decomposition

$$\frac{x^2 + 9x + 2}{(x-1)^2(x+3)} = \frac{2}{x-1} + \frac{3}{(x-1)^2} - \frac{1}{x+3}.$$